Geometric Modeling and Reasoning of Human-Centered Freeform Products

Charlie C. L. Wang

Geometric Modeling and Reasoning of Human-Centered Freeform Products

Springer

Charlie C. L. Wang
The Chinese University of Hong Kong
Hong Kong
China

ISBN 978-1-4471-4359-8 ISBN 978-1-4471-4360-4 (eBook)
DOI 10.1007/978-1-4471-4360-4
Springer London Heidelberg New York Dordrecht

Library of Congress Control Number: 2012943377

Printed on acid-free paper

Springer is part of Springer Science+Business Media (www.springer.com)

Dedicated to my family for their endless love, forbearance and encouragement

Preface

This book covers the content on using geometric approaches to design the freeform shape of products that are worn by human bodies. The recent trend about user-customized product design requires that the shape of such products be automatically adjusted according to the human bodies' shape, so that people will feel more comfortable when wearing them.

The motivation of this book is to introduce readers to the newly developed geometric solution for the design automation of customized freeform objects, which can greatly improve the efficiency of design processes in various industries involving customized products (e.g., garment design, toy design, jewel design, shoe design, and design of medial devices, etc.). The products in the above industries are usually composed of very complex geometry shape (represented by free-form surfaces), and are not driven by a parameter table but a digital human model with freeform shapes or part of human bodies (e.g., wrist, foot, and head models). After carefully designing a product around one particular human model, it is desirable to have an automated tool for "grading" this product to other shape-changed human bodies while retaining the original spatial relationship between the product and human models.

The techniques introduced in this book are based on my research conducted in this area in the past decade, which cover the algorithms of human body reconstruction, freeform product modeling, constraining and reconstructing freeform products, and shape optimization for improving the manufacturability of freeform products. Based on these techniques, the design automation problem for human-centered freeform products can be fundamentally solved. This book can be used as a reference book for researchers and developers working on problems of automatic designing of products customized for individuals, and it can also be used as a reference book for courses in computer-aided product design at the graduate level.

My research conducted in this area was initiated during my graduate study. I would like to thank Prof. Matthew M. F. Yuen, Prof. Kai Tang, Prof. Shana Smith, Prof. Qifu Wang, Dr. Jin Fan, Dr. Zhuang Wu and Prof. Jiansong Deng for their encouragement and valuable advice at the early stage of my work. Acknowledgments also go to my collaborators, students, and research staff, in particular, my collaborators—Prof. Kin-Chuen Hui, Prof. Xiaogang Jin, Prof. Chih-Hsing Chu, Prof. Alexandre Kung and Dr. Terry Chang, my students and research staff—Yunbo Zhang, Tsz-Ho Kwok, Siu-Ping Mok, Shengjun Liu, Yuwei Meng, Kwan-Chung Chan, Alan Yeung, Samuel Li, Hoi Sheung, Yuen-Shan Leung, and Ya-Tien Tsai, and all staff at the Department of Mechanical and Automation Engineering, The Chinese University of Hong Kong (CUHK).

The bulk of this research was supported by the Hong Kong Research Grants Council under grants CUHK/412405, CUHK/416307, and CUHK/417109, the Innovation and Technology Commission under the grant of Innovation Technology Fund ITS/026/07, the Shun Hing Institute of Advanced Engineering (SHIAE) under grant CUHK/8115022, the industrial sponsors—TPC (HK) Limited and Tak Wai Model Clothes Rack Co., and the direct research grants of CUHK.

Shatin, Hong Kong, May 2012 Charlie C. L. Wang

Contents

Chapter 1
Introduction

1.1 Design Automation in Commercial CAD/CAM Systems

As techniques developed since 1960s, *computer-aided design and manufacturing* (CAD/CAM) plays an important role in product design and manufacturing in many industries. CAD/CAM systems have been demanded by the design of new aircrafts and automobiles, which are mainly assembled from mechanical components. At the beginning, CAD software was used to replace the conventional engineering drawing board so that designers can modify drawings in an easier and more economical manner in the computer system. The later development of CAD techniques make a CAD system have components far beyond this primary function [7]. One of the major components is *parametric modeling* [1, 4, 5] (or called *parametric design* [8]), which helps the design automation of many industrial products like mechanical parts and architecture.

Parametric modeling uses parameters (such as dimensions) that can be modified later to define a model. Other than the dimensions to construct models, the constraints between geometric entities are also explicitly specified in many cases. For example, plane A must be parallel to plane B, and the distance between them should be greater than 100 mm. The constraints serve as bridges to link the dimensions of a model together; therefore, the relationships between them are specified. All the dimensions of a designed model can then be coupled by these geometric constraints to a few key parameters. Changing the values of these parameters varies the shape of the designed model, which leads to a family of variational models derived from the original design (such as described in [10]). An important function of design automation is to adjust the shape of a product automatically according to some design purposes (such as reducing the maximal stress in the product under a fixed loading). In many cases, the parametric modeling is based on the technique of constraint solver (see [3] for an example).

Figure 1.1 shows the progressive steps to build a gear model in a commercial CAD/CAM system by parametric modeling, where the thickness and the number of teeth are chosen as key parameters. In the later phases of the design cycle, a

C. C. L. Wang, *Geometric Modeling and Reasoning of Human-Centered Freeform Products*, DOI: 10.1007/978-1-4471-4360-4_1, © Springer-Verlag London 2013

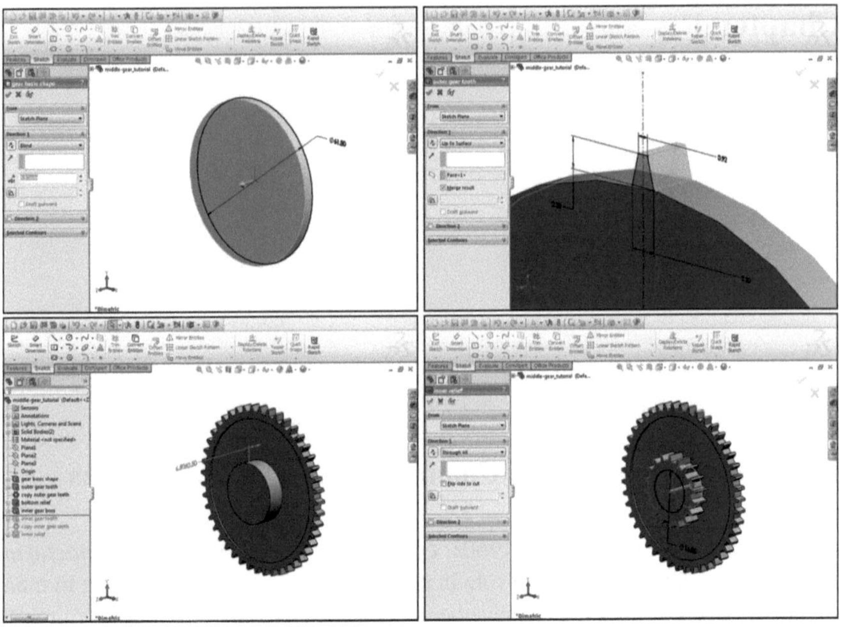

Fig. 1.1 Steps for constructing a gear model and specifying constraints between geometric entities

designer may wish to change the thickness of the gear. Based on the techniques of parametric modeling, instead of going through the modeling steps one by one again from the very beginning, he only needs to specify a new thickness. Then, a new gear model is automatically generated according to this new thickness while retaining the constraints specified in the original design (see Fig. 1.2 for an example). The shape variations of the gear can also be produced automatically according to the number of teeth (see Fig. 1.3). For a model with a complicated structure, a parameter table is extracted from the constraints and serves as the input of design automation of a product. The parameters for designing an optimal bicycle frame are given in [12] as an example.

1.2 Human-Centered Freeform Products

The parametric modeling technique focuses on models with relatively regular shapes (such as mechanical part). It has been widely used in different industries like automobile industry, aircraft industry, and ship industry, where variations in products are mainly driven by dimensions. There are other industrial applications, such as garment design, toy design, jewel design, shoe design, and glasses design. The products in these industries are usually composed of complex geometric shapes (represented

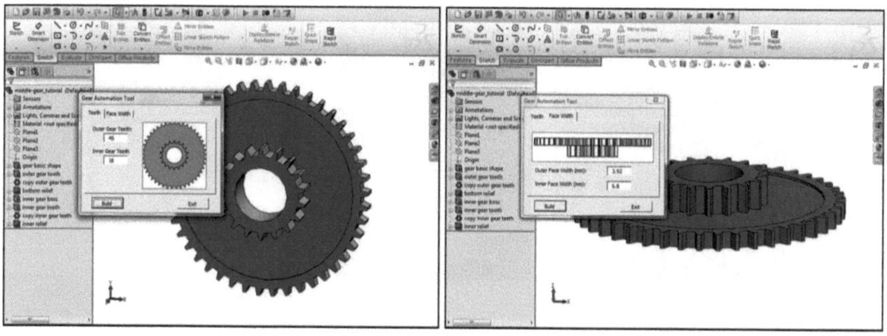

Fig. 1.2 A new gear model can be automatically generated by using a new thickness

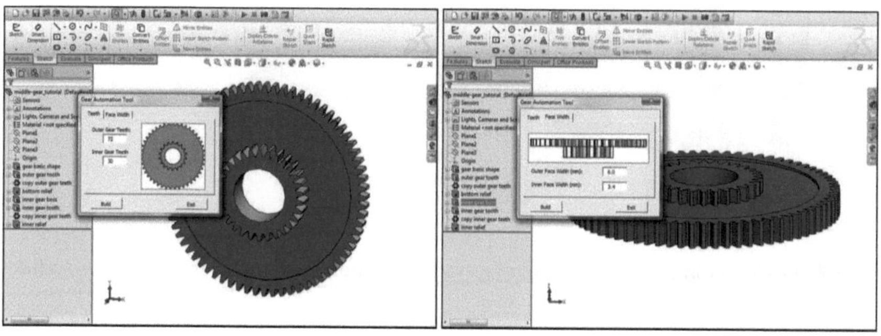

Fig. 1.3 Shape variation of the gear model shown in Fig. 1.1 by choosing different numbers of teeth

by freeform surfaces), and are not driven by a table of parameters but a part of human body with freeform shapes (such as mannequin, toy, wrist, foot, and head models). For products that are human body oriented, the reconstruction and processing methods for human bodies are presented in Chap. 2, where the techniques in four aspects are introduced, including surface reconstruction, feature point extraction, cross-parameterization, and shape space analysis.

Unlike the freeform surfaces in conventional geometric modeling techniques [9] where surfaces are usually differentiable and represented by parametric curved surfaces (such as Bézier and NURBS surfaces and their trimmed forms), the freeform surfaces for designing human-centered products are in a complex geometric form (for example, the jeans pants, the fabric toys, the sofa, the wetsuit and the shoes shown in Fig. 1.4). Piecewise linear mesh surfaces (especially triangular mesh surfaces) are usually used to represent this kind of surfaces in a geometric modeling system. Moreover, 3D human models are also represented in the form of piecewise linear mesh surfaces. Detailed data structure for this representation and its manipulation can be found in Chap. 3.

A more difficult problem than representing such products is how to automatically vary the shape of a product designed around a mannequin having a standard shape

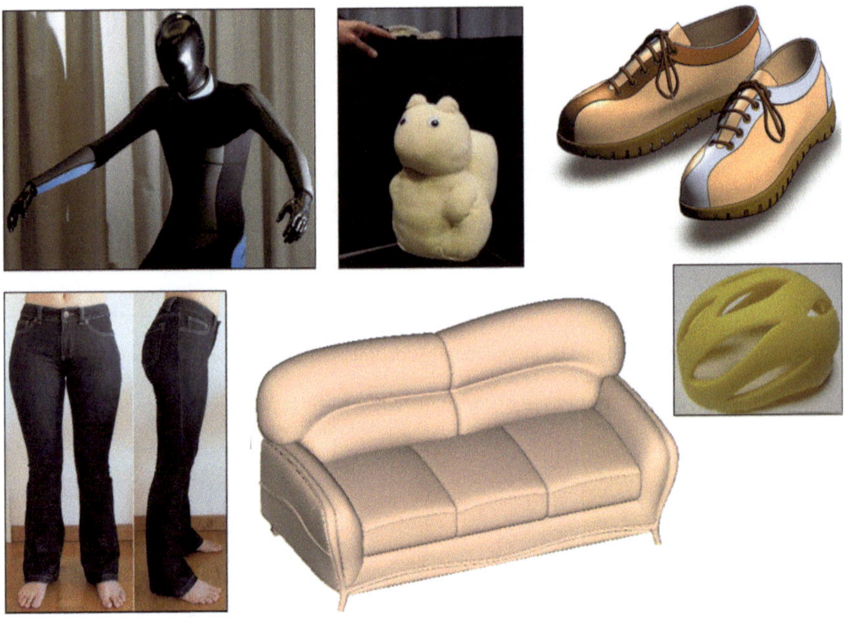

Fig. 1.4 Products formed by freeform surfaces that are hardly represented by continuous surfaces

to the shapes that 'fit' different individuals. In many industries, 'fitness' can be interpreted as maintaining the designed spatial relationship between a product and human bodies. Several techniques are introduced in Chap. 4 for this purpose, which include transformation technique, shape warping based on freeform deformation, an approach based on volumetric parameterization, and a method to realize flexible shape control.

1.3 Demanded New Technology for Design

The software and method to support the design automation of human-centered freeform product is a new research area of CAD/CAM. After carefully designing a freeform product around one particular mannequin by using the modern discrete geometric modeling and processing approaches (such as [2]), there is a demand for an automated tool to 'grade' the product to 'fit' other mannequins in different shapes. The spatial relationship between the product and mannequins is retained. Market available CAD/CAM systems, which are based on dimension-driven parametric modeling, cannot support design automation in this manner.

In the other aspect, a natural way for designers to create new fashion is to have a freehand drawing tool like drawing on a piece of paper with a pencil. This is a reason why a sketch-based interface becomes very popular in many 3D computer

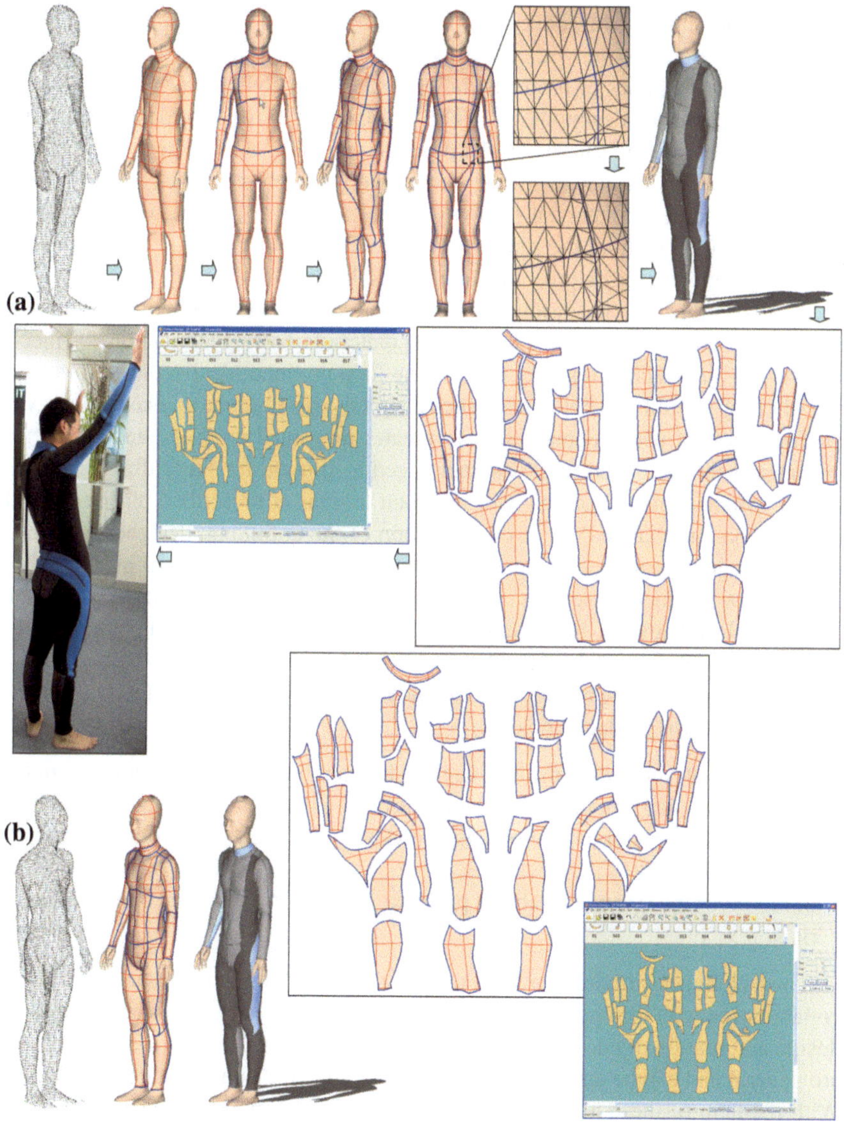

Fig. 1.5 An example of design automation in wetsuit design and fabrication. **a** Steps from styling design to fabricated wetsuit on a human model H_A: after using freehand drawing tools to get the styling design on the mesh surface generated from a scanned point cloud of human body, the surface of a wetsuit is trimmed into 3D pieces, unfolded into 2D pieces and then output to commercial 2D Garment CAD software in DXF file format. The real wetsuit customized for H_A is subsequently fabricated from these 2D patterns; **b** design transformation—the styling design of a wetsuit on H_A can be easily transferred to another human model H_B to generate planar pieces for industrial fabrication (Image taken from [11] with permission ©2010 IEEE)

graphics applications. Moreover, designers also want to have curve creating and editing tools, such as the widely used spline curves in CAD systems since these tools can define shapes of products more precisely. In some applications (e.g., the application of wetsuit design shown in Fig. 1.5a), the final shape of a designed product is specified by styling curves. Tight products skinning on the surface of human bodies are usually designed in this way. Although such tools have recently been provided in some commercial CAD systems (such as [6]), their functionality is greatly degraded without the support of the aforementioned design transformation and the manufacturability processing described below.

In many industries (for instance, the apparel industry and shoe industry), human-centered products are fabricated from planar materials. During fabrication, 2D pieces are warped and stitched together to build the final product. Ideally, the warping and the stitching should be stretch-free since a stretch will produce an elastic energy in the final product which debases fitness, creates unwanted wrinkles, and even leads to material fatigue. Therefore, designers need a solution for 'unfolding' given 3D surface patches into 2D pieces with invariant lengths at prescribed places (or with length variation bounded). As a dual-solution of the unfolding, designers may also request for geometric processing tools to modify the shape of a model into a slightly different form, which however is easier to be fabricated from planar materials. Details of approaches that solve this manufacturability problem for products fabricated by planar materials can be found in Chap. 5.

In summary, the new techniques demanded for the design automation of human-centered freeform products include:

- Modeling of three-dimensional digital human bodies which supports point-to-point mapping among individual bodies;
- Design transformation for both tight and loose-fit freeform products worn by human bodies;
- Solution for the manufacturability problem of products fabricated by planar materials.

The techniques introduced in this book are based on the author's research work and collaboration with the apparel industry in the past decade, which fill the niche of the design automation of human-centered freeform products. The presented approaches are general and can be applied to other industries having human-centered products. With the help of these new techniques, personalized design and manufacturing of human-centered freeform products can be realized in CAD/CAM systems.

References

1. Anderl, R., Mendgen, R.: Modelling with constraints: theoretical foundation and application. Comput. Aided Des. **28**(3), 155–168 (1996)
2. Botsch, M., Kobbelt, L., Pauly, M., Alliez, P., Lévy, B.: Polygonal Mesh Processing. A K Peters, Watertown (2010)

3. Gao, X.S., Lin, Q., Zhang, G.: A C-tree decomposition algorithm for 2D and 3D geometric constraint solving. Comput. Aided Des. **38**(1), 1–13 (2006)
4. Hoffmann, C., Kim, K.J.: Towards valid parametric CAD models. Comput. Aided Des. **33**(1), 81–90 (2001)
5. Hoffmann, C.M., Robert, J.A.: Parametric modeling. In: Farin, G., Hoschek, J., Kim, M.-S. (eds.) Handbook of Computer Aided Geometric Design, pp. 519–542. North Holland, Amsterdam (2002)
6. Lectra: Design Concept Furniture. http://www.lectra.com/en/index.html (2009)
7. Lee, K.: Principles of CAD/CAM/CAE Systems. Addison Wesley, Reading (1999)
8. Monedero, J.: Parametric design: a review and some experiences. Autom. Constr. **9**(4), 369–377 (2000)
9. Mortenson, M.E.: Geometric Modeling. Wiley, New York (1997)
10. van der Meiden, H.A., Bronsvoort, W.F.: Tracking topological changes in parametric models. Comput. Aided Geom. Des. **27**(3), 281–293 (2010)
11. Wang, C.C.L., Zhang, Y., Sheung, H.: From designing products to fabricating them from planar materials. IEEE Comput. Graph. Appl. **20**, 74–85 (2010)
12. Yoshihuku, Y., Herzog, W.: Optimal design parameters of the bicycle-rider system for maximal muscle power output. J. Biomech. **23**(10), 1069–1079 (1990)

Chapter 2
Digital Human Body

2.1 Reconstruction of Human Bodies for Individuals

This section starts by analyzing the current shape acquisition methods, where advantages and drawbacks of different surface reconstruction methods are reviewed. Based on the general defects in scanned data, two techniques are introduced to help improve the results of surface reconstruction. The first one is a robust method that assigns correct orientations to input points, which plays an important role in the functional-computation based surface reconstruction methods. The second one is an iterative consolidation framework which is developed for 'massaging' a set of scanned points in the presence of noise, outliers, and non-uniformities. The consolidation framework removes outliers, adds new samples to sparse regions, and extrapolates the boundaries of holes by an iterative down-sampling and up-sampling procedure.

2.1.1 Shape Acquisition

At present, there are several techniques and devices available for capturing the 3D shape of a real object. Basically, they can be grouped into three categories: (1) laser based (such as [8, 43]), (2) structure light based (such as [39, 59]), and (3) consumer camera based (such as [62]).

Among these techniques, the laser-based scanners can always capture 3D shape with a high accuracy but at a low speed as the scanners need to move and sweep laser lines across the surface of objects to be scanned. 3D positions of the points located on the lines are calculated by the stereo-viewing principle (see Fig. 2.1 for an illustration). The structure light based 3D shape acquisition runs much faster than laser scanners but provides results with a lower accuracy. The consumer camera based method can help capture 3D shape by low-cost equipment (such as digital camera or webcam); however, it generates 3D points in a much lower accuracy.

C. C. L. Wang, *Geometric Modeling and Reasoning of Human-Centered Freeform Products*, DOI: 10.1007/978-1-4471-4360-4_2, © Springer-Verlag London 2013

Fig. 2.1 Principles of a laser triangulation system, where the projected laser line is imaged by a charge-coupled device (CCD) camera

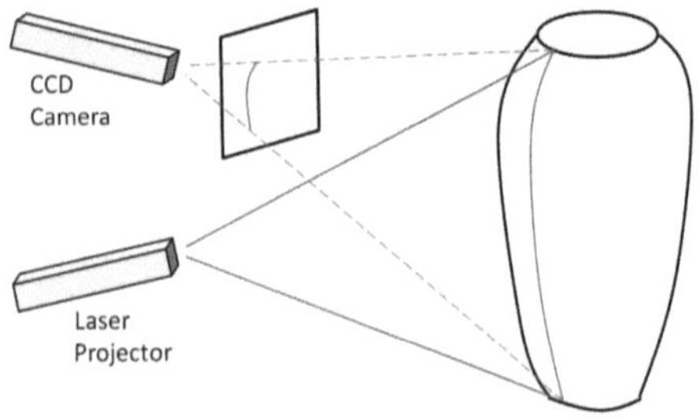

Table 2.1 Abbreviations of the algorithms

Abbreviations	Algorithm	References
POS	Poisson method of surface reconstruction	[34]
RBF	Radial basis function based surface reconstruction	[58]
INT	Integrating meshing method	[57]
CON	Consolidation of unorganized point clouds	[29]
ORT	Orienting unorganized points	[49]
ICO	Iterative consolidation	[48]

Although optical scanners are the most economical and efficient acquisition devices to obtain the 3D digital model from a real object, they always produce incomplete and noisy point clouds due to occlusions and physical limitations of the scanners. In regions that are invisible to cameras (such as deep cavities and bifurcations), the surface of the scanned model is not covered by sample points. The under-sampled or completely missed regions on the scanned point cloud of a real-world geometry will lead to an imperfect shape on the surface reconstructed by most reconstruction algorithms (see Fig. 2.2). The abbreviations of surface reconstruction and point processing algorithms discussed in this chapter are listed in Table 2.1.

2.1.2 Methods for Surface Reconstruction

Existing approaches for reconstructing surfaces from scattered sample points can be classified into two major groups: (1) approaches based on computational geometry and (2) techniques based on volumetric fitting.

Computational geometry approaches are usually based on the Voronoi diagram (VD) of a given point cloud and reconstruct a mesh surface by directly linking the input samples. Normal information is not required. Amenta et al. [4] gave a provable guarantee of reconstructing a correct model given a minimum sampling density dependent on the local feature size. The approach was further extended to be able to handle noisy inputs [51]. However, as they did not remove outliers, the quality of the resultant meshes was not good. Several variations of [4] are available in [5, 14, 15, 38]. When applying these algorithms to practical data sets, there are two difficulties. First, both memory and time costs to compute VD are expensive. Second,

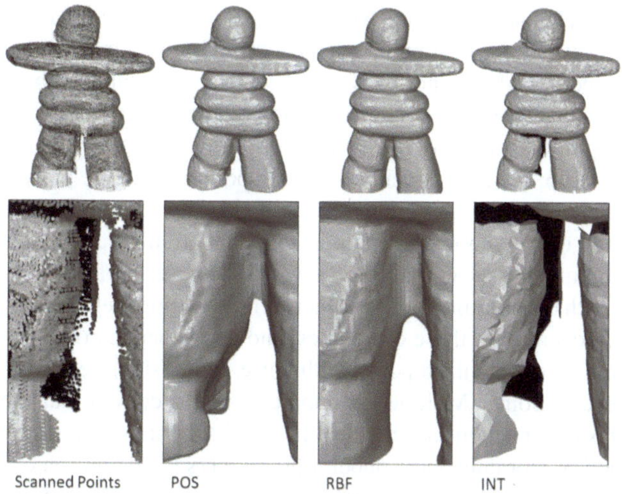

Fig. 2.2 The point cloud of an Inukshuk model obtained from a 3D scanner with incomplete sample points. Reconstructed surfaces generated by various algorithms in the literature are poor in regions with imperfect input samples (Image taken from [49] with permission)

Fig. 2.3 A computational geometry based surface reconstruction approach is not robust to noisy inputs: the scanned raw data (left), (middle) the result from Tight CoCone [15], and the successful reconstruction obtained by ORT [49] + POS [34] (right)

such approaches request input points to satisfy the d-covering requirement—i.e., the point set S sampled from a model H has any point $\mathbf{p}$ on H that can find a point $\mathbf{q} \in S$ such that $\|\mathbf{p} - \mathbf{q}\| \leq d$, where d is less than the smallest feature's size on H. This is hard to be satisfied, especially in regions with highly sparse points and with noise embedded (see the example shown in Fig. 2.3 where the Tight CoCone approach [15] fails). To the best of our knowledge, the integrating approach presented in [57] is a good approach that can handle the above difficulties while does not require the input points to be equipped with normal vectors. Nevertheless, their algorithm does not preserve the connectivity of underlying surfaces in regions with few points.

Volumetric reconstruction techniques attempt to build a signed implicit function that interpolates or approximates point-cloud samples (Refs. [11, 16, 26, 34, 56, 58, 76, 83]), and then extract its zero level set using one of the contouring methods—such as the Marching cubes (MC) algorithm [50], particle based polygonization method [9], or the dual contouring (DC) [32]. Compared with the approaches based on computational geometry, the reconstruction methods in this category in general can bear a higher level of noise embedded in input point sets (see the right of Fig. 2.3 as an example). However, computation of a signed implicit function requires point-cloud samples to be equipped with normal vectors, which can hardly be obtained directly from shape acquisition devices. The estimation of consistently oriented normals on given cloudy points is actually one of the most critical steps in the reconstruction pipeline (Refs. [26, 52])—especially when the points are in the presence of noise, holes, sharp features, or thin structures. The most widely adopted approach to obtain the consistently oriented normal vectors is the orientation propagation algorithm in [26] using minimal spanning tree (MST), which however performs very poorly on point clouds with the aforementioned defects. The orienting method presented below in this section overcomes these difficulties, and can generate consistently oriented normal vectors on unorganized points with noise, holes, sharp features, and regions with highly sparse points. Moreover, compared with the conventional approach that uses MST, this new method for orienting unorganized points is much faster.

The method of Hornung and Kobbelt [28] requires the definition of a watertight voxel crust in which the unknown surface is supposed to lie. To complete the crust, the authors used flood-fill and dilation operators. Sharf et al. [70] evolved an explicit mesh in a scalar field guided by the local feature size in a coarse-to-fine manner to avoid local optimum and capture details. The method also requires a volumetric grid to evaluate distance transformation, and topological changes have to be tracked. The computational implementation can be quite intricate (especially when varying the topology on two-manifold mesh surfaces). In [71], Sharf et al. interactively reconstruct a surface using only the positions of raw scanned data, where users define the general in/out orientation and assist the interpretation of data in topologically unstable regions that can be automatically detected.

2.1.3 Orienting Unorganized Points for Surface Reconstruction

To orient unorganized points effectively and efficiently, an algorithm (ORT) is developed here by extending the integrating approach for meshing scattered point data [57]. First, a modified scheme is proposed to generate adaptive spherical cover (ASC) for unorganized points by adding a sphere splitting step based on eigenvalue analysis. With this step, our approach can preserve the connectivity of surface in highly sparse regions. After getting the spherical cover for input points, the triangulation and topology cleaning procedure (Ref. [57]) can generate a triangular mesh surface M roughly presenting the underlying surface S. This mesh M may not be a good approximation of S; however, it gives a very robust evidence for assigning the orientation of input points. It is a straightforward thinking to find the closest point $\mathbf{c}_p$ on M for each input point $\mathbf{p}$, then assign the normal vector $\mathbf{n}_{\mathbf{c}_p}$ of $\mathbf{c}_p$ as the normal vector of $\mathbf{p}$. Nevertheless, as M is an inaccurate approximation of S, such normal vectors give inaccurate surface information to the downstream mesh reconstruction algorithm. Therefore, instead of assigning $\mathbf{n}_{\mathbf{c}_p}$ to $\mathbf{p}$, we only let $\mathbf{p}$ hold the orientation of $\mathbf{n}_{\mathbf{c}_p}$—thus, this method is named as *orienting* approach (ORT) [49]. An orientation-aware principal component analysis (PCA) step is adopted to assign correct and consistently oriented normal vectors to unorganized points. The most recent work presented in [29] tries to address a similar problem. However, their projection and down-sampling based strategy may destroy some small features of the underlying surface when the points are sparse, whereas the approach introduced below does not have such a defect.

2.1.3.1 Adaptive Spherical Cover

The ASC generated in [57] can successfully work on a noisy point cloud $P = \{\mathbf{p}_1, \ldots, \mathbf{p}_n\}$ with n scattered points, and output a set of covering spheres which are employed to construct triangular meshes by linking the auxiliary points in the spheres (Ref. [57]). To compensate the variations in point density on P, every point is assigned with a weight

$$w_i = \frac{1}{k} \sum_{j=1}^{k} \|\mathbf{p}_i - \mathbf{p}_j\|^2 \tag{2.1}$$

where $\{\mathbf{p}_j\}_{j=1}^{k} \subset P$ are the k-nearest neighbors of $\mathbf{p}_i$. Usually, $k = 10$ is selected, which balances the speed and quality well. In addition, unit normals $\{\mathbf{n}_1, \ldots, \mathbf{n}_n\}$ at the points of P are evaluated by their k-nearest neighbors using a standard covariance-based technique [26]. Normal vectors are not necessary to be consistently oriented at this moment. Only their directions are useful for the evaluation of quadric-error function below.

At the beginning of the spherical covering algorithm, all points in P are assigned as *uncovered*. Then, the algorithm generates m spheres centered at $\{\mathbf{c}_1, \ldots, \mathbf{c}_m\} \subset P$ with the radii $\{r_1, \ldots, r_m\}$ chosen adaptively by repeating the following steps until no point is found to be uncovered.

1. Randomly select a point $\mathbf{c}_i$ from the set of uncovered points to serve as the center of a new sphere.
2. For each sphere, if its radius r is known, a quadric-error function centered at $\mathbf{c}_i$ is defined as

$$Q_{\mathbf{c}_i,r}(\mathbf{x}) = \sum_j w_j G_\sigma(\|\mathbf{p}_j - \mathbf{c}_j\|)(\mathbf{n}_j \cdot (\mathbf{x} - \mathbf{p}_j))^2 \tag{2.2}$$

with $G_\sigma(\rho)$ being a compactly supported Gaussian-like function

$$G_\sigma(\rho) = \begin{cases} \exp(-8(\rho/\sigma)^2), & |\rho| \in [0, \sigma/2] \\ 16(1 - \rho/\sigma)^4/e^2, & |\rho| \in (\sigma/2, \sigma] \\ 0, & |\rho| \in (\sigma, \infty]. \end{cases}$$

The function $Q_{\mathbf{c}_i,r}(\mathbf{x})$ is locally defined on all the scattered points $\mathbf{p}_j$ falling in the range of $\|\mathbf{p}_j - \mathbf{c}_j\| \le \sigma$. A practical choice for the support size is $\sigma = 2r$. The position $\mathbf{x}_{\min}$ which minimizes the value of $Q_{\mathbf{c}_i,r}(\mathbf{x})$ can be determined robustly through singular value decomposition (SVD) of the linear equations

$$\partial Q_{\mathbf{c}_i,r}(\mathbf{x})/\partial \mathbf{x} = 0.$$

The value of r is determined by solving the following nonlinear equation

$$Q_{\mathbf{c}_i,r}(\mathbf{x}_{\min}) = (\varepsilon L)^2, \tag{2.3}$$

where L is the length of the main diagonal of the bounding box of the point set P, and ε is an error control threshold. $\varepsilon = 10^{-5}$ is employed here.

3. After obtaining r, whether the point $\mathbf{x}_{\min}$ lies in the sphere $\|\mathbf{x} - \mathbf{c}_i\| \leq r$ is checked. If it does, $\mathbf{x}_{\min}$ serves as an *auxiliary point* of the sphere; otherwise, the sphere's center $\mathbf{c}_i$ is simply assigned as the auxiliary point.
4. Project the set of points in the sphere onto a tangent plane at $(\mathbf{x} - \mathbf{c}_i) \cdot \mathbf{c}_i = 0$, and compute the 2D convex hull of the projections. Points that are not on the boundary of the 2D convex hull are then labeled as covered.

The ASC generated by these steps is triangulated into mesh surfaces. Stimulated by the nerve complex associated with a family of balls [64], a triangle is created for every three spheres if and only if there exists two intersection points of them and at least one of the intersection points is *not* inside other spheres (except these three). The triangle links the auxiliary points in the spheres. Although non-manifold entities may be generated, they can be removed by a mesh cleaning step (details can be found in [57]). The mesh cleaning step also propagates a consistent orientation along the resultant two-manifold mesh surface M.

2.1.3.2 Modified ASC Scheme

The weights defined on points in Eq. (2.1) are based on the average distance to their k-nearest neighbors of $\mathbf{p}_i$. This strategy works according to the non-uniform density of input points. However, such weights do not reflect the distribution of the points around a point $\mathbf{p}_i$. For this reason, the spheres generated by the above ASC scheme are problematic in regions with sparse points where the sparseness is *anisotropic*.

An illustration of such a case is shown in Fig. 2.4a, where only one single line of the spheres is constructed by the ASC scheme. No triangles can be generated around such spheres. As a result, the surfaces in the left and the right regions are separated on the resultant mesh M. The regions are further separated after the cleaning step which removes hanged faces and edges. Figure 2.6a shows such a region on the feet of a human model from scanned raw data. The connectivity for linking these regions is important as breaking the connectivity may make two originally connected regions have opposite orientations; that is, one is flipped whereas another is not. Such an orientation flip fails the downstream surface reconstruction algorithm. To prevent the orientation flip, the best way is to avoid breaking the connectivity of spheres of ASC in anisotropic sparse regions.

First, the spheres in the anisotropic sparse regions are detected. For a sphere $s_{\mathbf{c}_i}$ centered at $\mathbf{c}_i$, the spheres centered at $\mathbf{c}_j$ intersecting with $s_{\mathbf{c}_i}$ form its neighboring sphere set. The neighbor-voting tensor for the distribution of spheres around $s_{\mathbf{c}_i}$ can be computed by

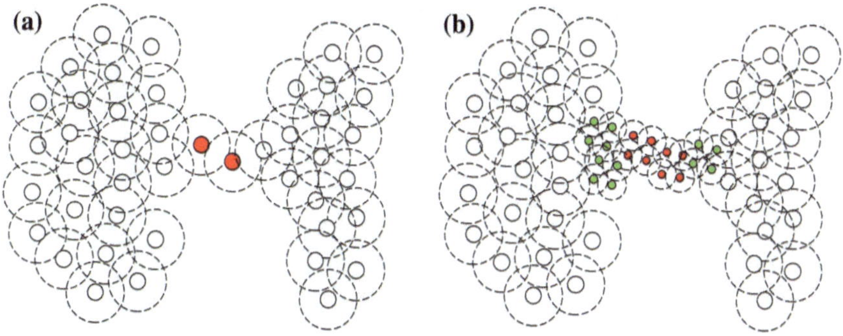

Fig. 2.4 An illustration of spheres generated in anisotropic sparse regions: **a** problematic spheres generated by the ASC scheme [57]—the *red points* and their corresponding spheres, and **b** the problem can be solved by our sphere splitting scheme. In **b**, the *red spheres* are split from the spheres in the anisotropic sparse regions, and the *green* ones are split from the neighboring spheres of the detected spheres

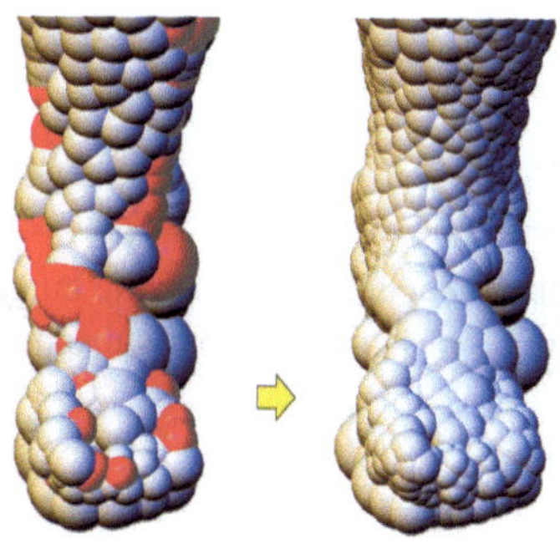

Fig. 2.5 Spheres in anisotropic regions are detected and displayed in *red*. These spheres are split into subspheres to improve robustness (Image taken from [49] with permission)

$$F_{\mathbf{c}_i} = \sum_j (\mathbf{c}_j - \mathbf{c}_i)(\mathbf{c}_j - \mathbf{c}_i)^T, \tag{2.4}$$

which is a 3×3 matrix. After computing the eigenvalues ($|\lambda_1| \geq |\lambda_2| \geq |\lambda_3|$) of the neighbor-voting tensor $F_{\mathbf{c}_i}$, if the other two eigenvalues are trivial compared with the one with the greatest magnitude, the distribution of spheres around $s_{\mathbf{c}_i}$ should be anisotropic. In practical implementation, spheres in anisotropic distributed regions can be detected by checking if $|\lambda_1| > \mu|\lambda_2|$ with $\mu \in [2.0, 5.0]$. $\mu = 3.0$ is employed for all examples shown here. Figure 2.5 shows the sphere detection on a foot example of Fig. 2.6.

In the anisotropic sparse regions, spheres that cannot preserve the connectivity of the regions on their two sides during triangulation and mesh cleaning, are modified by a splitting scheme. Every such sphere is split into four subspheres. For a sphere $s_{\mathbf{c}_i}$ in the anisotropic region, its center $\mathbf{c}_i$ and its unit normal vector $\mathbf{n}_i$ at the center are defined. If ν_1 is the eigenvector of its neighbor-voting tensor $F_{\mathbf{c}_i}$ corresponding to the largest eigenvalue $|\lambda_1|$, ν_1 is first projected onto the tangent plane $(\mathbf{x}-\mathbf{c}_i)\cdot\mathbf{n}_i = 0$ at $\mathbf{c}_i$.

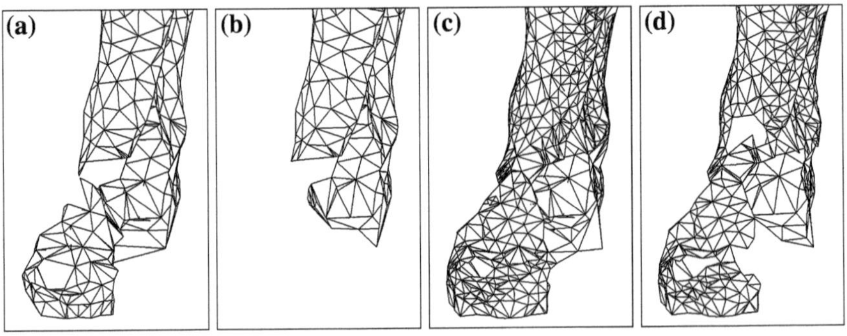

Fig. 2.6 A comparison of mesh surfaces generated by the original and the modified ASC schemes: **a** a triangular mesh generated by ASC [57], **b** the connectivity on separated regions is damaged after applying the cleaning step, **c** the triangular mesh generated by the modified adaptive spherical cover, and **d** the connectivity on separated regions is preserved after the cleaning step (Image taken from [49] with permission)

$$\mathbf{v}_a = \frac{\nu_1 - (\nu_1 \cdot \mathbf{n}_i)\mathbf{n}_i}{\|\nu_1 - (\nu_1 \cdot \mathbf{n}_i)\mathbf{n}_i\|} \tag{2.5}$$

Its orthogonal vector on the tangent plane can then be found by the cross-product as $\mathbf{v}_b = \mathbf{n}_i \times \mathbf{v}_a$. The four subspheres split from $s_{\mathbf{c}_i}$ are located at

$$\mathbf{c}_i \pm \frac{r}{2}\mathbf{v}_a \pm \frac{r}{2}\mathbf{v}_b + \varepsilon \tag{2.6}$$

with r being the radius of $s_{\mathbf{c}_i}$ and ε being a small random perturbation for each sphere. $\frac{r}{2}$ is assigned as the radius of these subspheres. To enhance the robustness of this splitting scheme, we also split the spheres adjacent to (that is, intersecting) the detected spheres in the anisotropic distributed regions. An illustration is given in Fig. 2.4b.

The points of a given point cloud covered by the sphere $s_{\mathbf{c}_i}$ are checked to see if they are covered by a new subsphere. The auxiliary point in a new sphere is assigned by finding a point minimizing the quadratic-error function defined in Eq. (2.2). In addition, the new $\mathbf{x}_{\min}$ must be detected to see whether it is inside the corresponding sphere. If it is not, the sphere's center will simply be used as the auxiliary point. Note that, there is no guarantee that all the points originally covered by $s_{\mathbf{c}_i}$ will be covered by the new spheres. However, this does not affect the later meshing and normal estimation steps of this algorithm. From the foot model example shown in Fig. 2.6c, d, we can easily find that the modified ASC scheme with sphere splitting step can better preserve the original shape of the underlying surface.

2.1.3.3 Orienting Unorganized Points

After triangulating auxiliary points in the modified adaptive spherical cover into a triangular mesh and cleaning its topology, a rough mesh surface M is obtained for approximating the underlying surface represented by the input scattered data point P. Although M does not accurately approximate the shape of the underlying surface S, it gives a very good estimation of S's topology. Therefore, a good estimation of the orientation on S for the points in P can be found from M.

Basically, there are three ways to determine the normal vectors of the points in the given set $P = \{\mathbf{p}_1, \ldots, \mathbf{p}_n\}$. All of them need to first search for the closest point $\mathbf{c}_{p_i}$ on the rough mesh surface M to every scattered point $\mathbf{p}_i$. The normal vector of $\mathbf{c}_{p_i}$ on M, $\mathbf{n}_{c_{p_i}}$, can be obtained at the same time.

1. *Direct Transfer*. Simply assign $\mathbf{n}_i = \mathbf{n}_{c_{p_i}}$.
2. *Direct Flipping*. With the precomputed unit normal vector $\mathbf{n}_i$ using PCA analysis, if $\mathbf{n}_{c_{p_i}} \cdot \mathbf{n}_i < 0$, let $\mathbf{n}_i = -\mathbf{n}_i$.
3. *Orientation-aware PCA*. For the k neighboring points $\mathbf{p}_j$ around $\mathbf{p}_i$, only the points which satisfy $\mathbf{n}_{c_{p_j}} \cdot \mathbf{n}_{c_{p_i}} \geq 0$ are included to be applied with a PCA to calculate a new normal vector $\mathbf{n}_i^{new}$ on $\mathbf{p}_i$.

By experimental tests, it is found that the orientation-aware PCA scheme gives better results in reconstruction (e.g., by using the RBF-based approach [58]).

The only issue left unsolved now is how to efficiently find the closest point on M to the scattered point $\mathbf{p}_i$. A general solution for closest point search is to establish a bounding volume hierarchy (BVH) (such as the swept sphere volume hierarchy (SSVH) presented in [42]). However, as the closest point search problem has its own characteristic, a faster approach should be developed rather than using some general methods like SSVH.

Every vertex on the rough mesh M is always covered by spheres of the ASC. Moreover, most scattered points in S must also be covered by the spheres from ASC though there are few escapees after the adaptive sphere splitting. Therefore, for a point $\mathbf{p}_i$ from S, if it is covered by a sphere ϱ, its closest point is searched on the triangles of a set T_{set} that T_{set} includes all the triangles with at least one vertex enclosed by ϱ. If the vertex enclosed by ϱ is removed during the mesh cleaning process, the triangles adjacent to the vertices enclosed by ϱ_{neigh} are included in T_{set}. Here, ϱ_{neigh} are spheres intersecting ϱ. If the point $\mathbf{p}_i$ is not covered by any sphere, the sphere which covers the closest neighbor of $\mathbf{p}_i$ is selected as ϱ to determine the set of triangles, T_{set}. In the experimental tests shown in [49], the search conducted by SSVH has been compared with this local search scheme. The results show that this local search scheme is three to four times faster than SSVH as the construction of SSVH takes a lot of time while this local search only uses the ASC structure which has already been established.

The example shown in Fig. 2.7 is a human model captured by a laser-based human scanning system with only two columns of laser scanners. Different from the conventional system with four or six columns, the points are roughly two pieces—one is captured from the front and another from the back. Reconstructions from the Tight

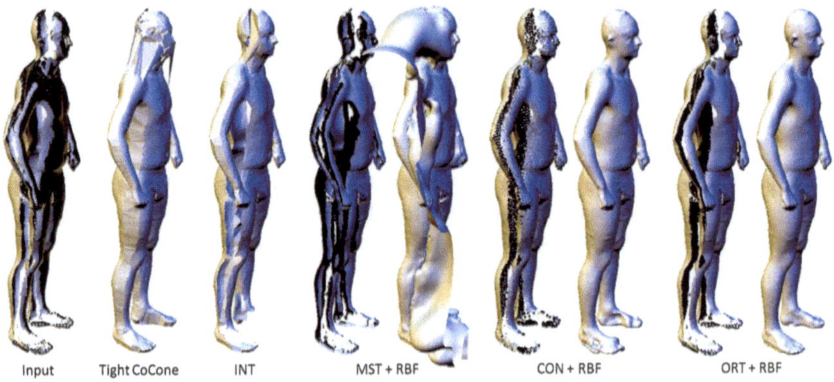

Fig. 2.7 Surface reconstruction results on a human model with 170 k points by different approaches. The oriented normals obtained from MST, CON and ORT are displayed in a shaded mode (Image taken from [49] with permission)

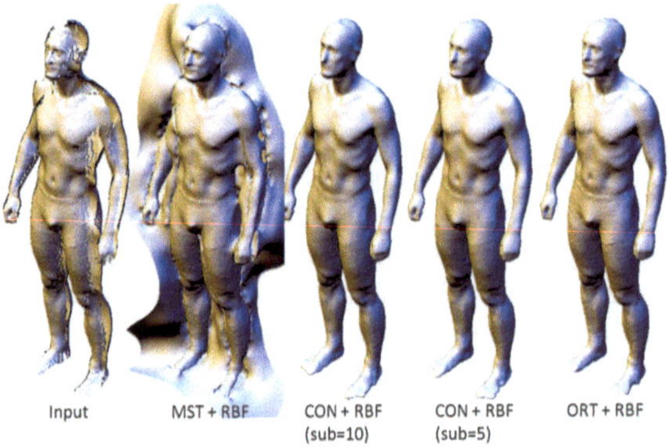

Fig. 2.8 Surface reconstruction results on a human model with 85.8 k points. For the consolidation based point processing (CON), 'sub' denotes the subsampling ratio (that is the number of input points each particle is sampled from). $\mu = 0.45$, $h = 2.68$ and 50 iterations are used in the tests of CON [29] (Image taken from [49] with permission)

CoCone [15] and the integrating approach (INT) [57] do not give a good surface approximation at places where no scan point is present. The conventional normal orienting method based on a MST [26] generates flipped orientation on the front and back parts of a human model, so the resultant implicit surface from an RBF-based reconstruction [58] has incorrect topology. The surface generated from the down-sampled particles and their normals from the consolidation scheme (CON) [29] gives a very poor shape to the right foot of the human body because the points are highly sparse in that region, and the down-sampling strategy from [29] further damages the already limited number of data points there. The parameter h used in the consolidation here is chosen by a trial-and-error procedure—several values are tried for h and the one with the best reconstruction result is displayed (with $h = 1.6$). Furthermore,

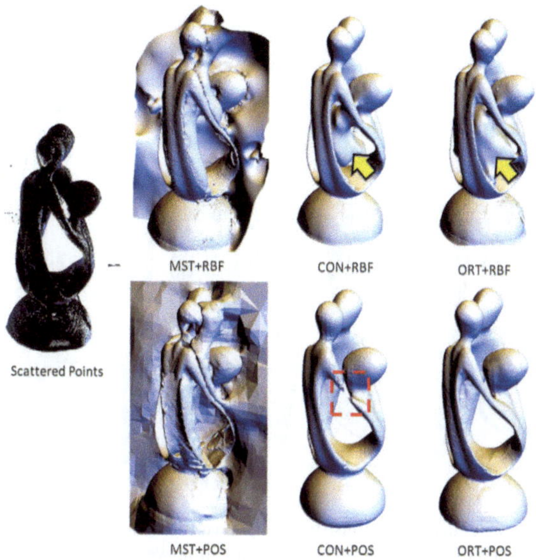

Fig. 2.9 Surface reconstruction results on incomplete scattered data of a model with outliers and structured noise. *Top row* Surfaces reconstructed by RBF-based reconstruction [58] where unreasonable surface parts (pointed by the *yellow arrow*) are generated by CON + RBF and ORT + RBF. *Bottom row* Surfaces generated by Poisson surface reconstruction [34] where only the result from ORT + POS is satisfactory (Image taken from [49] with permission)

when testing the consolidation approach [29], in order to estimate normals correctly, the normal propagation and the orientation-aware PCA steps are iteratively applied for ten times. Details can be found in [29]. Another example for the reconstruction of a human body is shown in Fig. 2.8.

A major limitation of this ORT approach is that the outliers in the input are retained and assigned with normals. An incomplete model with outliers and structured noise is tested on a few approaches and shown in Fig. 2.9. It seems that none of the approaches can generate satisfactory results when using an RBF-based surface reconstruction. When applying the Poisson surface reconstruction approach, only the oriented normal vectors generated by our approach (ORT) can give a satisfactory result (see the bottom row of Fig. 2.9).

2.1.4 Iterative Consolidation of Unorganized Points

A point cloud processing framework—iterative consolidation (ICO) [48] is introduced in this section to improve the quality of point clouds and thus improve the quality of reconstructed surfaces. The input of this approach is an unorganized point cloud which may contain outliers, noise and non-uniformities. Based on the point

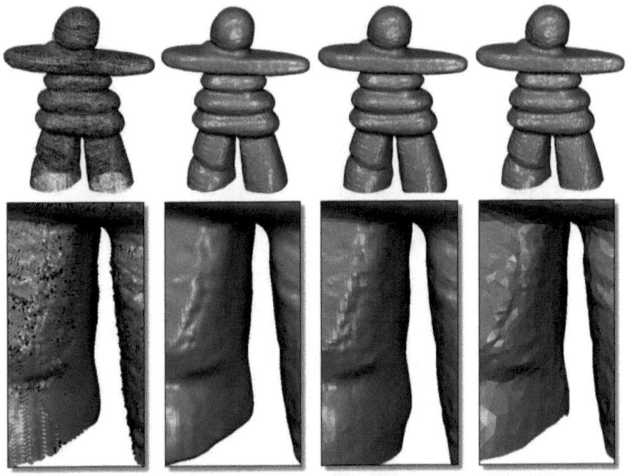

Fig. 2.10 The quality of surfaces reconstructed by various methods from a point cloud processed by the iterative consolidation approach (ICO) are all improved. The oriented normal vectors are generated by ORT (Sect. 2.1.3) for the surface reconstruction approaches that need consistently oriented normal vectors (Image taken from [48] with permission ©2012 IEEE)

positions alone, input points are processed to be evenly distributed by inserting samples into sparse regions using the down-sampling, up-sampling, and selection mechanism. Outliers can be removed by a mean-shift based particle selection operator. As the consequence of applying a novel particle repulsion operator in the framework, missed regions on a given point set can be extrapolated by the newly inserted sample points. The resultant point cloud processed by ICO and the surfaces reconstructed by POS, RBF, and INT are shown in Fig. 2.10.

Given an unorganized set $P = \{\mathbf{p}_j\} \subset \Re^3$ presenting noise, outliers and non-uniformities, surface reconstruction from these may cause significant misinterpretation of the data, thus leading to an erroneous surface. The iterative consolidation framework presented in this section aims to recover the structural information of P without losing the geometric details by inserting points in sparse regions to make the points evenly distributed, adding points into missed regions with a large area, and removing outliers that are far away from the up-sampled surface.

This point processing method is inspired by an image completion approach based on multi-resolution techniques [18]. Their method is based on the observation that a lower resolution representation of an image contains stronger structural information while a higher resolution representation contains more details. Therefore, structural information is recovered at a lower resolution. The structural and non-structural information on 3D models represented by a set of sample points is analogous. In the iterative consolidation framework, the points in P are first down-sampled into k particles, which are redistributed to the surface defined by the samples in P. The redistribution of the particles is performed by iteratively applying the weighted locally optimal

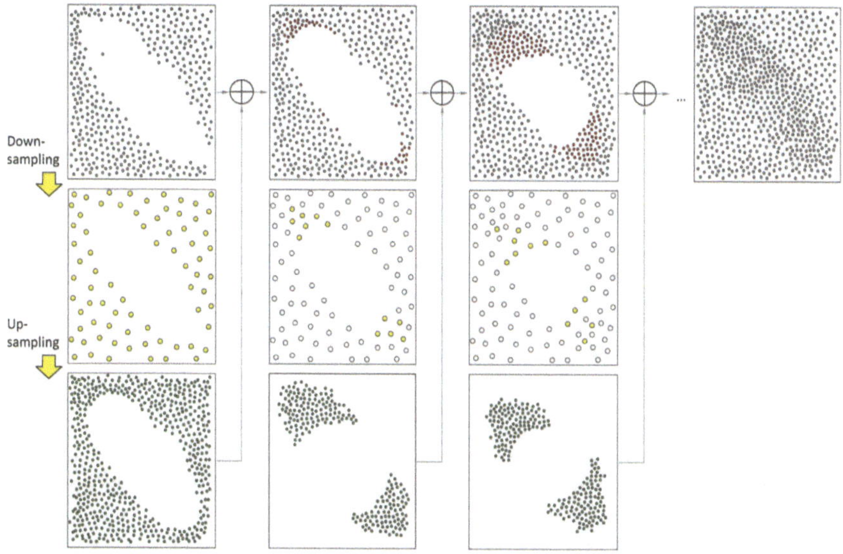

Fig. 2.11 Down-sampling and up-sampling are iteratively and adaptively taken in the framework to consolidate an input point set. The given unorganized points (in small *white dots*) are first down-sampled into particles (in *yellow*) and redistributed, and then up-sampled into a dense point set (in small *green dots*). Among the points generated by the up-sampling, the ones falling in regions that are lack of samples in the given point set are selected and added into the point set. The newly added points are illustrated by small *red dots*. In the next iteration, the points added from up-sampling are down-sampled into new particles (*yellow* ones in the second and the third columns), redistributed, and up-sampled into new points (the *green dots* in the pictures of the last row)

projection (WLOP) operator in [29]. Outliers are removed by applying a mean-shift based selection operator to the particles. The oriented normals of particles can be estimated on the cleaned particles by ORT or MST, and a new repulsion operator based on the algebraic point set surface (APSS) [23] is then applied to extrapolate the surface by pushing particles into the missed regions. After that, an up-sampling step is performed to refine redistributed particles into a smooth point set surface by a $\sqrt{3}$-like interpolatory refinement scheme [22]. The newly generated sample points are selectively merged into the given point set P, while the points in P are considered as outliers and are removed if they are far away from the up-sampled points (that is, a smooth surface interpolating the redistributed particles). The down-sampling, repulsion, up-sampling, and selection steps are repeatedly applied to the point set. The iteration stops when only a few new points are inserted into the point set P.

Nevertheless, the repeated application of down-sampling and up-sampling to the whole set of point samples wastes a lot of time in the regions that have been processed in the previous iteration steps. To reduce redundant computations, an adaptive up-sampling strategy is conducted here. As illustrated in Fig. 2.11, only the newly added points are down-sampled into particles while retaining the particles used in the previous iterations, and only the newly added particles are up-sampled into new points.

Algorithm 1 Iterative-Consolidation

$P^0 \Leftarrow P$ and $i \Leftarrow 0$;
Initialize a particle set X by down-sampling all points of P into m particles;
repeat
  $X \Leftarrow X \cup X^i$;
  Repeatedly move the particles in X^i by the WLOP operator[1];
  if $i = 0$ **then**
    Remove the outlier particles from X^0 by a mean-shift based selection operation;
  end if
  Estimate the orientation of particles by ORT (Sect. 2.1.3);
  Apply the repulsion operator based on APSS to all particles in X^i;
  Refine the points in X^i into a set of up-sampled points Υ^i;
  if $i = 0$ **then**
    Remove the outliers in P according to Υ^0;
  end if
  Select the points of Υ^i into a subset P^{i+1};
  $P \Leftarrow P \cup P^{i+1}$ and $i \Leftarrow i + 1$;
  Down-sample all points of P^i into X^i with m^i particles ($m^i = 2m|P^i|/|P^0|$ with $|\cdots|$ being the number of points);
until the terminal condition is reached
Estimate the consistently oriented normals on the sample points in P by ORT (Sect. 2.1.3);
return P;

[1]The computation will also use the particles in $(X \setminus X^i)$ although only particles in X^i are moved.

The pseudo-code of this point processing algorithm is listed in Algorithm 1. A hybrid terminal condition is employed for the iterations: (1) $\frac{|P^{i-1}|-|P^i|}{|P^{i-1}|} < 20\,\%$ or (2) more than ten iterations have been conducted, where $|\cdots|$ stands for the number of points in a set.

Details of each operator are presented below.

Position Update in WLOP

In WLOP, every particle $\mathbf{x}_i \in X^k$ is moved to a new position by the formula below. The update of position consists of two terms, where the first term attracts the particle to the given point set by the weighted local density

$$v_j = 1 + \sum_{\mathbf{p}_l \in (P \setminus \{\mathbf{p}_j\})} \theta(\|\mathbf{p}_j - \mathbf{p}_l\|)$$

and the second term repulses a particle away from other particles by the density

$$w_p = 1 + \sum_{\mathbf{x}_q \in (X \setminus \{\mathbf{x}_p\})} \theta(\|\mathbf{x}_p - \mathbf{x}_q\|).$$

The updated position of $\mathbf{x}_i$ is

$$\mathbf{x}_i = \sum_{\mathbf{p}_j \in P} \mathbf{p}_j \frac{\theta(\|\mathbf{x}_i - \mathbf{p}_j\|)/v_j \|\mathbf{x}_i - \mathbf{p}_j\|}{\sum_{\mathbf{p}_j \in P} \theta(\|\mathbf{x}_i - \mathbf{p}_j\|)/v_j \|\mathbf{x}_i - \mathbf{p}_j\|} + \mu \sum_{\mathbf{x}_q \in (X \setminus \{\mathbf{x}_i\})} (\mathbf{x}_i - \mathbf{x}_q) \frac{w_q \theta(\|\mathbf{x}_i - \mathbf{x}_q\|)/\|\mathbf{x}_i - \mathbf{x}_q\|}{W} \tag{2.7}$$

where $W = \sum_{\mathbf{x}_q \in (X \setminus \{\mathbf{x}_i\})} w_q \theta(\|\mathbf{x}_i - \mathbf{x}_q\|)/\|\mathbf{x}_i - \mathbf{x}_q\|$, $\| \cdots \|$ is the L^2-norm, and $\theta(r) = e^{-16r^2/h^2}$ is adopted as in [29]. $\theta(r)$ is a rapidly decreasing smooth weight function with the support radius h defining the size of the influenced neighborhood. $\mu \in [0, 0.5)$ and h are served as two parameters selected by users to tune the performance of the operator. During experimental tests in [48], the default values $\mu = 0.45$ and $h = 2L_{\text{avg}}$ work well on all models with L_{avg} being the average distance between particles and their k-nearest neighboring particles. $k = 20$ is usually chosen to balance speed and robustness.

2.1.4.1 Down-Sampling and Relaxation

For a given point set P^i, m^i points are randomly selected to form a set X^i. The points in X^i are called *particles*, and $m^i = \frac{2m|P^i|}{|P^0|}$ with m being a user specified parameter—$m = \frac{1}{20}|P^0| \sim \frac{1}{5}|P^0|$ is usually used. These m particles are then iteratively moved to a new position which is defined by two terms, where the first term attracts a particle to the given point set and the second term repulses a particle from other particles. Such a position update is called WLOP.

When $i \neq 0$, only the particles in X^i are moved. All other particles in X are involved in the computation but with their positions fixed. Another parameter used in this down-sampling and relaxation step is the number of particles, m. As studied in [29], using too few particles could easily damage the existing small features on the given point set, while a too large value of m will slow down the computation in relaxation as well as the following repulsion of particles. Therefore, it is suggested to use $m = \frac{1}{20}|P^0|$ for processing a dense point cloud and $m = \frac{1}{10}|P^0|$ for relatively sparse points (Ref. [48]).

2.1.4.2 Outlier Removal

Although the WLOP operator can efficiently filter out noise by the robust down-sampling and relaxation, it, however, cannot remove outliers that are far away from the real surface of a model—see the middle-left of Fig. 2.12 for an example. These unwanted particles will potentially affect the quality of consolidation in the down-stream operations and are considered to be removed. According to the nature that

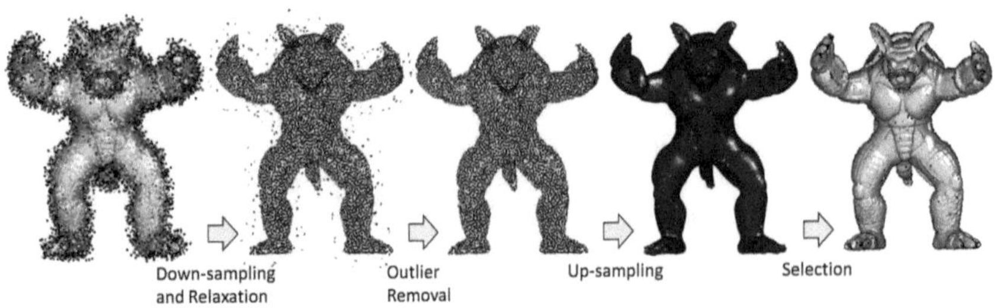

Fig. 2.12 Mean-shift based outlier removal on down-sampled and relaxed particles: an input point set with outliers (*leftmost*), the particles after down-sampling and relaxation (*middle-left*), outlier particles removed (*middle*), the up-sampling result (*middle-right*), after removing points that are far away from the up-sampled point set (*rightmost*)

these outliers are usually far away from the majority of their neighbors, outlier particles can be detected and removed by a mean-shift based method during the first iteration of consolidation.

Each particle $\mathbf{x}$ in X is iteratively shifted to the average position of its k-nearest neighboring particles ($k = 20$). Then, the distance from the shifted mean position to the original position of $\mathbf{x}$ is compared with the average distance between its shifted neighbors. If the difference between two distances is significant, $\mathbf{x}$ is considered as an outlier particle. Pseudo-code of the mean-shift based outlier removal algorithm can be found in Algorithm 2, by which the outlier particles are removed from X. This outlier removal step is only performed in the first iteration of the iterative consolidation algorithm.

Note that, the average distance between shifted particles is computed locally so that it is adaptive to the non-uniform distribution of particles. Based on experimental tests, choosing $b = 3$ and $s = 3$ for Algorithm 2 gives a good balance between speed and quality. See Fig. 2.12 for an example. The mean-shift iteration converges very fast; therefore, using a larger b may not change the result of outlier identification and removal.

Simply deleting the outlier particles cannot successfully remove the outliers from the input point set P since the outlier particles are only a small subset of the points in P. A simple selection step can be performed to remove outliers from P. After obtaining a set of cleaned particles and then up-sampling them back into points in Υ^0 as the samples of a smooth surface, the points in P which are far away from the surface represented by Υ^0 are considered as outliers. In short, the criterion is:

- $\forall \mathbf{p}_i \in P$, it is removed from P when $\|\mathbf{p}_i - \mathbf{q}_j\| > h$ $(\forall \mathbf{q}_j \in \Upsilon^0)$.

This simple selection operation can effectively remove outliers embedded in the given point cloud (see the rightmost of Fig. 2.12 for an example).

Algorithm 2 Outlier-Particle-Removal

for all $\mathbf{x}_i \in X$ **do**
  Initialize $\mathbf{x}'_i \Leftarrow \mathbf{x}_i$
  for $j = 1$ to b **do**
    Search in X the k-nearest neighbors of $\mathbf{x}'_i$ and let them be the subset K;
    $\mathbf{x}'_i \Leftarrow \frac{1}{|K|} \sum_{\forall \mathbf{k}_i \in K} \mathbf{k}_i$;{Mean-shift step}
  end for
end for
for all $\mathbf{x}_i \in X$ **do**
  Search in X the k-nearest neighbors of $\mathbf{x}'_i$ and let them be the subset K;
  $\bar{d}_{\mathbf{x}'_i} \Leftarrow \sum_{\forall \mathbf{k}_i \in K} \frac{\|\mathbf{k}_i - \mathbf{x}'_i\|}{|K|}$;{The average distance}
  if $\|\mathbf{x}'_i - \mathbf{x}_i\| > s\bar{d}_{\mathbf{x}'_i}$ **then**
    $\mathbf{x}_i$ is considered as an outlier particle;
  end if
end for
Remove all outlier particles from X;
return X;

APSS Fitting

Using $\mathbf{u} = [u_0, \ldots, u_4]$ as the vector of scalar coefficients describing a general algebraic sphere in $\Re^3$, the solution of algebraic sphere fitting at a given point $\mathbf{x} \in \Re^3$ can be evaluated by

$$\mathbf{u} = \arg \min_{\mathbf{u}, \mathbf{u} \neq \mathbf{0}} \|\mathbf{W}^{\frac{1}{2}}(\mathbf{x})\mathbf{u}\|^2. \tag{2.8}$$

The solution of $\mathbf{u}$ can be found by solving the following linear equation system (details can be found in [23]).

$$\mathbf{W}^{\frac{1}{2}}(\mathbf{x})\mathbf{D}\mathbf{u} = \mathbf{W}^{\frac{1}{2}}(\mathbf{x})\mathbf{b} \tag{2.9}$$

The coefficient matrices have $4n$ rows where n is the number of particles.

$$\mathbf{W}(\mathbf{x}) = \begin{bmatrix} \ddots & & & & & \\ & w_i(\mathbf{x}) & & & & \\ & & \beta w_i(\mathbf{x}) & & & \\ & & & \beta w_i(\mathbf{x}) & & \\ & & & & \beta w_i(\mathbf{x}) & \\ & & & & & \ddots \end{bmatrix}, \mathbf{D} = \begin{bmatrix} \vdots & \vdots & \vdots \\ 1 & \mathbf{x}_i^T & \mathbf{x}_i^T \mathbf{x}_i \\ 0 & \mathbf{e}_0^T & 2\mathbf{e}_0^T \mathbf{x}_i \\ 0 & \mathbf{e}_1^T & 2\mathbf{e}_1^T \mathbf{x}_i \\ 0 & \mathbf{e}_2^T & 2\mathbf{e}_2^T \mathbf{x}_i \\ \vdots & \vdots & \vdots \end{bmatrix} \text{ and}$$

$$\mathbf{b} = \begin{bmatrix} \vdots \\ 0 \\ \mathbf{e}_0^T \mathbf{n}_{\mathbf{x}_i} \\ \mathbf{e}_1^T \mathbf{n}_{\mathbf{x}_i} \\ \mathbf{e}_2^T \mathbf{n}_{\mathbf{x}_i} \\ \vdots \end{bmatrix}$$

Here, the weight function $w_i(\mathbf{x}) = \phi(\frac{\|\mathbf{x}-\mathbf{x}_i\|}{h})$ describes the weight of the particle $\mathbf{x}_i$ for the local evaluation of the APSS at the position $\mathbf{x}$ with

$$\phi(r) = \begin{cases} (1 - r^2)^4 & |r| < 1 \\ 0 & |r| \geq 1 \end{cases}$$

h is the support size of the repulsion where the same value is chosen as the orientation-aware WLOP above and $\{\mathbf{e}_k\}$ represents the unit basis vector of the coordinate system. $\beta = 10^6 h^2$ is adopted as suggested in [23] to compensate the variance of scaling. Note that, in practical computations, only the particles whose distance to $\mathbf{x}$ is less than h are employed to determine the coefficients in $\mathbf{u}$ since the weight function $w_i(\mathbf{x})$ only shows nonzero values on these particles. After computing the $\mathbf{u}$ vector, the center $\mathbf{c}$ and the radius of the algebraic sphere can be calculated as

$$\mathbf{c} = -\frac{1}{2u_4}[u_1, u_2, u_3]^T \text{ and } r = \sqrt{\mathbf{c}^T\mathbf{c} - u_0/u_4}.$$

2.1.4.3 Repulsion of Particles

The WLOP operator can evenly redistribute particles along the surface defined by a given point set P. However, the movement of particles driven by WLOP stops at the boundary of missed regions with a large area (such as the missed region on thighs in Fig. 2.13). The same observation has also been reported in [29]. The repulsion operator introduced below can overcome this drawback of WLOP by moving the particles into the missed regions. The observation shows that the normals of missed regions are always nearly perpendicular to the viewing direction of scanners. The missed regions can be filled by extrapolating the surface (defined by the existing particles) along its tangential direction.

As the particles in X after the WLOP based relaxation and the mean-shift based outlier removal have become more uniform and less noisy, consistently oriented normal vectors can be estimated on the particles by ORT (or even MST in [26]).

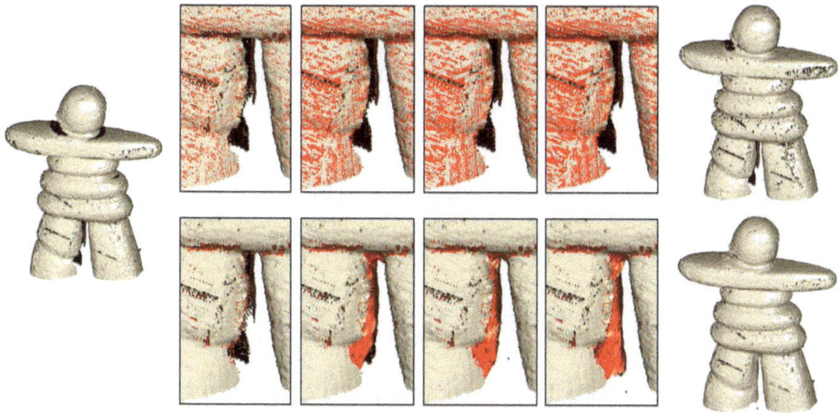

Fig. 2.13 Progressive results without (*top row*) versus with (*bottom row*) using the repulsion operator. The iterative consolidation framework can progressively improve the quality of the point cloud (see the *bottom row*), where the up-sampled points are displayed in *red*

Equipped with the oriented normals, an APSS [23] can be defined by the particles, $\forall \mathbf{x}_i \in X$, and their normal vectors, $\mathbf{n}_{\mathbf{x}_i}$. APSS is a kind of moving least-square surface (MLS). Instead of plane fitting, APSS directly fits higher order algebraic spheres. The advantage is that APSS yields more stable results than planar MLS does at regions with a high curvature (that is, thin and sharp features).

By the APSS defined on $\forall \mathbf{x}_i \in X$, the particles can then be moved along the APSS in a repulsive manner. The movement consists of two components: the tangential component and the projective component. The tangential component of a particle $\mathbf{x}_i$ is determined by rotating it along an axis $\mathbf{r}_{\mathbf{x}_i}$ that passes through the center $\mathbf{c}_{\mathbf{x}_i}$ of the algebraic sphere corresponding to $\mathbf{x}_i$.

- The tangential component of repulsion is derived from the second term of WLOP as

$$\mathbf{l} = \mu \sum_{\mathbf{x}_p \in (X \setminus \{\mathbf{x}_i\})} (\mathbf{x}_i - \mathbf{x}_p) \frac{w_p \rho}{\sum_{\mathbf{x}_p \in (X \setminus \{\mathbf{x}_i\})} w_p \rho} - \mathbf{x}_i \tag{2.10}$$

  with $\rho = \theta(\|\mathbf{x}_i - \mathbf{x}_p\|)/\|\mathbf{x}_i - \mathbf{x}_p\|$. Generally speaking, $\mathbf{l}$ is not perpendicular to the normal vector $\mathbf{n}_{\mathbf{x}_i}$ at $\mathbf{x}_i$. Thus, the corresponding vector orthogonal to $\mathbf{n}_{\mathbf{x}_i}$ is computed by

$$\mathbf{l}' = \mathbf{l} - (\mathbf{l} \cdot \mathbf{n}_{\mathbf{x}_i})\mathbf{n}_{\mathbf{x}_i}. \tag{2.11}$$

- The rotation axis of the tangential component is

$$\mathbf{r}_{\mathbf{x}_i} = \frac{(\mathbf{x}_i - \mathbf{c}_{\mathbf{x}_i}) \times \mathbf{l}'}{\|(\mathbf{x}_i - \mathbf{c}_{\mathbf{x}_i}) \times \mathbf{l}'\|}, \tag{2.12}$$

  and the rotation angle is obtained by

$$\varpi = \frac{\|\mathbf{l}'\|}{2\pi \|\mathbf{x}_i - \mathbf{c}_{\mathbf{x}_i}\|}.$$

The tangential component of the movement is then defined by

$$\mathbf{R}_{\mathbf{r}_{\mathbf{x}_i}}(\varpi)(\mathbf{x}_i - \mathbf{c}_{\mathbf{x}_i}) + \mathbf{c}_{\mathbf{x}_i} - \mathbf{x}_i, \tag{2.13}$$

with $\mathbf{R}_{\mathbf{r}_{\mathbf{x}_i}}(\varpi)$ being the rotation matrix around the axis $\mathbf{r}_{\mathbf{x}_i}$. In order to improve the stability of the particle movement, the rotation angle is restricted by

$$\varpi = \min\left\{\frac{\|\mathbf{l}'\|}{2\pi \|\mathbf{x}_i - \mathbf{c}_{\mathbf{x}_i}\|}, \frac{L_{\text{avg}}}{2\pi \|\mathbf{x}_i - \mathbf{c}_{\mathbf{x}_i}\|}, \frac{\pi}{4}\right\}, \tag{2.14}$$

where L_{avg} is the average distance between particles and their k-nearest neighboring particles (with $k = 20$).

After applying the tangential component on a particle $\mathbf{x}_i$, the ICO algorithm consecutively applies the following projective component on it three times to retain the moved particles on the APSS defined by the particles equipped with normal vectors.

$$\mathbf{c}_{\mathbf{x}_i} + r(\mathbf{x}_i)\frac{\mathbf{x}_i - \mathbf{c}_{\mathbf{x}_i}}{\|\mathbf{x}_i - \mathbf{c}_{\mathbf{x}_i}\|} - \mathbf{x}_i \tag{2.15}$$

The resultant particles obtained by this repulsion operator are distributed into the missed regions. Note that, applying this repulsion once only cannot fill up the whole missed regions with a large area. These regions need to be filled iteratively (see the bottom row of Fig. 2.13 for an example).

2.1.4.4 Up-Sampling and Selection

An up-sampling step is conducted in ICO to generate more sample points on the surface that interpolates the particles in X^i. The up-sampled point set Υ^k is expected to regularize the scattered samples and converge on a smooth surface interpolating the particles and their normals. A good candidate up-sampling scheme that satisfies these requirements is the interpolatory refinement method presented in [22], which is therefore used in the ICO framework for generating Υ^k from X^i.

To fit in with the adaptive down-sampling/up-sampling strategy, only the 'alive' particles (that is, particles that can be moved in the repulsion) and the points up-sampled from the 'alive' particles are used as the centers to generate refined points. Refinement at a center is prevented if it is too close to the neighbors (for instance, less than $1/3$ of the support size h of particles used above). Up-sampling is stopped when no refinement on any center is allowed.

In fact, the repulsion operator and the up-sampling operator work together to generate sample points on a smooth surface extrapolating the particles generated by WLOP. Such an extrapolation helps to improve the uniformity of sample points in highly sparse regions. The points in Υ^k are selected to merge into the point set P^k

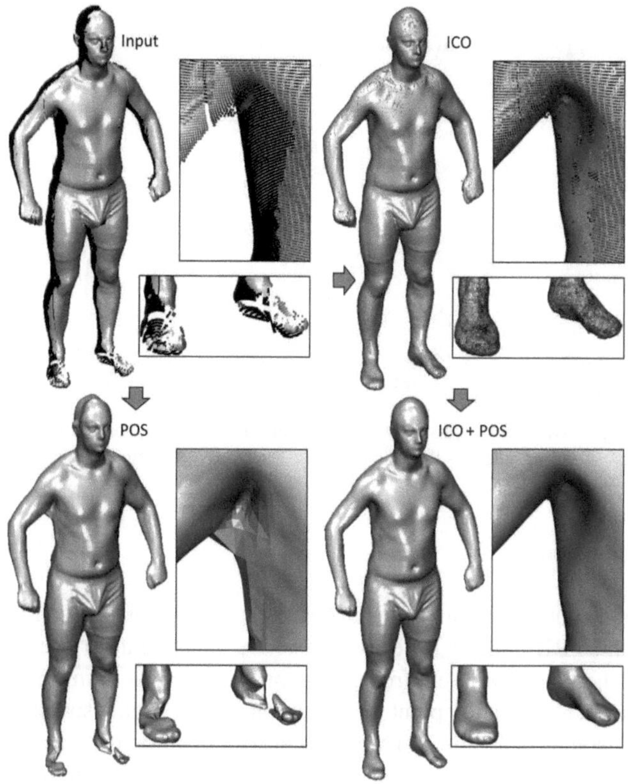

Fig. 2.14 Missed regions on a scanned human model can be repaired by the iterative consolidation framework (ICO)—see the armpit and the feet. Without filling sample points into these regions, the surface reconstruction result (for example, by Poisson method [34]) is poor in highly sparse regions. After being processed by ICO, the quality of reconstruction is significantly improved (Image taken from [48] with permission ©2012 IEEE)

to form a new point set P^{k+1}. In the next iteration of point processing, only the particles down-sampled from these newly inserted points are 'alive' to be moved in the WLOP based relaxation. Therefore, simply adding all the points of Υ^k into P^{k+1} leads to too many particles in the later iterations, which significantly slows down the computation. However, adding too few up-sampled points results in too few 'alive' particles in the next iteration, which also slows down the extrapolation generated by particle repulsion. Υ^k and P^k can be merged by the criterion that:

- $\forall \mathbf{q}_j \in \Upsilon^k, \mathbf{q}_j$ should be excluded from P^{k+1} if $\exists \mathbf{p}_i \in P^k$ with $\|\mathbf{q}_j - \mathbf{p}_i\| < \frac{1}{2} L^k_{\text{avg}}$,

where L^k_{avg} is the average of distances between all samples in P^k to their nearest neighbors. Note that, according to the experimental tests of Liu et al. [48], the quality of processed point sets by using different values for $\frac{1}{2} L^k_{\text{avg}}$ does not have significant differences. The value of $\frac{1}{2} L^k_{\text{avg}}$ mainly affects the speed of computation.

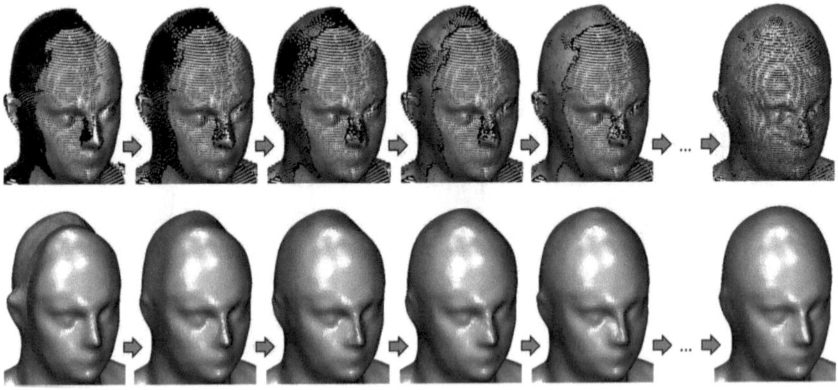

Fig. 2.15 Progressive results show how the points are filled into the missed regions by ICO (*top row*). The corresponding surfaces that can be reconstructed by POS are given in the *bottom row*

2.1.4.5 Discussion

For some point data which are embedded with small holes only, reconstruction algorithms (such as RBF and POS) can successfully generate a surface without being affected by the holes. Different from the reconstruction algorithms, the aim of ICO is to construct a high-quality point set which can recover highly sparse regions while preserving geometric details. High-quality surfaces can be reconstructed from the processed point clouds by a variety of surface reconstruction algorithms (see the example shown in Fig. 2.10). More examples can be found in [48]. For human modeling purpose, ICO has also been tested on some raw data of scanned human bodies. As shown in Figs. 2.14 and 2.15, the missed regions between the front and the back scans can be repaired by ICO. The unoriented raw scans are rendered using the normals computed by the method of Liu and Wang in [49].

Another interesting study is about the robustness of ICO to noise in different levels. As shown in Fig. 2.16, ICO is applied to an Armadillo model embedded with 5, 10, and 20 % noisy points. It is easy to find that the surface reconstruction method generates mesh models that are highly affected by noise without processing the given noisy points. However, after processing the points by ICO, the same surface reconstruction method produces smooth models with geometric details preserved.

2.2 Feature Point Extraction

This section describes the techniques for extracting feature points on human bodies which are reconstructed from scanned points and represented by triangular meshes. The feature points on human bodies are always defined according to applications instead of shape extremities (like the shoulder point), and they serve as anchor

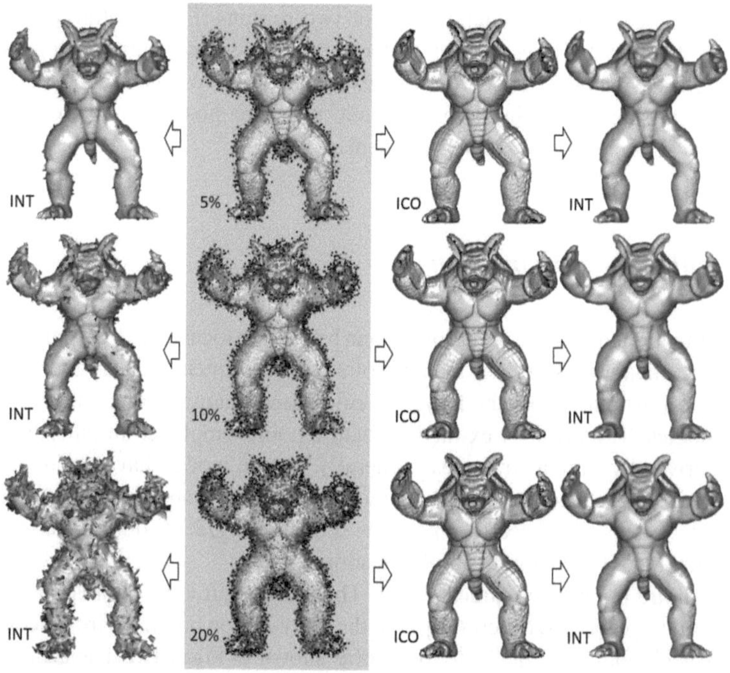

Fig. 2.16 Examples to show the performance of ICO on an Armadillo model with different amounts of noise—noisy points are randomly distributed in the range of 2 % of the bounding box's diagonal length. To demonstrate how the noise affects the mesh reconstruction method, the mesh surfaces generated by the integrating meshing method (INT) [57] are also shown (Image taken from [48] with permission ©2012 IEEE)

points for computing cross-parameterization in Sect. 2.3 to establish correspondences between human bodies. After giving the problem definition of feature point extraction, two approaches are presented in this section: (1) a rule-based extraction method, and (2) an algorithm based on bending-invariant matching. The rule-based method is easy to be implemented while the matching-based algorithm is more robust to the pose variations of scanned human bodies.

2.2.1 Problem Definition

Given a template 3D human model T represented by a triangular mesh surface $M_T \in \Re^3$ with a set of predefined feature points, G_T, the corresponding feature points, G_H, on the surface M_H of an input human model H need to be extracted. Without loss of generality, M_H is also represented by a triangular mesh in $\Re^3$ and both M_T and M_H have surface normals facing outwards.

Automatic extraction is challenging for two reasons. First, feature points on a human body are not always located at shape extremities; therefore, the local shape-matching based methods cannot robustly give satisfactory results. Second, the robustness of local shape matching is problematic when the poses of human bodies are varied (that is, when the 3D bodies are bent).

2.2.2 Rule-Based Extraction

As aforementioned, feature points on human bodies are located in different places for different applications. Here, the feature points in the garment industry are used as an example to illustrate the methodology of rule-based feature point extraction on 3D human bodies. The rules for extracting feature points employed in other industries can be derived by similar strategies. Generally, human models have a wide range of variations in body shape and size. To extract feature points robustly, it is assumed that each model stands in a prescribed pose (as shown in Fig. 2.14) during the scanning, and a fuzzy logic based approach [80] is adopted here.

Key feature points are extracted first. There are eight key features on a human model: the armpit points, the crotch point, the belly-button point, the front neck point, the back neck point, and the busty points (as shown in Fig. 2.17). The basic idea of using the rule-based method to extract these key feature points is as follows.

- Using a set of horizontal planes to slice a 3D human body H into 2D contours;
- Uniformly sampling the 2D contour into a sequence of points $\{\mathbf{p}_i\}$;
- Searching for turning points on the 2D contours by the 'sharp' angles along the contours[1];
- Extracting the key feature points with the help of turning points.

Other feature points can be found from the key features by proportion rules used in the garment industry [73, 75].

Crotch Point. A human body H is sliced downwards from half of its height until the intersection contour breaks into two circles. The crotch point is located at the center of the bounding box of the two legs' contours [80].

Neck Points. The right view of the human body is projected to obtain its silhouette (see Fig. 2.17). The turning point $\mathbf{p}_1$ indicates the position of the front neck point while $\mathbf{p}_2$, which is closest to $\mathbf{p}_1$ on the right boundary of the silhouette, indicates the location of the back neck point.

Belly-button Points. The height from the back neck point to the highest point of the scan is the head's height, h_{head}. The height of belly-button is usually about the height of crotch point, h_{crotch}, plus h_{head}. Thus, by cutting several horizontal planes around the height of $h_{\text{crotch}} + h_{\text{head}}$ (see the right lower part of Fig. 2.17), if a turning point (which is marked as $\mathbf{p}_3$ in Fig. 2.17) is found, it is the belly-button point.

[1] On a 2D polygon, if the positions of three adjacent points $\mathbf{p}_{i-1}$, $\mathbf{p}_i$, and $\mathbf{p}_{i+1}$ make the value of angle $\angle \mathbf{p}_{i-1}\mathbf{p}_i\mathbf{p}_{i+1}$ smaller than a threshold, $\mathbf{p}_i$ is considered as a turning point.

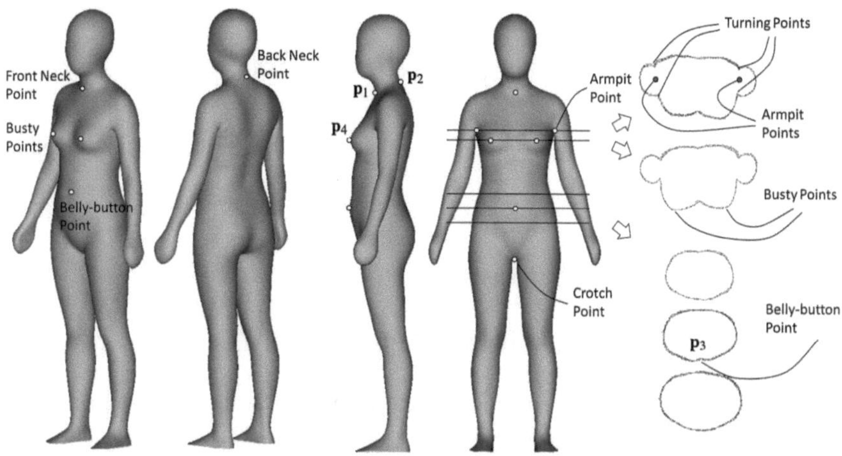

Fig. 2.17 A rule-based method to determine key feature points on a human model

Busty Points. From the turning point $\mathbf{p}_4$ on the silhouette of the right view, the height of busty points, h_{busty}, can be obtained. To determine the exact busty points, the model is sliced at the height of h_{busty} to obtain a contour as shown in Fig. 2.17. The left and the right lowest points on the contour are the two busty points.

Armpit Points After the height of the belly-button point, h_{belly}, and the height of the back neck point, h_{beckneck}, are determined, the height of the underarm point, h_{armpit}, can be computed by an anthropometrical equation:

$$h_{\text{armpit}} = h_{\text{belly}} + 0.55\left(h_{\text{beckneck}} - h_{\text{belly}}\right). \tag{2.16}$$

Slicing the human model horizontally at h_{armpit}, four turning points can be found on the contour. Two Armpit points are in the middle of the front and the back turning points on the cutting plane (see the right upper corner of Fig. 2.17).

Based on the above methods, the key feature points can be automatically determined on more than 85 % of examples. For the examples that fail the automatic extraction, users need to manually locate the key feature points on the scans (which is similar to [2]).

2.2.3 Bending-Invariant Matching Algorithm

This subsection introduces a bending-invariant matching algorithm [44], which is in fact a global deformation based fitting method, to automatically find the location of feature points on an input human body H according to a template human body T. H and T are represented by oriented triangular mesh surfaces, M_T and M_H. As a global alignment technique, a mapping Υ is sought to minimize the distortion function E

as

$$E(\Upsilon) = \int \|M_H - \Upsilon(M_T)\|^2 ds \tag{2.17}$$

with $\| \cdots \|$ being the L^2-norm in $\Re^3$. In other words, with the optimal mapping function

$$\Upsilon = \arg\min E(\Upsilon), \tag{2.18}$$

the feature points can be determined by

$$G_H = \{\mathbf{g} \mid \mathbf{g} = \Upsilon(\mathbf{q}), \forall \mathbf{q} \in G_T\}. \tag{2.19}$$

To have a refined matching, some important points in G_H should have local shape distributions similar to their corresponding points in G_T. This serves as constraints for the minimization problem defined in Eq. (2.18).

This bending-invariant matching algorithm integrates *Global Surface Alignment* and *Feature-Based Matching* techniques, which are found to be two major types of techniques in shape-matching studies. The integration of these two techniques inherits their advantages. The matching algorithm has three major steps: (1) pose alignment, (2) surface fitting and (3) feature matching refinement.

- First, the *pose alignment* step transforms the template model to the input model non-rigidly based on the control point mapping defined by the models' similar isometric signatures—the multidimensional scaling (MDS) embedding. The MDS embedding of a given model is defined in a k-dimensional domain according to the relative distribution of surface points on the model. The robustness of finding good initial correspondences according to the MDS embedding is guaranteed by the observation that (1) shapes of the same human body in different poses are nearly isometric to each other and (2) isometric shapes have the same MDS embedding.
- Second, the *surface fitting* step refines the surface of the transformed template by optimizing the fitness and the smoothness iteratively. Two main processes, *surface fitness optimization* and *surface smoothing*, are repeatedly applied until changes on the surface converge to a limited amount. The surface fitting procedure employs a bidirectional mapping concept and an orientation-aware movement to improve the fitting quality of the template model.
- Third, the *feature matching refinement* step further refines the correspondences by adopting feature descriptor constraints on particular surface regions. At this stage, the descriptor is encoded on a surface point with curvature distribution information on the surface around it. The concept is similar to the *Curvature Maps* presented by Gatzke et al. [21] but in a constrained manner. Hence, the predefined feature points on the template model can be mapped to the input model according to the feature-aligned models.

Unlike the approaches that require input models to be two-manifold (such as the recently developed method by Möbius transformations [47]), this algorithm is based on sampling points and spatial transformation.

In the rest of this section, the methods for computing the MDS embedding and solving the sign-flip problem are presented first. After that, the three major steps of this bending-invariant matching algorithm are detailed.

2.2.3.1 MDS Transformation

To robustly establish the initial correspondences between the template human model T and the input model H, their MDS embeddings T_{MDS} and H_{MDS} are computed via the classical MDS transformation, which involves a computationally expensive step—eigenvector analysis. In order to simplify and speed up the computation, their surface models M_T and M_H are sampled into m points, $\tilde{M}_T = \{\mathbf{t}_1, \ldots, \mathbf{t}_m\}$ and $\tilde{M}_H = \{\mathbf{h}_1, \ldots, \mathbf{h}_m\}$, by the farthest point sampling (FPS) method [53]. According to this simplified shape representation, when the samples of $\tilde{M}_T$ are mapped to new positions, for example, $\tilde{M}_T^* = \{\mathbf{t}_1^*, \ldots, \mathbf{t}_m^*\}$, the newly mapped (or warped) shape of M_T can be determined by a warping function based on radial basis function (RBF) (see [81, 83] for reference).

To compute MDS transformation, the geodesic distances between the sample points on a human body are calculated by the fast Marching algorithm on triangulated domains [36]. The result is stored as an $m \times m$ symmetric and zero diagonal matrix $\mathbf{D}$

$$\mathbf{D} = \{d_{ij}\} = \{\xi^2(i, j)\} \tag{2.20}$$

where $\xi(i, j)$ evaluates the geodesic distance between the sample points $\mathbf{t}_i$ and $\mathbf{t}_j$ on M_T (or between $\mathbf{h}_i$ and $\mathbf{h}_j$ on M_H). The MDS process is an important step to align a pair of given models into the same orientation, scale and pose. Here, the Gaussian affinity matrix $\mathbf{A}$ is used as the input of MDS transformation which is similar to [30] as

$$\mathbf{A} = \{a_{ij}\} = \{1 - e^{-\frac{d_{ij}}{2\delta}}\}. \tag{2.21}$$

Shape Warping Based on RBF

Given a set of points $\{\mathbf{p}_1, \ldots, \mathbf{p}_n\}$, each of which is mapped to a point $\Re^3$ as $\mathbf{p}_j \mapsto \mathbf{q}_j$ $(j = 1, \ldots, n)$, a continuous warping function $\Psi(\mathbf{x})$ can be established based on RBF in $\Re^3$ while interpolating the above mapping constraints. Without loss of generality, an RBF is represented in a piecewise form

$$\Psi(\mathbf{x}) = \mathbf{a}_n + [\mathbf{a}_{n+1}, \mathbf{a}_{n+2}, \mathbf{a}_{n+3}]\mathbf{x} + \sum_{i=1}^{n} \mathbf{a}_i g(\|\mathbf{x} - \mathbf{p}_i\|), \tag{2.22}$$

where the vector coefficients $\mathbf{a}_n, \ldots, \mathbf{a}_{n+3} \in \Re^3$ define the affine transformation of the point $\mathbf{x}$, $\mathbf{a}_i$s define the weights of the point $\mathbf{x}$ to the knots $\mathbf{p}_i$, and the basis function $g(r)$ is chosen as the Gaussian, the thin-plate spline $g(r) = r^2 \ln(r)$, or the polyharmonic spline $g(r) = r^k$ $(k = 1, 3, 5, \ldots)$. The warping function $\Psi(\mathbf{x})$ can be computed by determining the coefficients $\mathbf{a}_j$ using the constraints

$$\Psi(\mathbf{p}_i) = \mathbf{q}_i \ (i = 1, \ldots, n)$$

together with the compatibility conditions

$$\sum_i \mathbf{a}_i = \sum_i \mathbf{a}_i \mathbf{p}_i^x = \sum_i \mathbf{a}_i \mathbf{p}_i^y = \sum_i \mathbf{a}_i \mathbf{p}_i^z = \begin{bmatrix} 0 \\ 0 \\ 0 \end{bmatrix}.$$

This leads to the following linear equation system

$$\begin{bmatrix} \mathbf{G} & \mathbf{P}^T \\ \mathbf{P} & \mathbf{0} \end{bmatrix} [\mathbf{a}_i] = \begin{bmatrix} \mathbf{Q} \\ \mathbf{0} \end{bmatrix}, \tag{2.23}$$

where $\mathbf{G} = [g_{ij}]$ with $g_{ij} = g(\|\mathbf{p}_i - \mathbf{p}_j\|)$, $\mathbf{P}^T = [1 \ \mathbf{p}_i^x \ \mathbf{p}_i^y \ \mathbf{p}_i^z]$, and $\mathbf{Q} = [\mathbf{q}_i^x \ \mathbf{q}_i^y \ \mathbf{q}_i^z]$. The system is symmetric and positive definite unless all $\mathbf{p}_i$ are coplanar, which rarely happens in practice and can be easily detected. In some scenarios, the interpolation constraints can be released to the approximation constraints $\Psi(\mathbf{p}_i) \approx \mathbf{q}_i$ by changing the above linear equation system to

$$\begin{bmatrix} (\mathbf{G} - \alpha \mathbf{I}) & \mathbf{P}^T \\ \mathbf{P} & \mathbf{0} \end{bmatrix} [\mathbf{a}_i] = \begin{bmatrix} \mathbf{Q} \\ \mathbf{0} \end{bmatrix} \tag{2.24}$$

with α being a regularization (smoothing) parameter of the warping function [83]. The greater α is, the stronger its smoothing effect is. Notice that, the x-, y-, and z-components of the vector coefficients, $\mathbf{a}_i$s, can be determined independently. In other words, the dimension of linear system in Eqs. (2.23) and (2.24) is $(n + 4)$ instead of $3(n + 4)$.

The maximum value among the elements of $\mathbf{D}$ is chosen as the width, δ, of the Gaussian kernel. The positions of m sample points in the k-dimensional MDS embedding domain can then be computed by building and decomposing the inner product matrix

$$\mathbf{B} = -\frac{1}{2}\mathbf{JAJ} \tag{2.25}$$

where $\mathbf{J} = \mathbf{I} - \frac{1}{m}\mathbf{1}\mathbf{1}^T$ and $\mathbf{1}_{1\times m} = [1, 1, \ldots, 1]^T$. First, k most dominated eigenvalues of $\mathbf{B}$, $\lambda_1 > \lambda_2 > \ldots > \lambda_k \geq 0$, and their corresponding eigenvectors are calculated

Fig. 2.18 Examples of two human models after the MDS transformation. The Euclidean coordinates in $\Re^3$ and the RGB colors represent the normalized coordinates in the six-dimensional MDS domain (Image taken from [44] with permission ©2011 IEEE)

by the power method. The k eigenvectors each of which with m components are listed in the matrix $\mathbf{V}_{m\times k}$. Lastly, the resultant coordinates of the sampling points in the MDS domain can be determined by

$$\mathbf{X}_{m\times k} = \mathbf{V}_{m\times k}\Lambda_{k\times k}^{\frac{1}{2}} \tag{2.26}$$

with $\Lambda_{k\times k} = \mathrm{diag}(\lambda_1, \lambda_2, \ldots, \lambda_k)$. Each row of $\mathbf{X}$ represents a point coordinate in the k-dimensional MDS domain.

The value of k directly affects the robustness of initial shape matching in this algorithm. By experimental tests, $k = 6$ can always give satisfactory results while keeping an acceptable computational speed. Therefore, in all the figures shown in this section, the first three components of a point in the MDS domain are displayed as the Euclidean coordinates in $\Re^3$ and the next three components are displayed by the RGB colors. See Fig. 2.18 for an example.

2.2.3.2 Sign-Flip Correction

The shapes of human bodies in the MDS domain are quite similar. Ideally, the correspondences between the points $\mathbf{t}_i^* \in \tilde{M}_T^*$ and $\mathbf{h}_i^* \in \tilde{M}_H^*$ can be determined by the closest point search. However, such a mapping between $\tilde{M}_T^*$ and $\tilde{M}_H^*$ is neither bijective nor robust. The study in [44] finds that one of the the main challenges comes from the random selection of the sign of eigenvalues (therefore the direction of eigenvectors) in the MDS analysis. Thus, the shapes of $\tilde{M}_T^*$ and $\tilde{M}_H^*$ can greatly differ in terms of their axis directions called *sign-flip* (see Fig. 2.19 for an example).

Under the exhaustive search framework proposed by Shapiro and Brady [69], the alignment of the MDS embedding can be achieved by finding the combination of axes swapping which minimizes a shape difference metric. Specifically, there are $2^6 = 64$ sign-flipping combinations for six-dimensional MDS embeddings. Therefore, 2^d different sign-flipping functions can be defined as

$$\forall \mathbf{t}^* \in \tilde{M}_T^*, \ \phi_{k\ (k=1,\ldots,2^d)}(\mathbf{t}^*) = \mathbf{S}_{d\times d}^k \mathbf{t}^* \tag{2.27}$$

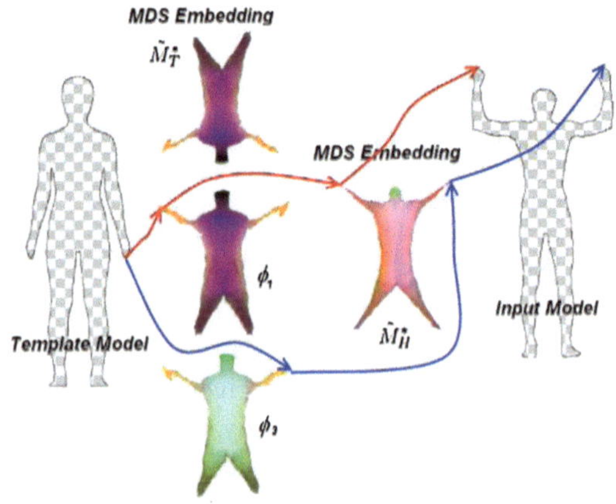

Fig. 2.19 The MDS embeddings (*colored*) of two human bodies (*checkerboard*) could be significantly different due to the sign-flip, with which the direct closest point search will give wrong correspondences between $\tilde{M}_T^*$ and $\tilde{M}_H^*$

with $\mathbf{S}_{d\times d}^k = \mathrm{diag}(\ldots, (-1)^{k \bmod i}, \ldots)$ $(i = 1, \ldots, d)$. Among them, the one giving the minimal cost on a shape error metric is selected as the sign-flip corrected MDS embedding.

According to the closest point search in the MDS space, every flipped point $\phi_k(\mathbf{t}_i^*) \in \tilde{M}_T^*$ can find its closest point $\mathbf{h}_i^* \in \tilde{M}_H^*$, which is its corresponding point. By transferring these correspondences into the spatial domain, a mapping function $\mathbf{f}(\cdots)$ can be obtained to map a point $\mathbf{t}_i \in \tilde{M}_T$ to a point $\mathbf{h}_i \in \tilde{M}_H$. Considering the sign-flipping function $\phi_k(\mathbf{t}^*)$ applied to the points on MDS embedding, the mapping becomes a functional to ϕ_k (i.e., $\mathbf{f}_{\phi_k}(\cdots)$). A continuous mapping function can be established by the deformation that is based on RBF according to the interpolation constraints that

$$\mathbf{f}_{\phi_k}(\mathbf{t}_i) \equiv \mathbf{h}_i. \quad (i = 1, \ldots, n) \tag{2.28}$$

Here, a small number of sample points are used to avoid being trapped by local optimum at the beginning of the corresponding matching algorithm. $n = 20$ is chosen in the pose alignment step, and a large value of α, 10^8, is adopted to obtain a correct sign-flip function.

In [69], a cost function based on the length variations in sample points is employed to score the mapping according to different sign-flipping functions as

$$E_{\mathrm{rr}}(\phi_k) = \sum_{i=0}^{m}\sum_{j=0}^{m} \big|\|\mathbf{t}_i - \mathbf{t}_j\|^2 - \|\mathbf{f}_{\phi_k}(\mathbf{t}_i) - \mathbf{f}_{\phi_k}(\mathbf{t}_j)\|^2\big|, \tag{2.29}$$

which actually evaluates

$E_{\mathrm{rr}}(\phi_k) = \sum_{i=0}^{m}\sum_{j=0}^{m} \big|\|\mathbf{t}_i - \mathbf{t}_j\|^2 - \|\mathbf{h}_i - \mathbf{h}_j\|^2\big|.$

A major drawback of this cost function is that it only measures the distortion in shape but does not consider the swap of surface orientation. In order to solve this problem, a modified cost function is proposed in [44] as

$$E_{\mathrm{rr}}(\phi_k) = \sum_{i=0}^{m}\sum_{j=0}^{m} \left| \|(\mathbf{t}_i+\omega\mathbf{n}_{\mathbf{t}_i})-(\mathbf{t}_j+\omega\mathbf{n}_{\mathbf{t}_j})\|^2 - \|\mathbf{f}_{\phi_k}(\mathbf{t}_i+\omega\mathbf{n}_{\mathbf{t}_i})-\mathbf{f}_{\phi_k}(\mathbf{t}_j+\omega\mathbf{n}_{\mathbf{t}_j})\|^2 \right| \tag{2.30}$$

with $\mathbf{n}_{\mathbf{t}_i}$ being the unit normal vector on the template model M_T. A very small number should be selected for the offset value ω. As suggested by Li et al. [44], $\omega = 0.5\,\mathrm{cm}$ is employed for the scanned human models. Then, an optimal sign-flip function can be found by

$$\phi^* = \arg\min E_{\mathrm{rr}}(\phi_k) \tag{2.31}$$

2.2.3.3 Pose Alignment

Given the sample points $\mathbf{t}_i^* \in \phi^*(\tilde{M}_T^*)$ with sign-flip corrected, the approximate-nearest-neighbor (ANN) search [55] is employed to find the closest points $\mathbf{c}^*(\mathbf{t}_i^*) \in \tilde{M}_H^*$ of $\mathbf{t}_i^*$. Using these correspondences, the mapping of samples in $\Re^3$ from M_T to M_H is defined as $\mathbf{t}_i \mapsto \mathbf{c}(\mathbf{t}_i) \in M_H$. Then, a transformation function $\mathbf{f}_{\mathrm{pose}}$ can be defined on n such correspondences by the RBF-based thin-plate spline transformation as

$$\mathbf{f}_{\mathrm{pose}}(\mathbf{p}) = \mathbf{a}_n + [\mathbf{a}_{n+1}, \mathbf{a}_{n+2}, \mathbf{a}_{n+3}]\mathbf{p} + \sum_{i=1}^{n} \mathbf{a}_i g(\|\mathbf{p} - \mathbf{c}(\mathbf{t}_i)\|). \tag{2.32}$$

By $\mathbf{f}_{\mathrm{pose}}(M_T)$, the new sign-flip corrected MDS embedding of $\mathbf{f}_{\mathrm{pose}}(M_T)$ can be computed. The new correspondences and the new transformation function $\mathbf{f}_{\mathrm{pose}}^1$ can thus be determined. By repeatedly applying this corresponding step, the transformation function $\mathbf{f}_{\mathrm{pose}}^j(M_T)$ $(j = 1, 2, 3, \ldots)$ is iteratively updated to align with the pose of M_H. During the iteration, the value of the regularization parameter, α, is decreased by about one tenth gradually after each loop. According to the experiments, the changes of $\mathbf{f}_{\mathrm{pose}}^j(M_T)$ would converge to a small value (such as 10^{-5}) within ten iterations (see [44] for reference). Then, it is assumed that the template model has been aligned with an optimal pose $\mathbf{f}_{\mathrm{pose}}^*(M_T)$ according to the input human model M_H (Fig. 2.20).

2.2.3.4 Surface Fitting

Given a pose-aligned template model $\mathbf{f}_{\mathrm{pose}}^*(M_T)$ and an input model M_H, a surface fitting process can further increase the accuracy of the correspondence mapping Υ defined in Eq. (2.18). However, to simplify the evaluation of the distortion function E in Eq. (2.17), a discrete version of the function is evaluated on m sample points. For the examples shown in this section, the vertices of the template model are used as the sample points. For each sample point $\mathbf{t}_i \in M_T$, its closest point is found as $\mathbf{c}(\mathbf{t}_i) \in \{\mathbf{h}_j | \forall \mathbf{h}_j \in M_H\}$. The mapping function Υ is then updated by moving the

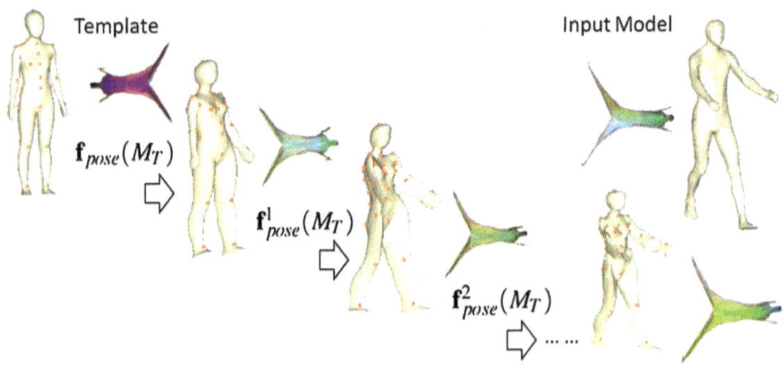

Fig. 2.20 The flow of the pose alignment step and the relationship between the spatial and the MDS domains of the template and the input models. The correspondences between the sample points in $\Re^3$ are established by the closest point search conducted in the MDS domain

vertices $\mathbf{t}_i \in M_T$ according to $\mathbf{c}(\mathbf{t}_i)$. However, simply using the closest point $\mathbf{c}(\mathbf{t}_i)$ to establish the mapping function has two defects. First, wrong mappings appear on the near but opposite surfaces. Second, the surface smoothness is disturbed in regions with dense meshes.

To overcome these defects, several modifications are added to enhance the surface fitting. First, vertices on the dynamic mesh (i.e., the warped template model) are not moved to their closest points on the model to fit. Instead, the vertices are moved along their own normal vectors like [82]. Second, the mapping established by the closest point match is validated by checking two constraints: (1) normal compatibility, that is, whether the angle between $\mathbf{n}_{\mathbf{t}_i}$ and $\mathbf{n}_{\xi(\mathbf{t}_i)}$ is greater than a threshold σ, and (2) relaxation power τ that controls the maximal allowed Euclidean distance between $\mathbf{t}_i$ and its mapped position. Last, a smoothing step is added to the fitting procedure to enhance the smoothness of the fitting result.

The pseudo-code of the surface fitting algorithm is listed in Algorithm 3. After applying this algorithm, the correspondence mapping Υ has been updated to $\Upsilon : M_T \mapsto M'_T$.

2.2.3.5 Feature Matching Refinement

After determining an updated mapping Υ by the surface fitting of M_T to M_H, we can extract the feature point set G_H by Υ as Eq. (2.19). However, the previous fitting steps are based on global alignment which do not consider the local shape distribution, and thus the locations of feature points may not be accurate enough. A feature matching refinement is conducted as the last step of this algorithm to further adjust the locations of the selected feature points.

To perform shape matching on the surfaces of human bodies, a local shape descriptor is employed. Basically, the shape descriptor needs to be invariant to the differences

Algorithm 3 TemplateSurfaceFitting

Update the template model M_T to M'_T by the optimized posture alignment function $\mathbf{f}^*_{\text{pose}}(\cdots)$;
$\tau \Leftarrow 0.5$ and $\sigma \Leftarrow 0.9$;
repeat
  $\forall \mathbf{t}_i \in M'_T$, find the closest point $\mathbf{c}(\mathbf{t}_i) \in M_H$;
  Establish a new mapping function ξ by $\mathbf{t}_i \mapsto \mathbf{c}(\mathbf{t}_i)$;
  for all $\mathbf{t}_i \in M'_T$ **do**
    $\mathbf{p}_i \Leftarrow \xi(\mathbf{t}_i)$;
    if $(\mathbf{n}_{\mathbf{t}_i} \cdot \mathbf{n}_{\mathbf{p}_i}) > \sigma$ **then**
      if $\|\mathbf{t}_i + 0.5\tau\mathbf{n}_{\mathbf{t}_i} - \mathbf{p}_i\| < \|\mathbf{t}_i - 0.5\tau\mathbf{n}_{\mathbf{t}_i} - \mathbf{p}_i\|$ **then**
        $\mathbf{t}_i \Leftarrow \mathbf{t}_i + 0.5\tau\mathbf{n}_{\mathbf{t}_i}$;
      else
        $\mathbf{t}_i \Leftarrow \mathbf{t}_i - 0.5\tau\mathbf{n}_{\mathbf{t}_i}$;
      end if
    end if
  end for
  $\sigma \Leftarrow 0.9\sigma$ and $\tau \Leftarrow 1.1\tau$;
  Smoothing M'_T by a Gaussian filter;
until $\sigma < \sigma_{\min}$ and $\tau > \tau_{\max}$
return;

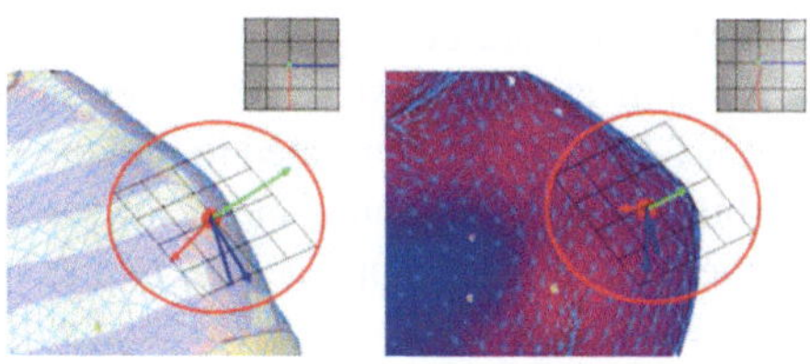

Fig. 2.21 The construction of a feature descriptor around a point on a template model (*left*) and on an input human model (*right*). The Gaussian curvature map is generated within the *red circle*. For illustration purpose, a Gaussian curvature map with 4×4 resolution is displayed. The three axes of local frames are displayed by *red*, *green* and *blue arrows* (Image taken from [44] with permission ©2011 IEEE)

in scale, orientation, and topology between the template model M_T and the input model M_H. Given a point $\mathbf{v}$ on a triangular mesh surface M, its feature descriptor $F_r(\mathbf{v})$ with a support size r is constructed as follows.

- First, a local frame $[\hat{\mathbf{t}}_1, \hat{\mathbf{t}}_2, \hat{\mathbf{t}}_3]$ at $\mathbf{v}$ is established by letting $\hat{\mathbf{t}}_2$ be along the surface normal at $\mathbf{v}$, $\hat{\mathbf{t}}_1$ be an arbitrary unit vector on the tangent plane at $\mathbf{v}$ and $\hat{\mathbf{t}}_3 = \hat{\mathbf{t}}_1 \times \hat{\mathbf{t}}_2$.
- Second, the points around $\mathbf{v}$ within a radius r are searched and assigned to a point set V_r. The Gaussian curvatures κ_G of these points are evaluated by the method of Alliez et al. [3], and the values of the Gaussian curvatures are normalized from $[\kappa_G^{\min}, \kappa_G^{\max}]$ to $[-1, 1]$, where $\kappa_G^{\min}$ and $\kappa_G^{\max}$ are the minimal and maximal Gaussian curvatures, respectively, among all the points in V_r.
- Last, the normalized Gaussian curvatures of the points in V_r are projected onto the tangent plane of $\mathbf{v}$ to form a Gaussian curvature image with 10×10 pixels—that is the feature descriptor, $F_r(\mathbf{v})$, used here.

Based on the experimental tests of Li et al. [44], selecting r as ten times of the average edge length on M_T is a good tradeoff between robustness and speed. Figure 2.21 shows an example of the feature descriptor at a local convex region.

Once the feature descriptor scheme has been developed and the feature points on the template model have been defined, the mapping Υ is refined iteratively by realigning the feature mapping between the template model M_T and the input model M_H once at a time.

For simplicity, the feature matching algorithm focuses on a single vertex $\mathbf{t}_a$ in G_T during each iteration. The correspondence $\mathbf{h}_a$ of $\mathbf{t}_a$ must be found on the surface of the input human model M_H, so that the cost of the feature descriptor

$$C_F(\mathbf{t}_a, \mathbf{h}_a) = \|F(\mathbf{t}_a) - F(\mathbf{h}_a)\| \tag{2.33}$$

is minimized as

$$\mathbf{h}_a = \arg\min C_F(\mathbf{t}_a, \mathbf{h}_a). \tag{2.34}$$

The search for an optimal $\mathbf{h}_a$ starts from $\mathbf{h}_a = \Upsilon(\mathbf{t}_a)$. A search window with a radius r is established to include all sampled surface points on M_H of a distance less than r from $\mathbf{t}_a$. Then, the minimal feature descriptor cost C_F between $\mathbf{t}_a$ and all these surface points can be found by an exhaustive search. Notice that, during the search, the local frames on the surface samples are rotated to find the best match as the axis $\hat{\mathbf{t}}_1$ of a local frame is arbitrary on the tangent plane of the surface point.

Figure 2.22 shows the results of this approach on two examples of real human bodies with different poses. All the results are generated automatically. The template model with predefined features is shown in the first column of Fig. 2.22. The computation can be completed in one minute on a PC with moderate configuration.

Another interesting study is about the number of sample points used in the pose alignment step and its effect on the final matching result. According to the experimental tests reported in [44], the pose alignment results do not lead to a satisfactory matching result if a small number of sample points are used. However, it does not mean that the more sample points are used, the more accurate result can be obtained. When further increasing the number of sample points, the computation of surface fitting may be stuck at some local optima. In addition, those testing examples with a moderate level of deformation show satisfactory results in the experimental tests. However, one limitation of this approach is its restriction on the deformation effect between the models in local regions—specifically, isometric deformation is assumed. Therefore, a particular highly stretched area, a dense-point-distributed region or a twisted surface may invalidate this algorithm.

2.3 Optimal Cross-Parameterization

The correspondences established between human bodies by locating feature points are not continuous. In order to support the design automation of human-centered freeform product, a bijective mapping between human models must be constructed.

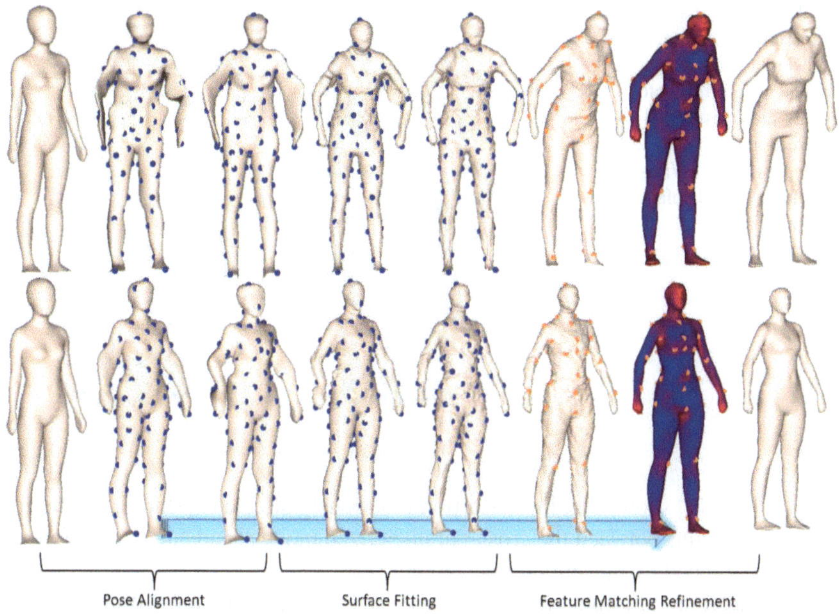

Fig. 2.22 Results of applying the matching-based feature extraction approach to two examples (in the last column) by using the template human model shown in the first column of the figure. The progressive results in different steps are shown in different columns of the figure. The feature points are displayed in *orange* (Image taken from [44] with permission ©2011 IEEE)

Computations of a bijective mapping among different models are useful for many geometry processing applications, such as texture mapping, morphing, pair-wise model editing, detail transfer, model completion, shape analysis, and model database preparation. A general solution for constructing such a mapping can be computed through the global parameterization approach (see [35, 46, 63]). However, for applications like morphing and database preparation, parameterization must be constrained by semantic features, which are correspondingly specified as anchor points on the surface of input models. A method called *cross-parameterization* [37] was proposed to solve this problem by constructing consistent domains on a pair of models, and Schreiner et al. [67] presented an approach named as *intersurface mapping* for a similar purpose.

Without loss of generality, models that need to be cross-parameterized usually have similar features, and the correspondences between anchor points should respect such similarities (for example, hands are mapped to hands as mentioned in [37]). In addition, the constructed cross-parameterization should preserve the shape of the models as much as possible. Similar to a 3D-to-2D surface parameterization problem (such as [19]), shape preservation is usually achieved by minimizing distortion occurring in the bijective mapping. A relaxation-based smoothing step was introduced by Kraevoy and Sheffer in [37] for this purpose. Nevertheless, the recent

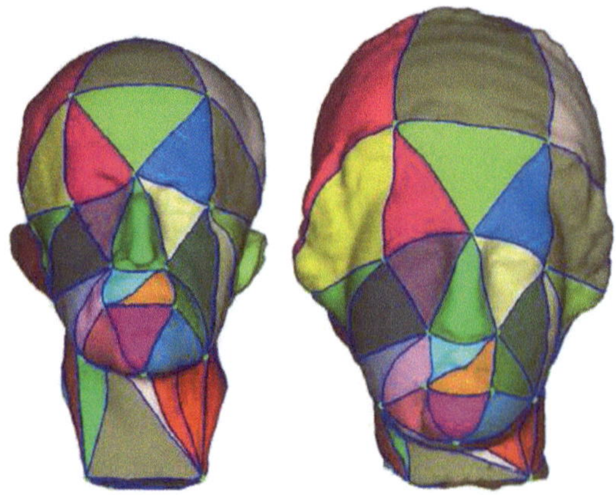

Fig. 2.23 Patch-based cross-parameterization: consistent patch layouts on two head models, where the patch layouts are constrained by feature points (displayed by *green dots*). The corresponding patches are displayed in the same colors

study in [40, 41] shows that the distortion is seriously affected by the shape dissimilarity of domains between the cross-parameterized models, and such distortion can hardly be reduced by the smoothing step in [37].

This section presents the techniques of computing an optimal cross-parameterization between a pair of human models. The method for computing a patch-based cross-parameterization is introduced first. Then, the content mainly focuses on how to construct common base domains constrained by anchor points and how to optimize the shape of the common base domains to reduce distortion in the cross-parameterization. The optimization is also constrained by the anchor points.

2.3.1 Patch-Based Cross-Parameterization

Given two surface meshes, the source model M_s and the target model M_t, which have the corresponding sets of anchor points G_s and G_t prescribed, linking the anchor points on both models in a consistent way (such as by using the algorithms in [37, 61]) can construct two triangular patch layouts P_s and P_t having the same connectivity. As an example, Fig. 2.23 shows the consistent patch layouts on two head models, where the patch layouts are constrained by the feature points. Following [61], a polygonal mesh M is represented by a pair (V, K), where V is a set of vertices and K is an abstract simplicial complex that contains all the topological information (such as adjacency). For convenience, a triangular patch layout is treated as a special mesh with anchor points being the vertices and curved patch boundaries being the edges. Assume that each patch P^i in a patch layout can find a corresponding planar domain B^i (by [19, 20]), the 3D-to-2D mapping is obtained as

$$\Gamma_s : P_s^i \Rightarrow B_s^i \text{ and } \Gamma_t : P_t^i \Rightarrow B_t^i.$$

Fig. 2.24 Star-shaped polygon around $\mathbf{v}_0$

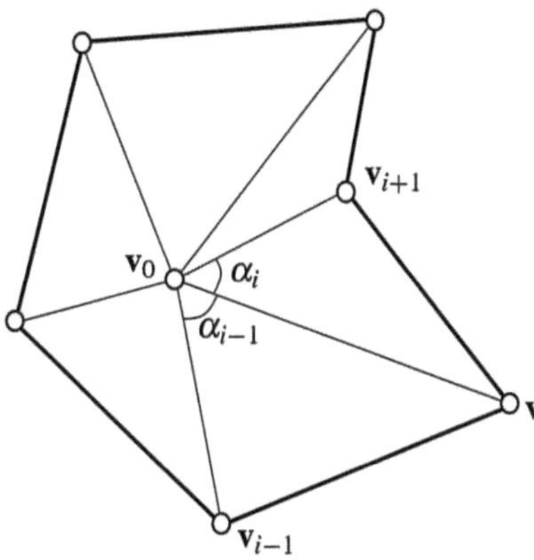

After establishing the mapping between 2D domains as $\Gamma_{\rm st} : B_s^i \Rightarrow B_t^i$ (for example, using mean-value coordinates [20]), one can establish the cross-parameterization

$$\Gamma = \Gamma_s \cdot \Gamma_{\rm st} \cdot \Gamma_t^{-1} \tag{2.35}$$

to map M_s to M_t.

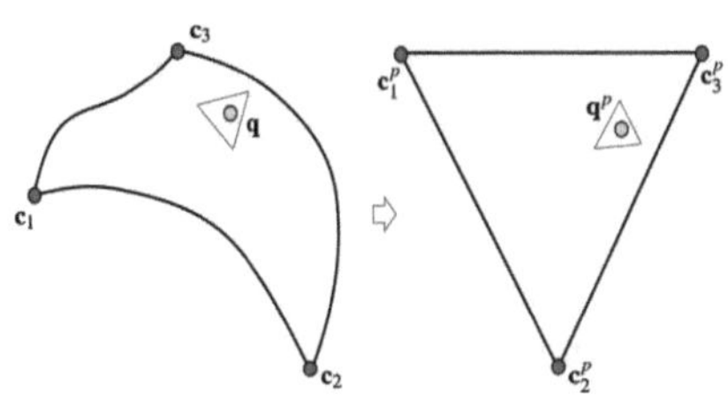

Patch Parameterization Based on Mean-Value Coordinates

Without loss of generality, let a freeform surface patch P be represented by a two-manifold piecewise linear surface composed of triangles. If the triangular patch P in $\Re^3$ has three corner vertices $\mathbf{c}_1$, $\mathbf{c}_2$ and $\mathbf{c}_3$, a 3D-to-2D mapping with small distortion can be found to convert its triangles into a planar triangular domain as follows.

First, the planar coordinates, $\mathbf{c}_1^p$, $\mathbf{c}_2^p$ and $\mathbf{c}_3^p$, of the corner vertices, are determined. There are many ways to do so and where the simplest one is to put them into the corners of a regular triangle. Second, the vertices on the boundary curves of the patch $P \in \Re^3$ are located on the edges of the planar regular triangle to make the 2D lengths of boundary edges proportional to their 3D lengths. Last, the planar coordinates of the interior vertices are determined by the mean-value coordinates.

For a vertex $\mathbf{v}_0$ on P whose one-ring neighboring vertices $\mathbf{v}_1, \ldots, \mathbf{v}_k$ are arranged in anticlockwise order around $\mathbf{v}_0$ (Fig. 2.24), the mean value coordinates assume that mapping the one-ring neighboring vertices onto a plane forms a star-shaped polygon $\mathbf{v}_1^p, \ldots, \mathbf{v}_k^p$ with $\mathbf{v}_0^p$ in its kernel. The position of $\mathbf{v}_0^p$ is a linear combination of $\mathbf{v}_1^p, \ldots, \mathbf{v}_k^p$ as

$$\mathbf{v}_0^p = \sum_{i=1}^{k} \lambda_i \mathbf{v}_i^p, \quad \sum_{i=1}^{k} \lambda_i = 1. \tag{2.36}$$

By Floater's analysis [20], the weights

$$\lambda_i = \frac{w_i}{\sum_{j=1}^{k} w_j}, \qquad w_i = \frac{\tan(\alpha_{i-1}/2) + \tan(\alpha_i/2)}{\|\mathbf{v}_0\mathbf{v}_i\|} \tag{2.37}$$

are served as coordinates for $\mathbf{v}_0$ with respect to $\mathbf{v}_1, \ldots, \mathbf{v}_k$. Notice that, the positions and angles in Eq. (2.37) are evaluated in $\Re^3$. Substituting Eq. (2.37) into (2.36), a linear equation system can be obtained as

$$\begin{bmatrix} \mathbf{M} & \mathbf{0} \\ \mathbf{0} & \mathbf{I} \end{bmatrix} \begin{bmatrix} \mathbf{U}^{\text{int}} \\ \mathbf{U}^{\text{bnd}} \end{bmatrix} = \begin{bmatrix} \mathbf{0} \\ \mathbf{U}^{\text{bnd}} \end{bmatrix} \tag{2.38}$$

with $\mathbf{M}$ being the matrix according to the linear combination in Eq. (2.36), $\mathbf{U}^{\text{int}}$ being the vector containing the planar coordinates of interior vertices and $\mathbf{U}^{\text{bnd}}$ denoting the planar coordinates of boundary vertices which have been fixed by the aforementioned method.

The planar coordinates of a point $\mathbf{q} \in \Re^3$ that is not a vertex but is inside a triangle $\Delta \in P$ can be determined by first calculating its barycentric coordinates in Δ. Applying the barycentric coordinates to the planar coordinates of Δ's three vertices can give the planar coordinates $\mathbf{q}^p$. The mapping from any $\mathbf{q} \in P$ to $\mathbf{q}^p$ gives the patch parameterization, and has been proved to be bijective (see [20] for reference). In fact, this method can also be generalized to any n-sided patches.

In this approach, the patch parameterization based on mean-value coordinates is employed to establish the bijective mapping between patches. It can be applied to the 3D-to-2D mapping as well as the 2D-to-2D mapping if one of the 2D patch is considered as a 3D patch to compute the mean-value coordinates. By the mapping Γ in Eq. (2.35), the mesh connectivity on the template human body T can be transformed to any other individuals H to obtain the mesh surfaces with consistent connectivity. Then, shape blending can be easily conducted by blending the positions of the corresponding vertices (see Fig. 2.25 for an example).

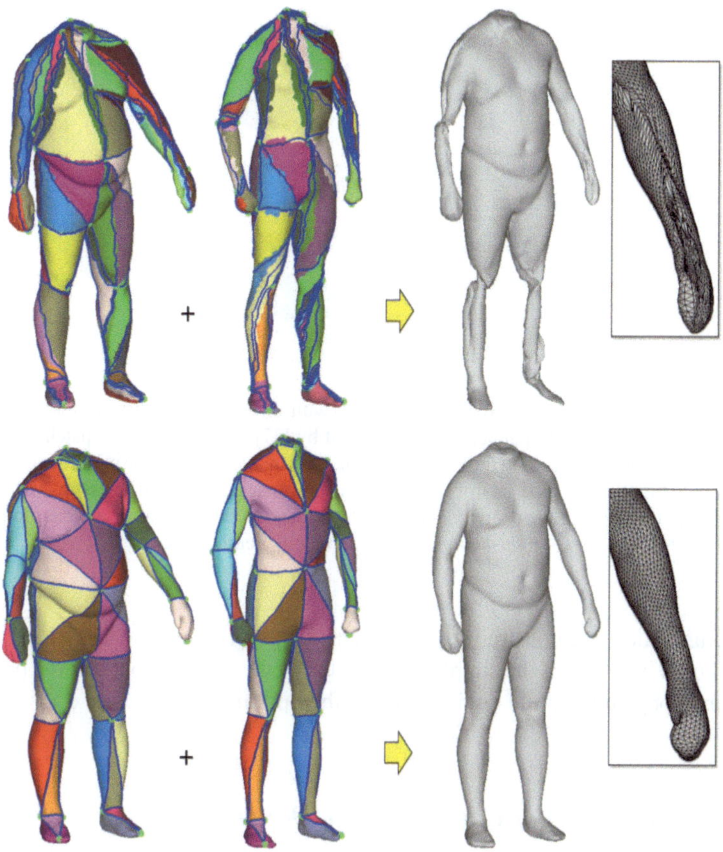

Fig. 2.25 The quality of common base domains (for example, the patches) greatly affects the result of cross-parameterization and therefore affects the shape blending result. *Top row* The cross-parameterization generated by [37]—notice that the boundary of the patches is not consistent with the paths linking the anchor points after applying the relaxation-based smoothing step, the purpose of which is to reduce distortion in the cross-parameterization. *Bottom row* The result generated by the method introduced in this section. Significant distortion occurs on the arms and the legs of the blending result according to cross-parameterization computed by [37]

Basically, the mapping in cross-parameterization is rarely stretch-free (that is, isometric). However, applications like human model synthesis and shape blending wish to minimize such distortion as it leads to unwanted shape distortion on the blending results—an example is shown in Fig. 2.25. Many prior researchers have dedicated to minimizing distortion in 3D-to-2D mappings as well as in 2D-to-2D mappings. The recent study of Kwok et al. in [40, 41] finds that constructing common base domains with a good shape and optimizing the topology of common base domains in P_s and P_t can further reduce the stretch, and the research presented in the rest of this section mainly focuses on this aspect of cross-parameterization. The construction

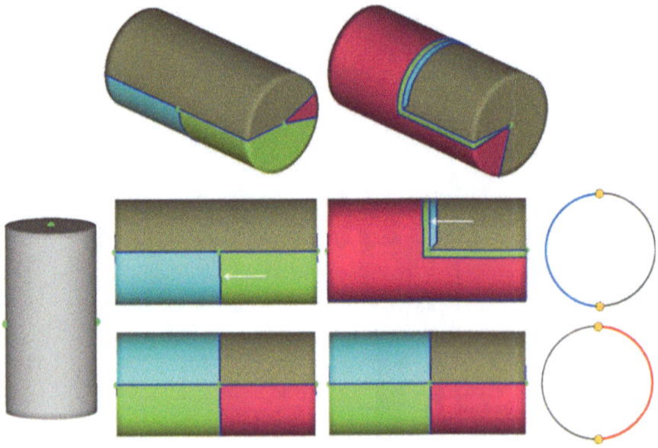

Fig. 2.26 Unwanted swirling: two input cylinders with different connectivity and four pairs of anchor points in *green* (*left*), patch layouts generated by [37] (*top-middle*), and patches generated by our method (*bottom-middle*). The path between the two anchor points in the middle could travel in two different directions due to a lack of uniqueness in the shortest path based heuristic. Analogously, two anchor points are placed at antipodal points of a circle, the shortest path may pass along either the *left* (*right-top*) or *right* (*right-bottom*) path around the circumference of the cylinder

and optimization are constrained by the anchor points. Specifically, every anchor point should be a part of the simplicial complex of the patch layout and the positions of the anchor points should be fixed during the optimization.

2.3.2 Construction of Common Base Domains

The research work in [17] presents that using the graph given by a VD to construct a base domain can reduce distortion in surface parameterization. Recent development in [40] tries to solve the problem of how to use VDs to construct common base domains for constrained cross-parameterization. Two challenging problems are:

1. VDs centered at anchor points may not be dual to a triangulation; therefore, the construction of base domains is not straightforward.
2. Although corresponding anchor points are given in two input models, the graphs of the two VDs may not be the same. Thus, the construction of consistent patch layouts is not guaranteed.

As mentioned by Kraevoy and Sheffer [37], models that need to be cross-parameterized usually have similar features. Such similarities are respected by correspondences between anchor points. Therefore, the VD centered at the anchor points is a good cue for the process, and the base domains construction based on such a global consideration is more reliable.

The strategy that uses VDs as cues to govern the construction of common base domains brings the following advantages:

Table 2.2 Local versus global approaches

Methods	Multi-paths	Swirling	Complexity
Local	Not supported	Tackled by postprocessing	Best case: $O(n_a n_e \log n_v)$
Global	Supported	Prevented by directional guidance	Worst case: $O(n_a n_e \log n_v)$

- Using VDs centered at anchor points to form Delaunay-like triangular layouts gives good-shape common base domains, thus generating less distortion in cross-parameterization.
- Computing the shortest paths within the corresponding clusters of triangles can speed up the whole process. The complexity of base domain construction algorithm is proved to be bounded by the prior method.
- With the help of multiple linkages, a valid triangulation can be formed on a VD which is not dual to a triangulation.
- The Voronoi tessellation gives directional guidance for path traveling when linking two anchor points. This prevents unwanted swirling when computing on a cylindrical shape, where prior methods need complicated operations to tackle this problem (see Fig. 2.26 for an illustration).

The smoothness of a domain boundary is affected by mesh quality. If the quality of an input mesh is poor (for example, having many irregular triangles and significantly non-uniform sizes of triangles), the shape of the base domain may be poor too. This problem can be solved by adding edge nodes adaptively to Dijkstra search. As the quality of parameterization between a set of patches highly depends on the shape of the patches, adding edge nodes can get rid of the influence of poor mesh quality and make the boundaries of meshes smooth.

2.3.2.1 Local Versus Global Approaches

Prior methods (such as [37, 67]) compute common base domains by the heuristic of having higher priority to link shorter paths. Patch layouts are constructed in a trial-and-error manner, which can be considered as a locally greedy search in the solution space and is therefore called *local* approach. In contrast to these local approaches, the work presented here proposes employing a VD centered at the anchor points as a governing field for the construction. The VD actually represents the distribution of anchor points as well as the topological structure of input models. This method for constructing common base domains in a *global* manner can tackle several problems in the local approaches, which are analyzed below. A summary of comparison is listed in Table 2.2.

Multiple Linkages

As mentioned by Kraevoy and Sheffer [37], the procedure for tracing paths may fail if there are not enough anchor points. The upper row in Fig. 2.27 displays the failure in base domain construction by [37] with six anchor points on two genus-one models. The major reason for the failure of the algorithm [37] in this case is that there are not

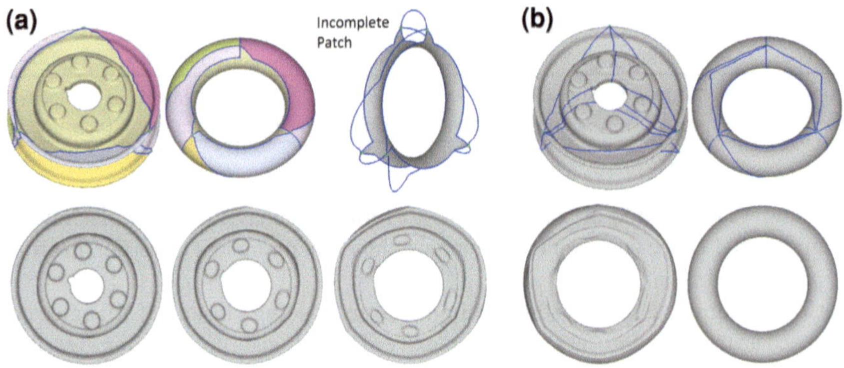

Fig. 2.27 A comparison between the algorithm [37] and this VD-based method in construction of common base domains on a pair of models with genus-one topology. Only six anchor points are specified. The algorithm [37] fails to construct a complete set of common base domains (**a**), as the patch in the inner tunnel loop has not been triangulated (that is incomplete patch). The patch is enclosed by all the six anchor points because a pair of anchor points alone cannot support multiple paths. On the contrary, the VD-based approach can build a valid set of common base domains (**b**). The *bottom row* shows the shape blending results obtained by cross-parameterization

enough anchor points to support new paths. Specifically, the anchor points cannot be linked multiple times because the local approach cannot identify the differences between two paths with the same endpoints. The authors of [67] tried to solve this problem by adding more anchor points automatically; however, it is not guaranteed that the newly added pairs of anchor points respect the corresponding similarities correctly.

In this global approach, anchor points are allowed to be connected for multiple times. This can be achieved by forcing the paths to travel through the corresponding cuts where the clusters connect each other. Benefited by VD, more paths can be supported with the same number of anchor points. Therefore, the VD-based approach can successfully generate common base domains for the example shown in Fig. 2.27. To address this feature, paths are identified by two anchor points and an index i indicating that it is the ith path linking these two anchor points.

Unwanted Swirling

Local approaches suffer from a lack of uniqueness in the shortest path problem; that is, they may have multiple solutions for the same inputs. This brings uncertainty to the local approaches for constructing base domains. An illustration is given in Fig. 2.26 where two points on a cylindrical surface can be linked in different ways. When there is no guidance to lead the corresponding paths generated in the same direction on two models, a corresponding path may travel downward on M_s while traveling upward on M_t; therefore, swirling happens. One real example containing unwanted swirling is shown in Fig. 2.25, where 37 anchor points are provided for cross-parameterization and the common base domains constructed by [37] lead to a highly distorted mapping caused by unwanted swirling (see the top row of Fig. 2.25). Thanks to the features of having directional guidance, the capability of supporting multiple linkages and the global connectivity information provided by VDs, the VD-based method can avoid involving complicated topological operations to handle swirling, and can thus produce a good result (see the bottom row of Fig. 2.25).

Computational Complexity

Local approaches (such as [37]) compute the shortest paths among all pairs of anchor points, and then construct the common base domains in a trial-and-error manner. Many invalid paths are computed unnecessarily. The procedures of searching the shortest paths are performed on the whole mesh surface. Using VDs as the underlying governing field, one knows exactly which paths need to be computed by the neighboring relationship of the diagrams. The VD-based approach can always limit the Dijkstra searching to be within two diagrams (that are local regions) of a VD centered at anchor points. Therefore, the computational complexity is greatly reduced and proved to be bounded by the local approaches (detail discussion is given at the end of Sect. 2.3.2.3).

Distortion in Parameterization

Last but not least, local approaches simply construct paths in a greedy way, which do not consider the shape of base domains. They just aim at building a valid common base domain. However, it is found that the distortion in cross-parameterization highly relates to the shape of base domains (see [41, 40] for a reference). As shown on the left arm of a human body in Fig. 2.25, the shape of the domain is very bad and the parameterization result is highly stretched even though swirling does not occur. Fortunately, VDs suggest the connectivity of the base domains so that the shape of the common base domains generated according to VDs is much better. With these base domains, a cross-parameterization with low distortion can be obtained.

2.3.2.2 Path Tracing

Before presenting the main part of the VD-based algorithm for constructing common base domains, a path tracing method which is a frequently used operator in the base domain construction algorithm is introduced here. Edge nodes are employed to improve the smoothness of paths (see Fig. 2.28) and a robust method (without using numerical predicates [25]) is exploited to detect intersections between paths.

The shortest path between two anchor points can be found by linking vertices and edge nodes so that it can travel across faces and give smoother boundaries for the constructed patches. The concept of adding edge nodes is similar to the strategy of adding Steiner vertices in [67]. Schreiner et al. [67] add Steiner vertices to ensure success in emanating new paths, whereas here aims at having smooth boundaries. Kraevoy and Sheffer [37] add face paths after edge paths to guarantee the completion of base domain; however, introducing face paths after edge paths may produce some poorly shaped patches. In this approach, new edge nodes are adaptively added to ensure that there are always enough nodes to pass through. Although the computation of the shortest path is slowed down if too many edge nodes are added, this algorithm

Fig. 2.28 The shortest paths are constructed to link the anchor points on an input cow model (*middle*). Only edge paths and face paths are used in [37] (*left*), and smoother curves can be obtained by using adaptively added edge nodes (*right*) (Image taken from [40] with permission)

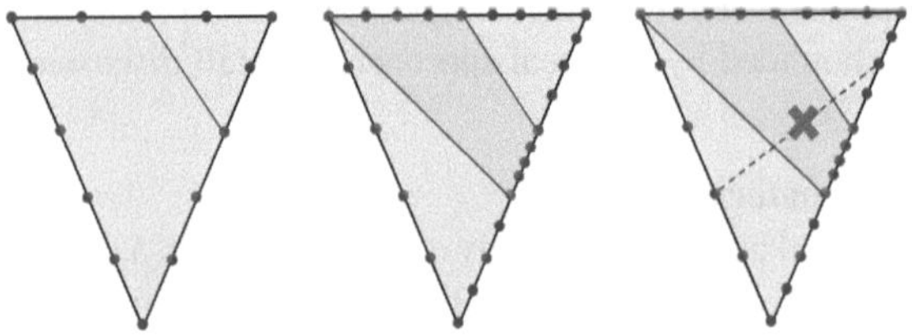

Fig. 2.29 An illustration of adaptive addition of edge nodes and robust intersection check: adding a new path across the face splits the face into two regions (*left*), new edge nodes are adaptively added and a new path further subdivides the face into three regions (*middle*), and a proposed new path (in *dash line*) intersecting the existing paths is prevented as its two ending nodes are located in different regions of the face (*right*)

is still fast enough as the computation of the shortest paths is limited in some local regions based on VDs.

Initially, each edge is associated with $e = \max(b, \frac{bl}{\bar{l}})$ edge nodes, where b is the minimal number of edge nodes on an e.g., l is the length of the current e.g., and $\bar{l}$ is the average length of its neighboring edges (in other words, adjacent edges of two end vertices of the current edge including the edge itself). This formula compares the length of the current edge with its neighbors', and adaptively adds more edge nodes if the edge is much longer than its neighbors. In the implementation (see [40]), $b = 3$ is used. When an edge is split by a path into two edges (as shown on the left and in the middle of Fig. 2.29), the number of edge nodes on the new edges must be adaptively adjusted based on the above formula.

Robust Detection of Intersections

Intersections between paths on a face must be prevented. However, using numerical predicates to detect such intersections suffers from robustness problems [25]. The intersections can be effectively and robustly detected by topological information in this algorithm. As illustrated in Fig. 2.29, adding a new path to a face subdivides the face into two regions. The edges nodes on the left of the newly added path can be easily assigned to one region and the edge nodes on the right side are assigned to another region. The same assignment is performed on the adaptively added edge

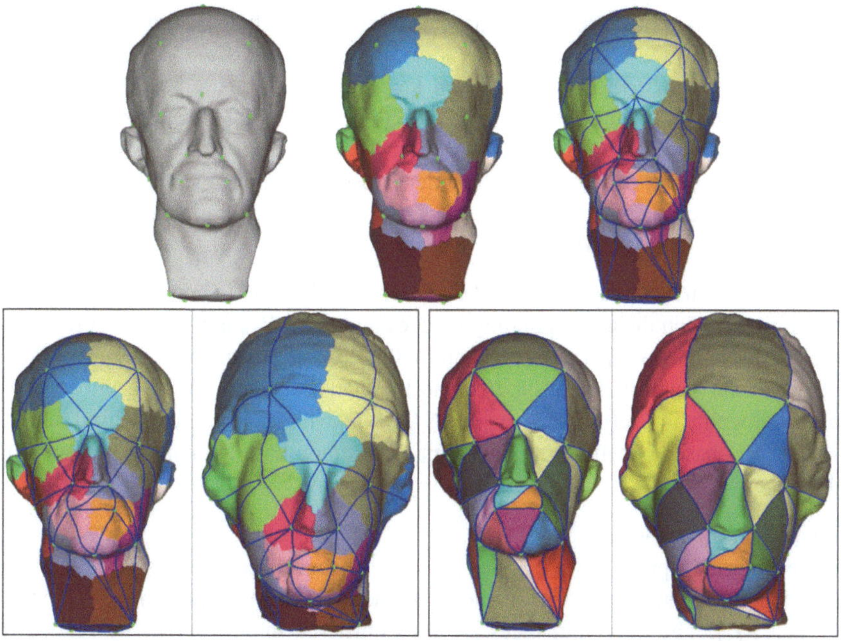

Fig. 2.30 An algorithm overview. *Top—from left to right* An input source model M_s with prescribed anchor points, computing the surface VD centered at the anchor points by face clustering, and building the base domain P_s on M_s. *Bottom—left* Constructing base domain P_t on M_t using the connectivity of P_s. *Bottom—right* Matching the corresponding patches in P_s and P_t that are shown in the same color

nodes. Another path can be further added if its two end-nodes are located in the same region of the face (see the middle of Fig. 2.29). The face is further subdivided into three regions by this new path, and the relevant edges nodes are assigned to the new regions. A path intersects other existing paths on a face when its two end-nodes fall in different regions of the face (see the proposed third path on the right of Fig. 2.29). By this method, intersection prevention can be robustly conducted in the path tracing operator without using any numerical predicates.

2.3.2.3 VD-Based Construction of Base Domains

The construction algorithm of common base domains on a pair of models, M_s and M_t, with prescribed anchor points consists of two stages (see Fig. 2.30 for an illustration).

- *Build base domains on source model M_s.* This stage finds out the connectivity of base domains constrained by anchor points on M_s. First, the VD of anchor points is computed by face clustering. Second, the neighboring relationship between the diagrams (that is, regions centered at the anchor points) is employed to establish the connectivity of base domains based on the anchor points. Last, triangulation is performed to produce a triangular patch layout P_s, which is stimulated by dual-graph of the VD.

- *Construct consistent patch layout on target model* M_t. After building the base domains on the source model M_s, the connectivity of P_s is transformed onto the target model, M_t. Again, the VD on M_t is first computed to speed up the process. Second, a proper linking order is used to produce single-linkage paths and multiple-linkage paths correctly. Last, a topologically consistent triangular layout P_t is constructed on M_t.

The result of this procedure is a set of common base domains in good shape. Bijective mapping between the corresponding patches in P_s and P_t can then be computed by the patch-based cross-parameterization method presented in Sect. 2.3.1. The shape of the planar patch is determined by the Euclidean distance between three corner points. Specifically, let $\|\mathbf{c}_i^p\mathbf{c}_j^p\| = \|\mathbf{c}_i\mathbf{c}_j\|$ $(i \neq j)$, which is the strategy used in [37].

The VD-based construction algorithm of common base domains is presented in detail below.

Base Domain on Source Model

For each anchor point in $\{\mathbf{g}_1, \ldots, \mathbf{g}_s\}$ on M_s, a face adjacent to it is first randomly selected. These selected faces are used as the sites of the approximate VD, $S = \{f_1, \ldots, f_s\}$. The triangular faces are then clustered into s regions by the multi-source Dijkstra's algorithm (see [13] for reference). As a result, every anchor point g_i is surrounded by a set of faces which are closer to g_i than to other anchor points. The region formed by these faces is the approximation of a Voronoi cell (called *cluster*, τ_i, in the rest of this section). Notice that the VD computed here is an approximation of the exact VD as it is based on clustering triangular faces into different regions (that is, the approximate cells of VD) centered at anchor points. The boundary of the clusters can be further smoothed by applying the following operations five to ten times:

- Locating a face $f_k \in \tau_i$ which is on the boundary of a cluster τ_i, that is, if any of its adjacent faces belong to a different cluster τ_j $(j \neq i)$.
- $f_k \in \tau_i$ is assigned to another cluster τ_j if two of its adjacent faces are part of τ_j.

Different from the approach of Eck et al. [17], the anchor points of the VD in this algorithm are hard constraints and cannot be moved. Therefore, it is easy to have the following observations.

Observation 2.1 When the sites (anchor points) are the hard constraints, the Voronoi diagram centered at them may not always be dual to a triangulation.

A *cut* on the clustered triangular mesh is a contiguous set of edges along which the same pair of clusters touch.

Observation 2.2 There are pairs of clusters that may share more than one *cut*.

Therefore, constructing the triangulation linking the anchor points as the dual-graph of the VD is not straightforward. A novel approach that allows multiple linkages between two clusters to form a valid triangulation is introduced below. Once the clusters centered at the anchor points are generated, the connections between clusters as well as the paths linking anchor points can be found by the following method.

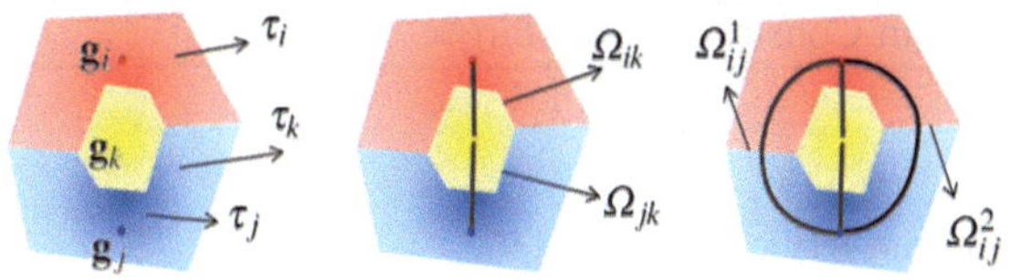

Fig. 2.31 An illustration of multiple linkages: after forming clusters, there are two cuts between τ_i and τ_j (*left*); a single linkage simply connects two anchor points of two neighboring clusters sharing only one cut (*middle*); and multiple linkages between a pair of anchor points can be constructed through the corresponding cuts indicated by a VD (*right*)

- *Initial connectivity*. For each cluster τ_i, its neighboring clusters, $\tau_j \in N(\tau_i)$, as well as the cut Ω_{ij} between τ_i and τ_j are found out first. When τ_i and τ_j share more than one cut, an index η is assigned to each of these cuts, where Ω^{η}_{ij} denotes the η-th cut shared by τ_i and τ_j (see Fig. 2.31 for an illustration).
- *Constructing paths*. After obtaining the initial connectivity of base domains, the boundary of base domains on M_s is constructed by tracing paths between anchor points. For example, if the clusters τ_i and τ_j share a cut Ω_{ij}, the shortest path across Ω_{ij} linking g_i and g_j is computed. Notice that, the search for the shortest path by the Dijkstra's algorithm has been narrowed to the regions occupied by τ_i and τ_j, which speeds up the computation significantly. For two clusters sharing more than one cut (that is $\eta > 1$), the anchor points must be linked by multiple paths (see Fig. 2.31 for an example). A path can be enforced to travel through a particular cut Ω^{η}_{ij} by removing the nodes and edge nodes (as well as the arcs linking them) located on other cuts (that is, Ω^{κ}_{ij} with $\kappa \neq \eta$).
- *Patch triangulation*. The paths generated in the above step sometimes do not subdivide the whole surface into triangular surface patches. In this case, triangular patches are created simply by adding paths between anchor points that are not adjacent to each other. This is similar to the triangulation method stated by Kraevoy and Sheffer [37]. Notice that, the search for the shortest path linking two anchor points by Dijkstra's algorithm is conducted locally on a non-triangular patch holding these two anchor points, and the computation is efficient.

This procedure ends with the base domains, P_s, in the form of triangular patch layouts on M_s.

Common Base Domain on Target Model

Having the connectivity of base domains P_s on M_s, the base domains P_t on the target model M_t need to be constructed, which should be topologically consistent with the connectivity of P_s. Similar to [61], *intersection*, *cyclical order* and *blocking* need to be checked in order to guarantee that the connectivity of P_s can be consistently transformed to P_t. Different from [61], which randomly picks paths to construct, this algorithm sorts out a proper order for linking anchor points multiple times. Otherwise, multiple paths cannot be constructed correctly.

First, clusters, $\{\tau_1^*, \ldots, \tau_t^*\}$, for the anchor points, $G_t = \{\mathbf{g}_1^*, \ldots, \mathbf{g}_t^*\} \in M_t$, are generated in the same way as that mentioned above. Second, paths corresponding to the ones in P_s are constructed one by one. Note that, the cyclical order of paths

Algorithm 4 Domain-Construction-on-Target

Input: P_s, M_t and G_t

Output: P_t

1: Construct $\{\tau_1^*, \ldots, \tau_t^*\}$ on M_t by G_t.

2: Obtain the neighboring relationship $\Omega_{ij}^{\eta *}$ on M_t.

3: Construct paths ξ_{ij}^* for $(\mathbf{g}_i^*, \mathbf{g}_j^*)$ if (τ_i^*, τ_j^*) share only one cut and there is only one path ξ_{ij} linking $(\mathbf{g}_i, \mathbf{g}_j)$ in P_s.

4: Construct paths ξ_{ij}^* if $(\mathbf{g}_i^*, \mathbf{g}_j^*)$ share no cuts but there is one path ξ_{ij} linking $(\mathbf{g}_i, \mathbf{g}_j)$ in P_s.

5: Construct multiple paths $\xi_{ij}^{\eta *}$ in their corresponding cuts $\Omega_{ij}^{\eta *}$ enforced by cyclical order defined in P_s. If the multiple paths are in the same sector at this moment, the construction of these paths is postponed.

6: Repeat step 5 if any construction of paths is postponed.

between anchor points must be enforced. In addition, once a candidate path is going to form a cycle, *non-blocking* check is performed to ensure that this cycle encloses the same set of anchor points on both M_s and M_t (details can be found in [37]). The paths are classified into different groups to be constructed in a proper order based on the configurations of the underlying VDs. First, the simple paths linking two anchor points, the clusters of which only share one cut, are constructed. Second, the paths between two anchor points located in non-neighboring clusters are generated. Last, the multiple paths linking the same pair of anchor points are traced on the surface of M_t. Pseudo-code is shown in Algorithm 4.

Complexity

Let n_v, n_e, n_f and n_a be the number of nodes on the mesh, the number of edges, the number of faces and the number of anchor points, respectively; the multi-source Dijkstra's algorithm that generates VD has complexity $O(n_e \log n_f)$ as the faces serve as 'nodes' on the graph and the neighboring faces have an 'edge' linking them on the graph. For a closed triangular mesh surface, $n_f \simeq 2n_v$; therefore, the complexity of the step generating VD is also $O(n_e \log n_v)$.

When computing the shortest path between two anchor points, the computational complexity is $O(n_e \log n_v)$. In practice, as the computation takes place in local regions (such as, two clusters holding two anchor points to be linked), the complexity can be reduced to $O\left(\frac{2n_e}{n_a} \log \frac{n_v}{n_a}\right)$. On a successfully constructed layout of triangular patches, the number of paths n_p is about $3n_a$. Therefore, constructing n_p paths leads to a computation with complexity $O\left(\frac{2n_p n_e}{n_a} \log \frac{n_v}{n_a}\right) \simeq O\left(n_e \log \frac{n_v}{n_a}\right)$. In fact, this is the optimal case. In the worst case, the numbers of edges and nodes are non-uniformly distributed in the clusters, so every path needs a computation in the complexity of $O(n_e \log n_v)$. Therefore, the overall computation has complexity $O(n_a n_e \log n_v)$. Since the number of edge nodes on each edge is a small constant, it is neglected in the analysis of computational complexity.

In summary, the upper bound of the computational complexity of the VD-based algorithm is $O(n_a n_e \log n_v)$ (which in fact rarely happens) while the lower bound is only $O(n_e \log n_v)$.

Considering the methods based on the heuristic of having higher priority to link the shortest paths (such as [37, 67]), they need to first compute the length of all the possible paths linking all anchor points, which leads to a computation with complexity $O(n_a(n_e \log n_v))$. The subsequent steps to add new paths in a trial-and-error manner are taken on the whole mesh surface (that is, may have an overall order of $O(n_a(n_e \log n_v))$ for a few runs). Although there is no theoretical analysis to find out the complexity of trial path generation, the best case of [37] gives the complexity $O(n_a(n_e \log n_v))$.

Remark 2.1 The speed of VD-based method is much higher on mesh surfaces with relatively uniform anchor points.

When more anchor points are given on the models, a more significant speedup can be achieved in this VD-based algorithm compared with prior methods. This can be verified by the experimental tests presented below.

2.3.2.4 Discussion

The VD-based approach can construct good common base domains in an efficient way, and it has been compared with the heuristic-based cross-parameteriztion method in [37]. To evaluate the distortion in cross-parameterization, the triangles of M_s are transferred onto M_t by the mapping, Γ, to obtain a new mesh surface M_{st}, which has the same connectivity as M_s but with a shape similar to M_t. Two error metrics are employed to measure the distortion in the mapping of cross-parameterization.

Angle Distortion:

$$E_{\mathrm{ang}} = \frac{1}{3n_f} \sum_{i=1}^{n_f} \sum_{j=1}^{3} \left(\frac{\phi_{i,j} - \theta_{i,j}}{\phi_{i,j}} \right)^2 \tag{2.39}$$

where n_f is the number of triangles on M_s, and $\phi_{i,j} \in M_s$ and $\theta_{i,j} \in M_{\mathrm{st}}$ are the corresponding angles in the jth angle of the ith triangle. If there is no angle distortion in Γ, $E_{\mathrm{ang}} = 0$.

L^2-stretch:

$$E_{L^2} = \sqrt{\frac{\sum_{f_i \in M_s} (L^2(f_i))^2 A_{\mathrm{st}}(f_i)}{\sum_{f_i \in M_s} A_s(f_i)}} \tag{2.40}$$

where $L^2(\ldots)$ computes the L^2-stretch of a triangle (see [65] for reference). For a triangle $f_s \in M_s$ and its corresponding shape defined by the mapping Γ as $\Gamma(f_s)$, the L^2-stretch is defined by computing the eigenvalues of the Jacobian formed by partial derivatives of a unique affine mapping between f_s and $\Gamma(f_s)$. To ease the computation, for f_s with three vertices $(\mathbf{p}_1, \mathbf{p}_2, \mathbf{p}_3)$ in $\Re^3$, a local planar frame is defined on $\Gamma(f_s)$ to obtain the planar coordinates of its three vertices as (s_k, t_k) $(k = 1, 2, 3)$. Then, the L^2-norm defined on the triangle f_s in the cross-parameterization is

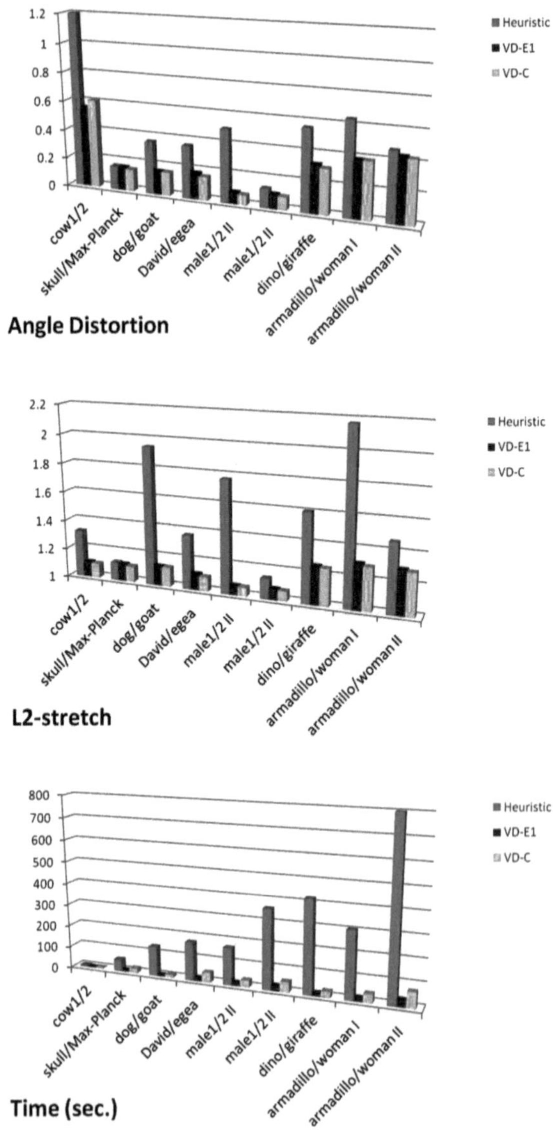

Fig. 2.32 Comparison charts of angle distortion, L^2-stretch and computing time between the VD-based construction approach [40] and the heuristic-based algorithm [37], where '-E1' stands for the VD-based algorithm using only one edge node, and '-C' is the complete version of the VD-based algorithm using an adaptive number of edge nodes. The computation time is evaluated on a standard PC with Intel Core 2 Duo CPU E6750 at 2.67GHz plus 2GB RAM

$$L^2(f_s) = \sqrt{(\varsigma_s \cdot \varsigma_s + \varsigma_t \cdot \varsigma_t)/2} \tag{2.41}$$

with

$$\varsigma_s = ((t_2 - t_3)\mathbf{p}_1 + (t_3 - t_1)\mathbf{p}_2 + (t_1 - t_2)\mathbf{p}_3)/(2A(\Gamma(f_s)))$$
$$\varsigma_t = ((s_2 - s_3)\mathbf{p}_1 + (s_3 - s_1)\mathbf{p}_2 + (s_1 - s_2)\mathbf{p}_3)/(2A(\Gamma(f_s)))$$

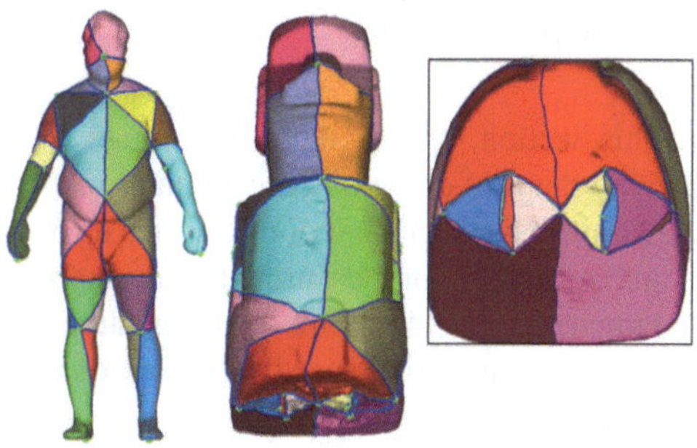

Fig. 2.33 The VD-based approach can successfully generate common base domains constrained by non-uniformly distributed anchor points. The two legs of a human model are mapped to the bottom surface of a Moai model

and $A(\cdots)$ defining the area of a triangle. In Eq. (2.40), the stretch values among triangles with different areas have been normalized by the factor $\sqrt{A_{st}(f_i)/A_s(f_i)}$ with $A_s(\ldots)$ and $A_{st}(\ldots)$ being the areas of a triangle on M_s and M_{st}, respectively. For a mapping with no distortion, the value of E_{L^2} is 1.0.

To conduct a fair comparison, when implementing [37], the postprocessing steps, flipping paths and smoothing, have been fully implemented to improve the quality of mapping. Moreover, the VD-based approach has also been tested by using only one edge node, which is similar to the face path used in [37]. The approaches have been tested and compared on several examples. The computational statistics and their corresponding charts are shown in Fig. 2.32. In these examples, the VD-based method outperforms [37] on both the quality of cross-parameterization (with less distortion) and the computation time. For the example of Armadillo/woman pair, when 45 anchor points are used, the VD-based method [40] is 18 times faster than [37]. After increasing the number of anchor points to 85, [40] is 25.4 times faster. The computational statistics also show that using more edge nodes could further reduce the distortion. Although the processing time is increased, it is still much faster compared with the prior method that uses the heuristic of shortest path.

One interesting observation from Fig. 2.32 is that the quality of results obtained from using one edge node is similar to that of using an adaptive number of edge nodes. Why do not simply use one edge node? Because if the triangles on input mesh surfaces are highly irregular, using one edge node will lead to a highly distorted map in cross-parameterization. The distortion in such cases can be greatly reduced by using an adaptive number of edge nodes.

Last but nor least, the VD-based approach can successfully construct common base domains constrained by anchor points that are distributed quite differently on M_s and M_t. As shown in Fig. 2.33, the anchor points are relatively uniform on the human model but are highly non-uniform on the Moai model. However, valid base domains in good shape can still be effectively constructed.

2.3.3 *Optimization*

In this section, an efficient algorithm is introduced to optimize the connectivity and the shape of common base domains in cross-parameterization. The optimization constrained by the anchor points is sped up by a new signature, length-preserved base domain (LPBD), for measuring the level of stretch between surface patches in cross-parameterization [41]. This new signature balances the accuracy of measurement and the computational speed well. Based on LPBD, a set of metrics are studied and compared. The best ones are employed in the domain optimization algorithm that consists of two major operators, *boundary swapping* and *patch merging*. After the optimization, every anchor point should still be a part of the simplicial complex of the patch layout. The positions of anchor points cannot be changed during the optimization as they have semantic meaning in applications (for example, digital human modeling).

2.3.3.1 Overview of Domain Optimization Algorithm

The domain optimization algorithm starts from the common base domains constructed by other methods (such as [37] or [40]). The three major steps are: boundary stretching, boundary swapping and patch merging. An example is shown in Fig. 2.34 to illustrate the function of each step.

Step 1: Boundary Stretching

This is a step to reduce the differences between the actual boundary shape of a surface patch in the patch layout and the shape of its corresponding planar domain. The stretching operator developed by Wang in [77] is applied to the boundary curves one by one. The principle of this curve stretching operator is to convert a curve into a geodesic curve locally by using an edge operator and a node operator to make the curve shorter dynamically—details can be found in [77]. After several iterations, the curve under stretching approximates a geodesic curve linking its two endpoints on the given mesh surface. Notice that, the stretched curve does not necessarily pass along the edges of the given mesh models. The intersection between a curve under stretching and other static curves must be prevented by the intersection avoidance method presented in [61]. The resultant patch layouts should have smooth boundaries for all patches. However, it is not difficult to find from the example shown in Fig. 2.34 that the distortion in the cross-parameterization is still very significant although stretching the boundary of patches can reduce the distortion slightly.

Step 2: Boundary Swapping

In this step of the optimization algorithm, the patch layouts on both models, M_s and M_t, are adjusted iteratively and locally to reduce the distortion in cross-parameterization. The *boundary swapping* operator is conducted to optimize the topology of the patch layouts. Without loss of generality, for two patches sharing a common boundary curve, applying the boundary swapping operator on them will convert the patches into two new patches by replacing the boundary curve with a new

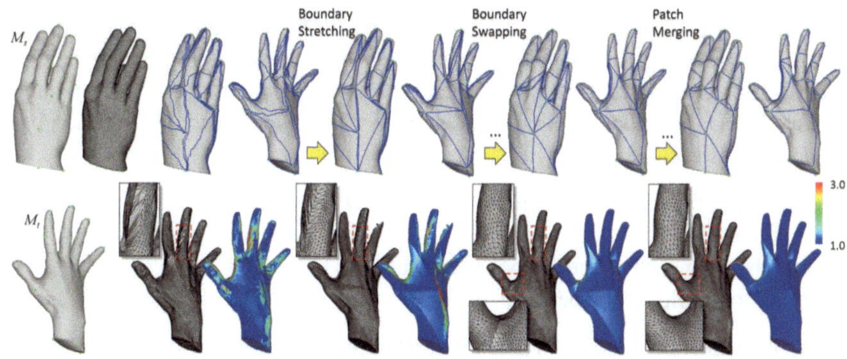

Fig. 2.34 An example for illustrating the overview of our domain optimization algorithm. The *left* most column shows two input hand models with prescribed anchor points (in *green*). *Top row* The topology and shape of common base domains on M_s and M_t are changed during the optimization. *Bottom row* A uniform mesh of the source model, M_s, is transferred onto the target model, M_t, using the established cross-parameterization where the transferred mesh on the target model should be smooth if the cross-parameterization has small distortion. The distortion of cross-parameterization is also visualized by the distribution of L^2-stretch [65] on the transferred mesh model (Image taken from [41] with permission ©2012 IEEE)

one linking its opposite anchor points in the triangular patches. The new boundary curve is constructed by first finding the shortest path along the mesh edges which links the opposite anchor points and then stretching the curve by the method of Wang in [77]. Again, the intersection between the new curve and other existing curves must be prevented. An alterative is the method of using adaptive edge nodes introduced in the above section. The priority list of applying the boundary swapping operator is built and maintained based on the metrics of the shape similarity of base domains, and serves as a signature to indicate the level of distortion in cross-parameterization. Detailed studies about the new signature, LPBD, are presented in Sect. 2.3.3.2.

Step 3: Patch Merging

A curved boundary shared by two patches will be removed in this step if such a removal helps reduce the distortion in the cross-parameterization. Apart from the reduction of distortion, other properties of the base domain (like the convexity and the flattenability) are also considered during patch merging to generate a valid and optimal result.

Following these main steps, the given mesh surfaces are trimmed by the boundary curves of the new patch layouts using a constrained delaunay triangulation (CDT) [45]. The newly constructed patch layouts lead to a cross-parameterization with less distortion (see Fig. 2.34 for an example).

2.3.3.2 LPBD as a Signature for Distortion

From the description of patch-based cross-parameterization in Sect. 2.3.1, it is known that a cross-parameterization is composed of three mapping functions: Γ_s, $\Gamma_{\rm st}$ and

Γ_t. Different from the domain optimization approach for a single model [60] that always employs regular triangles as the planar base domains in the mapping, the optimization algorithm of common base domains allows the shape of base domains to vary in 2D. Specifically, the planar domain B^i for a 3D patch P^i on the given model is constructed by

1. preserving the length of boundary curves on P^i (that is, having invariant length of boundaries),
2. straightening each of the 2D boundary curves,
3. trying to mimic the shape of P^i in 2D,

where the whole boundary of a patch is subdivided into curves by the anchor points. Obviously, the planar base domains generated in this way, named as LPBD, have different shapes and sizes. The dissimilarity between the planar base domains B_s^i and B_t^i is a good signature to indicate the cross-parameterization from P_s^i to P_t^i.

By retaining the length of boundary curves, the mapping from P^i to B^i is stretch-free along the boundaries. If the boundary curves on P^i are geodesic curves, flattening them into straight planar line segments results in a stretch-free mapping around/across the lines. If P_s^i and P_t^i are developable surfaces, the mappings Γ_s and Γ_t from P_s^i and P_t^i to the LPBD, B_s^i and B_t^i, are stretch-free too. As the cross-parameterization Γ from P_s^i to P_t^i is series-wound mappings, $\Gamma_s \cdot \Gamma_{\text{st}} \cdot \Gamma_t^{-1}$, the distortion of Γ is only introduced by the 2D-to-2D mapping $\Gamma_{\text{st}} : B_s^i \Rightarrow B_t^i$ in this scenario. In other words, the dissimilarity between B_s^i and B_t^i measures the distortion of cross-parameterization Γ.

A more interesting study relates to how significantly the distortion in Γ_{st} contributes to the overall distortion in the cross-parameterization Γ when the patches P_s^i and P_t^i are not developable. With a given planar shape B^i, it can be assumed that the distortion in the 3D-to-2D mapping from P^i to B^i has been minimized—for example, the parameterization based on mean-value coordinates [20] is used for this purpose. The mapping Γ_{st} is again established by mean-value coordinates [20]. The following studies how the developability of a 3D patch affects the robustness of using the signature, LPBD, to measure the stretch in cross-parameterization. As shown in Fig. 2.35, for an unchanged source model M_s, when the target model M_t becomes more and more non-developable while still keeping the same boundary, the stretch error of the cross-parameterization does not show a significant change. However, when keeping the geometric details (that is, the level of developability) of patches but changing their boundaries (that is, the shape of LPBD), the stretch errors of the cross-parameterization change significantly (see the changes in Fig. 2.35 by columns). In summary, the dissimilarity between the LPBDs, B_s^i and B_t^i, is a robust signature to indicate the stretch in the overall cross-parameterization. Different from the original cross-parameterization approach [37] that constructs base domains by linking anchor points with straight lines in 3D, mapping geodesic curves of the patch boundaries onto planar straight lines conveys more surface dissimilarity to the shape of 2D domains, B_s^i and B_t^i. This enhances the robustness of the LPBD signature.

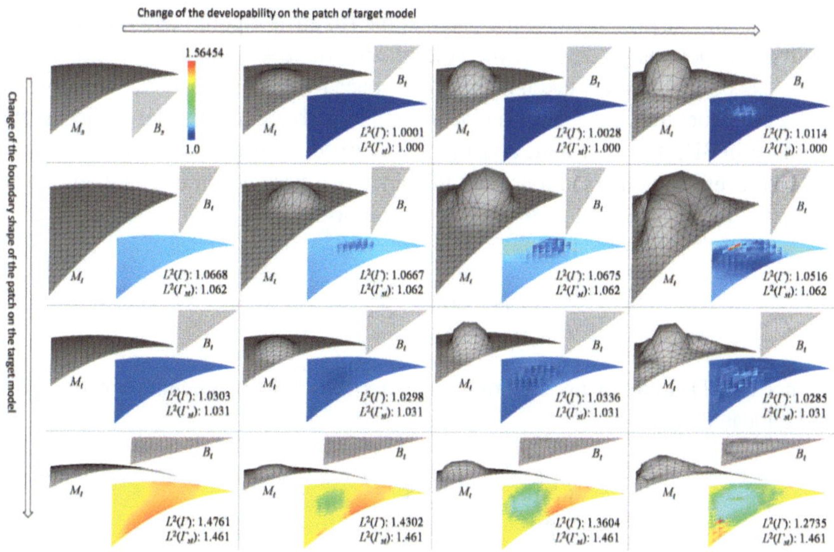

Fig. 2.35 A study of the robustness of using the dissimilarity in LPBDs to measure the distortion in cross-parameterization. For a given source model M_s shown in the upper left corner of the figure, the distortion between it and various target models M_t is tested, where the developability of patches keeps decreasing from left to right for patches in the same row and the boundaries of patches are changed for the patches in the same column. The L^2-stretches [65] in the cross-parameterization, $L^2(\Gamma)$, and the L^2-stretches on the planar base domains, $L^2(\Gamma_{st})$, are also listed in the figure. Notice that only the big triangles linking the anchor points are used to evaluate planar domain distortion, $L^2(\Gamma_{st})$. The color maps visualize the distribution of L^2-stretches generated by the cross-parameterization (Image taken from [41] with permission ©2012 IEEE)

Computation of LPBD

The LPBD, B^i, of a patch $P^i \in \Re^3$ on the given model can be easily computed by a constrained optimization framework similar to the computation of length-preserved free boundary (LPFB) in [79]. Assume that the boundary of P^i is determined by linking n anchor points with approximate geodesic curves, the optimal planar angles α_i on the ith anchor point are computed by solving a constrained optimization problem as

$$\begin{aligned} \min \sum_{i=1}^{n} \tfrac{1}{2}(\alpha_i - \theta_i)^2 & \\ s.t.\ n\pi - \sum_{i=1}^{n} \alpha_i &\equiv 2\pi \\ \sum_{i=1}^{n} l_i \cos\phi_i &\equiv 0 \\ \sum_{i=1}^{n} l_i \sin\phi_i &\equiv 0 \end{aligned} \tag{2.42}$$

where θ_i is the surface angle of the patch at the ith anchor point, and l_i denotes the length of the boundary curve between the ith and the $(i+1)$th anchor points on the patch. The computation is taken in the angle space by setting the angle variations as an objective function to generate a planar domain with a shape similar to the patch's boundary in 3D. The first hard constraint is derived from the closed-path theorem [54] to prevent self-intersections, and the latter two hard constraints are employed to ensure the generation of a closed boundary loop with

$$\phi_i = i\pi - \sum_{b=1}^{i} \alpha_i .$$

Once the 2D angles on anchor points are determined, the ith anchor point of LPBD can be located at

$$\left(\sum_{b=1}^{i} l_b \cos \phi_b , \sum_{b=1}^{i} l_b \sin \phi_b \right).$$

The LPBD can be computed efficiently because for a patch circled by n boundary curves, only n variables plus three constraints are involved in the computation (that is, a $(n+3) \times (n+3)$ linear equation system). This can be easily solved by the Quasi-Newton's method. Notice that the planar shape of a base domain can be directly computed by the lengths of its boundaries if the domain is triangular.

2.3.3.3 Boundary Swapping

Edge swapping (or called edge-flip) is a widely used operator to improve the quality of triangular meshes in mesh optimization (for instance, [10, 27, 74]). This optimization approach employs a similar concept to optimize the shape similarity of common base domains. Different from the edge swapping operator, performing swapping on the curved boundaries of base domains is more difficult because of the following four reasons:

- The shapes of domains on both the source and the target models need to be considered.
- The shapes of 2D domains after swapping a curved boundary cannot be predicted by that of 2D domains before swapping.
- The shape of a 2D domain after swapping could be degenerated, where the length of one boundary curve is longer than the sum of the other two, so the computed planar triangle could be an invalid LPBD.
- The computation of measuring the distortion in cross-parameterization based on dense meshes is time-consuming.

The purpose of the study below is to exploit an effective metric that can be efficiently evaluated.

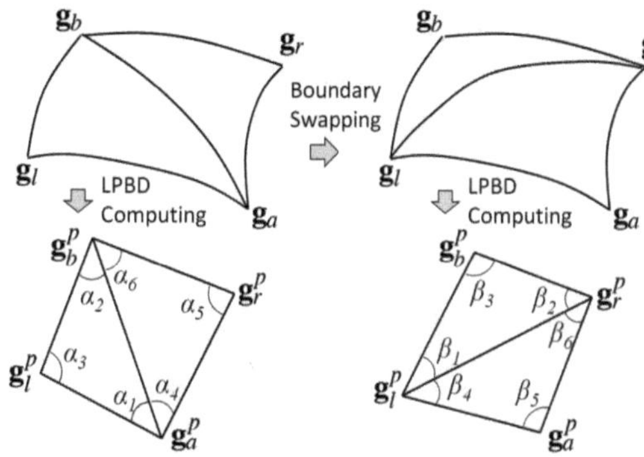

Fig. 2.36 *Boundary swapping* operator applied on a curved boundary edge $\widetilde{\mathbf{g}_a\mathbf{g}_b}$ shared by two patches $\widetilde{\Delta}\mathbf{g}_a\mathbf{g}_b\mathbf{g}_l$ and $\widetilde{\Delta}\mathbf{g}_b\mathbf{g}_a\mathbf{g}_r$ replaces $\widetilde{\mathbf{g}_a\mathbf{g}_b}$ by a new curve $\widetilde{\mathbf{g}_l\mathbf{g}_r}$ and therefore forms two new triangular patches $\widetilde{\Delta}\mathbf{g}_a\mathbf{g}_r\mathbf{g}_l$ and $\widetilde{\Delta}\mathbf{g}_b\mathbf{g}_l\mathbf{g}_r$ on the surface

Before going to the study of finding a good metric, let us assume that such a metric for boundary swapping has been well defined on every boundary curve $\widetilde{\mathbf{g}_a\mathbf{g}_b}$, as $\Upsilon(\widetilde{\mathbf{g}_a\mathbf{g}_b})$. For two patches $\widetilde{\Delta}\mathbf{g}_a\mathbf{g}_b\mathbf{g}_l$ and $\widetilde{\Delta}\mathbf{g}_a\mathbf{g}_b\mathbf{g}_r$ defined on four anchor points $\mathbf{g}_a$, $\mathbf{g}_b$, $\mathbf{g}_l$ and $\mathbf{g}_r$ and sharing a common boundary curve $\widetilde{\mathbf{g}_a\mathbf{g}_b}$, applying the boundary swapping operator can convert them into two new patches $\widetilde{\Delta}\mathbf{g}_a\mathbf{g}_r\mathbf{g}_l$ and $\widetilde{\Delta}\mathbf{g}_b\mathbf{g}_l\mathbf{g}_r$ (see Fig. 2.36). The metric $\Upsilon(\widetilde{\mathbf{g}_a\mathbf{g}_b})$ returns a value to indicate the level of distortion reduction in cross-parameterization by swapping $\widetilde{\mathbf{g}_a\mathbf{g}_b}$ to $\widetilde{\mathbf{g}_l\mathbf{g}_r}$, the value of which is the greater the better. Based on this metric, a greedy algorithm can be developed to conduct boundary swapping by using a maximum heap keyed on $\Upsilon(\cdots)$. Pseudo-code of the basic algorithm is given in the Algorithm 5. Notice that, when considering the new boundary $\widetilde{\mathbf{g}_l\mathbf{g}_r}$, the algorithm needs to prevent the paths from intersecting, blocking and being in wrong cyclical order. Moreover, the validity of LPBD needs to be verified. If invalid LPBDs will be generated after swapping, the operation must be prevented. In addition, if a boundary swapping can convert invalid LPBDs into valid ones, it must have the highest priority to be performed.

Metrics Based on Dense Meshes

A typical kind of metrics for $\Upsilon(\widetilde{\mathbf{g}_a\mathbf{g}_b})$ is one measured by the deformation between the dense meshes of $\widetilde{\Delta}\mathbf{g}_a\mathbf{g}_b\mathbf{g}_l$ and $\widetilde{\Delta}\mathbf{g}_a\mathbf{g}_r\mathbf{g}_b$ and their transformed shapes $\Gamma(\widetilde{\Delta}\mathbf{g}_a\mathbf{g}_b\mathbf{g}_l)$ and $\Gamma(\widetilde{\Delta}\mathbf{g}_a\mathbf{g}_r\mathbf{g}_b)$, which are defined by the cross-parameterization Γ. The metrics based on angle distortion or L^2-stretch on these dense meshes are presented below.

Angle Distortion: Borrowing the idea presented by Sheffer et al. [72], for any triangle $f_i \in \widetilde{\Delta}\mathbf{g}_a\mathbf{g}_b\mathbf{g}_l$ or $\widetilde{\Delta}\mathbf{g}_a\mathbf{g}_r\mathbf{g}_b$, angle distortion measures the change of its three angles $a_{i,1}, a_{i,2}$ and $a_{i,3}$ on the source model compared with the corresponding angles $a'_{i,1}, a'_{i,2}$ and $a'_{i,3}$ after applying the mapping Γ. If there are $|f|$ triangles in these two patches, the angle distortion on the patches can be calculated by

$$E_{\text{ang}}(\widetilde{\mathbf{g}_a\mathbf{g}_b}) = \frac{1}{3|f|}\sum_{i=1}^{|f|}\sum_{j=1}^{3}\left(\frac{a_{i,j} - a'_{i,j}}{a_{i,j}}\right)^2 \tag{2.43}$$

Algorithm 5 GreedyBndSwapping

Initialize an empty maximum heap H;
for every pair of $\widetilde{\mathbf{g}_a\mathbf{g}_b} \in M_s$ and $\widetilde{\mathbf{g}_a^*\mathbf{g}_b^*} \in M_t$ **do**
  Evaluate the metric $\Upsilon(\widetilde{\mathbf{g}_a\mathbf{g}_b})$ on them;
  Insert the curve $\widetilde{\mathbf{g}_a\mathbf{g}_b}$ into H when $\Upsilon(\widetilde{\mathbf{g}_a\mathbf{g}_b}) > 0$;
end for
while H is NOT empty **do**
  Remove a curved boundary edge $\widetilde{\mathbf{g}_a\mathbf{g}_b}$ from H;
  Apply the swapping operator on $\widetilde{\mathbf{g}_a\mathbf{g}_b}$;
  for any of $\widetilde{\mathbf{g}_a\mathbf{g}_l}$, $\widetilde{\mathbf{g}_a\mathbf{g}_r}$, $\widetilde{\mathbf{g}_b\mathbf{g}_l}$ and $\widetilde{\mathbf{g}_b\mathbf{g}_r}$ **do**
    Evaluate the metric $\Upsilon(\cdots)$ on the boundary curve;
    if the boundary curve is in H **then**
      Update its position in H;
    else
      Insert the curve into H when $\Upsilon(\cdots) > 0$;
    end if
  end for
end while

which has a lower bound of zero on examples without distortions. The boundary swapping metric based on angle distortion is then defined by the difference between the angle distortions $E_{\text{ang}}(\widetilde{\mathbf{g}_a\mathbf{g}_b})$ and $E_{\text{ang}}(\widetilde{\mathbf{g}_l\mathbf{g}_r})$.

$$\Upsilon_{\text{AD}}(\widetilde{\mathbf{g}_a\mathbf{g}_b}) = E_{\text{ang}}(\widetilde{\mathbf{g}_a\mathbf{g}_b}) - E_{\text{ang}}(\widetilde{\mathbf{g}_l\mathbf{g}_r}) \tag{2.44}$$

The angle distortion for the whole model, $E_{\text{ang}}(M_s)$, can be evaluated in a way similar to Eq. (2.43) by including all triangles on M_s, which in fact is introduced above in Eq. (2.39).

L^2-Stretch: Another widely used method for measuring the distortion of surface parameterization is the L^2-stretch introduced by Sander et al. in [65]. This is introduced above in Eqs. (2.40) and (2.41). The normalized L^2-stretch of all triangles in a particular region Ω is

$$E_{L^2}(\Omega) = \sqrt{\frac{\sum_{f_i \in \Omega} A(\Gamma(f_i)) \sum_{f_i \in \Omega}((L^2(f_i))^2 A(f_i))}{(\sum_{f_i \in \Omega} A(f_i))^2}}, \tag{2.45}$$

which has a lower bound of one. The boundary swapping metric based on the L^2-stretch on dense meshes is then defined on all triangles in the patches $\widetilde{\Delta}\mathbf{g}_a\mathbf{g}_b\mathbf{g}_l$, $\widetilde{\Delta}\mathbf{g}_a\mathbf{g}_r\mathbf{g}_b$, $\widetilde{\Delta}\mathbf{g}_a\mathbf{g}_r\mathbf{g}_l$ and $\widetilde{\Delta}\mathbf{g}_b\mathbf{g}_l\mathbf{g}_r$ as

$$\Upsilon_{L^2}(\widetilde{\mathbf{g}_a\mathbf{g}_b}) = E_{L^2}(\widetilde{\mathbf{g}_a\mathbf{g}_b}) - E_{L^2}(\widetilde{\mathbf{g}_l\mathbf{g}_r}). \tag{2.46}$$

Similarly, the L^2-stretch of the whole model, $E_{L^2}(M_s)$, can also be evaluated by including all triangles on M_s in the computation of Eq. (2.45), which in fact is the one used in Eq. (2.40).

Metrics Based on Shape of Domains

The main problem of directly applying the above metrics to the domain optimization for cross-parameterization is that the evaluation of them is generally time-consuming (especially when the given models have dense triangular meshes). Therefore, some other metrics, which are defined directly on the LPBD signature, have been studied. It is wished to find one among them that well balances the quality of results and the speed of computation. As the metrics, $\Upsilon_{\mathrm{AD}}(\widetilde{\mathbf{g}_a\mathbf{g}_b})$ and $\Upsilon_{L^2}(\widetilde{\mathbf{g}_a\mathbf{g}_b})$, are tightly coupled with the quality of cross-parameterization, the results of applying them to the Algorithm 5 are served as benchmarks for the selection of a good metric below.

Area similarity. The first tested metric is stimulated by Wang's analysis given in [79] that a surface parameterization with small distortion usually has small area variations (for example, flattening a developable mesh surface). Therefore, an area similarity metric is defined as

$$\Upsilon_{\mathrm{area}}(\widetilde{\mathbf{g}_a\mathbf{g}_b}) = E_{\mathrm{area}}(\widetilde{\mathbf{g}_l\mathbf{g}_r}) - E_{\mathrm{area}}(\widetilde{\mathbf{g}_a\mathbf{g}_b}) \tag{2.47}$$

where

$$E_{\mathrm{area}}(\widetilde{\mathbf{g}_c\mathbf{g}_d}) = (\psi(P_s^l) - \psi(P_t^l))^2 + (\psi(P_s^r) - \psi(P_t^r))^2$$

with P^l and P^r denoting the surface patches on the left and the right of the curved boundary $\widetilde{\mathbf{g}_c\mathbf{g}_d}$ respectively in the patch layout. The function $\psi(P^i)$ returns a value based on the area difference between a surface patch, P^i, and its LPBD, B^i.

$$\psi(P^i) = \begin{cases} A(P^i)/A(B^i) \ (A(P^i) \geq A(B^i)) \\ -A(B^i)/A(P^i) \ (A(P^i) < A(B^i)) \end{cases}$$

$A(B^i)$ is the area of a base domain, which can be directly calculated by *Heron's formula* using the lengths of boundary curves on P^i. However, the Heron's formula which is based on lengths may fail if the length of a boundary curve is larger than the sum of another two boundary curves' lengths—that is, a degenerated case happens. For a degenerated case, the area of B^i can be simply assigned with a small value (like 10^{-8}).

Domain angle distortion. In order to measure the similarity between triangular domains, one of the methods is to calculate the difference between the corresponding angles in the planar domains, B_s^i and B_t^i. Therefore, using the notation illustrated in Fig. 2.36 as the angles in LPBDs, the following metric can be defined to govern the swapping operator.

$$\Upsilon_{\mathrm{DAD}}(\widetilde{\mathbf{g}_a\mathbf{g}_b}) = \sum_{k=1}^{6} \left(\frac{\alpha_k^s - \alpha_k^t}{\alpha_k^s}\right)^2 - \sum_{k=1}^{6} \left(\frac{\beta_k^s - \beta_k^t}{\beta_k^s}\right)^2 \tag{2.48}$$

Again, a curved boundary edge $\widetilde{\mathbf{g}_a\mathbf{g}_b}$ is inserted into the priority list only when $\Upsilon_{\mathrm{DAD}}(\widetilde{\mathbf{g}_a\mathbf{g}_b}) > 0$.

Domain L^2-stretch. Another method that measures the similarity between the triangular domains is through the L^2-stretch which is used in Eq. (2.46). Differently, the L^2-stretch here measures the difference between a planar triangular domain of the source model and its corresponding domain of the target model. For example, $L^2(\Delta \mathbf{g}_i^P \mathbf{g}_j^P \mathbf{g}_k^P)$ gives the L^2-stretch between $\Delta \mathbf{g}_i^P \mathbf{g}_j^P \mathbf{g}_k^P$ and its corresponding triangular domain on the target model. For triangular domains adjacent to a curved boundary edge (as illustrated in Fig. 2.36), the metric of domain L^2-Stretch is defined as

$$\Upsilon_{\mathrm{DL}^2}(\widetilde{\mathbf{g}_a\mathbf{g}_b}) = E_{L^2}(\widetilde{\mathbf{g}_a\mathbf{g}_b}) - E_{L^2}(\widetilde{\mathbf{g}_l\mathbf{g}_r}) \tag{2.49}$$

with

$$E_{L^2}(\widetilde{\mathbf{g}_a\mathbf{g}_b}) = L^2(\Delta \mathbf{g}_a^P \mathbf{g}_b^P \mathbf{g}_l^P) + L^2(\Delta \mathbf{g}_b^P \mathbf{g}_a^P \mathbf{g}_r^P),$$
$$E_{L^2}(\widetilde{\mathbf{g}_l\mathbf{g}_r}) = L^2(\Delta \mathbf{g}_a^P \mathbf{g}_l^P \mathbf{g}_r^P) + L^2(\Delta \mathbf{g}_b^P \mathbf{g}_r^P \mathbf{g}_l^P).$$

Shorter diagonal length. By the experimental tests shown in [41], it is found that the distortion in cross-parameterization which is still high if the shapes of LPBDs of two models are similar (not exactly the same) to each other but are far from regular triangles. Therefore, the fourth metric tries to make the common base domains on both the source and the target models become 'regular' triangles. A rule employed in the mesh optimization to generate regular triangles is that a diagonal edge will be swapped if the swapped one is shorter. The metric Υ_{SDL} below is introduced for the same purpose.

$$\Upsilon_{\mathrm{SDL}}(\widetilde{\mathbf{g}_a\mathbf{g}_b}) = 2 - \left(\frac{\|\widetilde{\mathbf{g}_l\mathbf{g}_r}\|}{\|\widetilde{\mathbf{g}_a\mathbf{g}_b}\|} + \frac{\|\widetilde{\mathbf{g}_l^*\mathbf{g}_r^*}\|}{\|\widetilde{\mathbf{g}_a^*\mathbf{g}_b^*}\|} \right), \tag{2.50}$$

where $\|\widetilde{\mathbf{g}_a\mathbf{g}_b}\|$ denotes the length of the curve linking the anchor points $\mathbf{g}_a$ and $\mathbf{g}_b$, while $\|\widetilde{\mathbf{g}_a^*\mathbf{g}_b^*}\|$ denotes the length of the curve linking the anchor points $\mathbf{g}_a^*$ and $\mathbf{g}_b^*$. In this metric, the length changes of the boundary curves on both the source and the target models are considered. Swapping is prohibited if either side has become much worse after swapping—for example, when $\|\widetilde{\mathbf{g}_l\mathbf{g}_r}\| < \|\widetilde{\mathbf{g}_a\mathbf{g}_b}\|$ but $\|\widetilde{\mathbf{g}_l^*\mathbf{g}_r^*}\| > 2\|\widetilde{\mathbf{g}_a^*\mathbf{g}_b^*}\|$. Similarly, if the curves are elongated on both M_s and M_t, the swapping is also prevented since the value of Υ_{SDL} is negative. When the patches P_s^i and P_t^i for the common based domains on both models approach regular triangles, the shapes of their corresponding LPBDs, B_s^i and B_t^i, should be similar to each other as well (that is, the distortion in the mapping Γ_{st} is reduced).

Greatest angle reduction. In mesh optimization, another widely used rule which detects the possibility of applying the edge swapping operator is based on checking whether the greatest angle in two triangles adjacent to the edge is reduced. Assuming that α_k and β_k ($k = 1, \ldots, 6$) are the six angles in the two LPBD triangles adjacent to the edge $\widetilde{\mathbf{g}_a\mathbf{g}_b}$ and the swapped one $\widetilde{\mathbf{g}_l\mathbf{g}_r}$ respectively—see Fig. 2.36 for an illustration, we define the greatest angle reduction metric as

$$\Upsilon_{\mathrm{GAR}}(\widetilde{\mathbf{g}_a\mathbf{g}_b}) = \max_k\{\alpha_k^s, \alpha_k^t\} - \max_k\{\beta_k^s, \beta_k^t\}. \tag{2.51}$$

The swapping is applied only when $\Upsilon_{\mathrm{GAR}}(\widetilde{\mathbf{g}_a\mathbf{g}_b}) > 0$. Similar to using Υ_{SDL} to govern boundary swapping, the metric Υ_{GAR} also tends to make the shapes of the based domains on both models approach regular triangles. Therefore, it can reduce distortion in cross-parameterization. It should be remarked that all the angles of LPBDs in 2D are calculated by the length of the boundary curves directly using the *Law of Cosine*. For the degenerated cases (e.g., the length of one side is longer than the sum of the other two), the computed value ς for arccos by the *Law of Cosine* does not fall in the range $[-1, 1]$. For these cases, the angles are reassigned by the rules as $(-\varsigma)\pi$ ($\forall \varsigma \in (-\infty, -1)$) or $(1-\varsigma)\pi$ ($\forall \varsigma \in (1, \infty)$).

Analysis and Discussion of all Metrics

In order to select the 'best' metric for the *GreedyBndSwapping* algorithm whose purpose is to reduce distortion in cross-parameterization through optimizing the shape of common base domains, the algorithm has been applied to a variety of models by using all metrics mentioned above. The distortion in cross-parameterization based on the resultant base domains is measured by warping a source model M_s to the shape of a target model M_t and calculating the angle distortion $E_{\mathrm{ang}}(M_s)$ and the L^2-stretch $E_{L^2}(M_s)$ (by Eqs. (2.39) and (2.40)). Figure 2.37 shows the bar charts for the statistics of $E_{\mathrm{ang}}(M_s)$ and $E_{L^2}(M_s)$ on different models.

When analyzing the results, several interesting phenomena can be observed.

- First, the evaluation results obtained from using $E_{\mathrm{ang}}(M_s)$ and $E_{L^2}(M_s)$ to measure the distortion in cross-parameterization are not consistent with each other. Take David-egea models as an example, after using the angle distortion metric Υ_{AD} based on dense meshes to optimize the common base domains, the L^2-stretch, $E_{L^2}(M_s)$, becomes even worse than the cross-parameterization before domain optimization. Another example is the models of men. Using the L^2-stretch metric Υ_{L^2} to govern the boundary swapping turns out to have enlarged the global angle distortion, $E_{\mathrm{ang}}(M_s)$.
- Second, the results obtained from using the L^2-stretch metric Υ_{L^2} are better than that obtained from using the metric Υ_{AD} in terms of the L^2-stretch error, $E_{L^2}(M_s)$. However, using Υ_{AD} does not always give the result that a smaller value is on $E_{\mathrm{ang}}(M_s)$ than the results obtained by using Υ_{L^2} in the optimization.
- Third, when using the metrics defined by the dissimilarity of LPBDs to govern optimization, Υ_{DL^2} results in cross-parameterizations with less distortion than Υ_{area} and Υ_{DAD}—based on the evaluations of both $E_{L^2}(M_s)$ and $E_{\mathrm{ang}}(M_s)$.
- Last, after considering the regularity of the shape of base domains, the optimizations using the metrics Υ_{SDL} and Υ_{GAR} always give better results than one using the metrics, Υ_{area}, Υ_{DAD} and Υ_{DL^2}, which only consider the dissimilarity. Among them, using Υ_{SDL} generates better results in most cases.

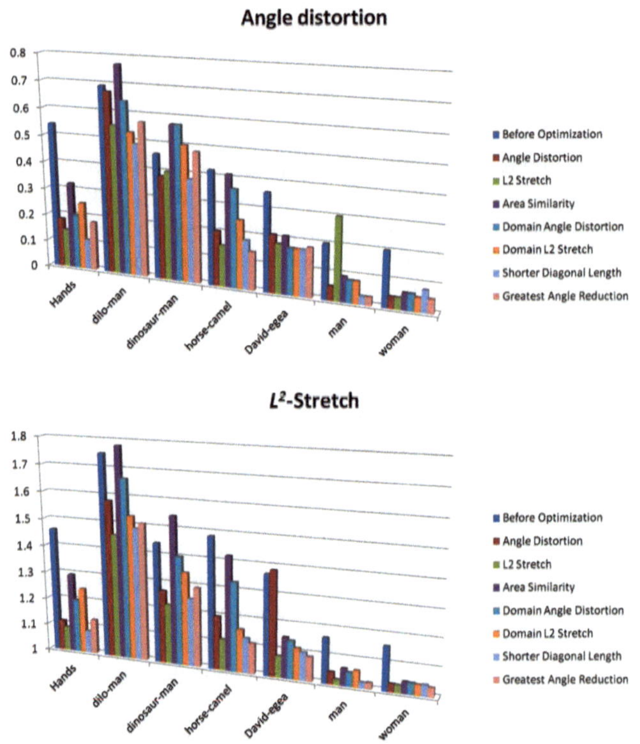

Fig. 2.37 Statistics of using different metrics in the greedy algorithm when applying the *boundary swapping* operator to improve the shape similarity of common base domains and thus reduce the distortion in cross-parameterization. The *top* bar chart shows the angle distortion $E_{\text{ang}}(M_s)$ while the *bottom* one gives the results of $E_{L^2}(M_s)$

Based on these observations, the metric Υ_{SDL} can be concluded as the best choice (based on the results of experimental tests) to reduce the distortion in cross-parameterization under the greedy boundary swapping algorithm. Furthermore, the metric Υ_{DL^2} gives better results in measuring the similarity of base domains than Υ_{DAD}.

Balanced Algorithm

Using the metrics defined by the shape of LPBDs instead of dense meshes can speed up the optimization procedure; however, a more significant speedup can be obtained if some of the swapping operations on boundary curves are skipped. Specifically, operations on those 'similar-enough' common base domains are skipped. This leads to a balanced algorithm for boundary curve swapping, which consists of three major steps.

- Step 1: The domain L^2-stretch, $E_{\text{DL}^2}(\widetilde{\mathbf{g}_a\mathbf{g}_b})$, defined on every curved boundary edge $\widetilde{\mathbf{g}_a\mathbf{g}_b}$ is first evaluated by Υ_{DL^2} in Eq. (2.49).

- Step 2: Among all curved boundaries, only those having $E_{\mathrm{DL}^2} > 2.0 + (1 - \varphi)\max\{E_{\mathrm{DL}^2} - 2.0\}$ are defined as active boundary curves. The golden ratio, 0.618, is selected for φ.
- Step 3: The greedy boundary swapping algorithm is only applied to the active boundary curves and the degenerated cases to optimize the common base domains by using the metric Υ_{SDL}.

This balanced algorithm gives a good tradeoff between the quality and the speed. Around 5–15 times speedup can be gained—more discussion on the computation time can be found in [41].

2.3.3.4 Patch Merging

The patch merging operator can be applied to further reduce the distortion in cross-parameterization. Without loss of generality, when applying the patch merging operator to a curved boundary edge $\widetilde{\mathbf{g}_a\mathbf{g}_b}$ adjacent to two triangular patches $\widetilde{\Delta}\mathbf{g}_a\mathbf{g}_b\mathbf{g}_l$ and $\widetilde{\Delta}\mathbf{g}_b\mathbf{g}_a\mathbf{g}_r$, the triangular patches are merged into a quadrilateral patch $\widetilde{\Diamond}\mathbf{g}_a\mathbf{g}_r\mathbf{g}_b\mathbf{g}_l$. Although merging patches into n-sided patches ($n > 4$) could possibly further reduce the distortion in cross-parameterization, it is easier for a polygon with $n > 4$ have concave corners—using which as the 2D domain may violate the correctness of cross-parameterization. Thus, in practice, it is rarely applied so as to ease the implementation and shorten the computation time.

The metric for governing the order of patch merging is defined based on the shape similarity of the LPBDs measured by angle distortion.

$$\Upsilon_{\mathrm{merge}}(\widetilde{\mathbf{g}_a\mathbf{g}_b}) = \frac{1}{6}\sum_{k=1}^{6}|\alpha_k^s - \alpha_k^t| - \frac{1}{4}\sum_{k=1}^{4}|\gamma_k^s - \gamma_k^t|, \tag{2.52}$$

where α_k and γ_k are angles on the LPBDs before and after merging respectively (as illustrated in Fig. 2.38). The metric is set to be $-\infty$ or removed from the heap to prevent patch merging operation when any of the following cases occurs.

- Any of the patches adjacent to the curved boundary $\widetilde{\mathbf{g}_a\mathbf{g}_b}$ is a quadrilateral patch.
- There is one edge of the to-be-merged quadrilateral patch longer than the sum of the other three—this will lead to a degenerated result when computing the LPBD of the patch.
- $\exists\gamma_k > \pi$ among $k = 1, \ldots, 4$. This case is prevented since any angle greater than π will make LPBD a concave polygon, which generates self-overlaps on the 3D-to-2D parameterization and such the bijective mapping for the cross-parameterization cannot be established. Detailed discussion about self-overlapping in mesh parameterizations can be found in [33].

Again, the greedy strategy is used with the metric $\Upsilon_{\mathrm{merge}}(\widetilde{\mathbf{g}_a\mathbf{g}_b})$ to perform patch merging. See the pseudo-code in Algorithm 6 for more information.

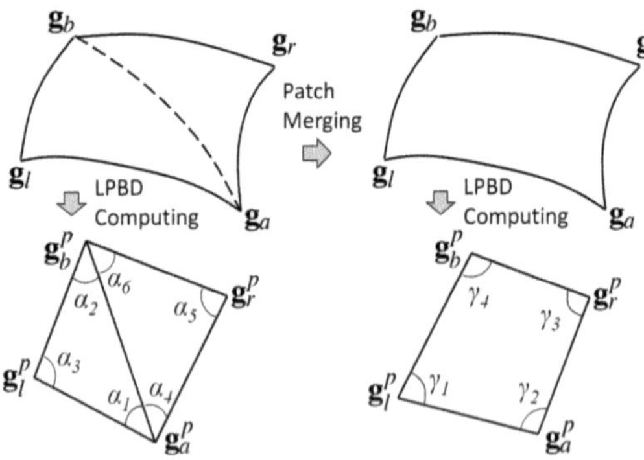

Fig. 2.38 *Patch merging* operator applied on a curved boundary edge $\widetilde{\mathbf{g}_a\mathbf{g}_b}$ (shown in *dash line*) adjacent to two triangular patches

Algorithm 6 GreedyPatchMerging

Initialize an empty maximum heap $H_{\max}$;
for every pair of $\widetilde{\mathbf{g}_a\mathbf{g}_b} \in M_s$ and $\widetilde{\mathbf{g}_a^*\mathbf{g}_b^*} \in M_t$ **do**
  Evaluate the metric $\Upsilon_{\text{merge}}(\widetilde{\mathbf{g}_a\mathbf{g}_b})$ on them;
  Insert $\widetilde{\mathbf{g}_a\mathbf{g}_b}$ into $H_{\max}$ when $\Upsilon_{\text{merge}}(\widetilde{\mathbf{g}_a\mathbf{g}_b}) > 0$;
end for
while $H_{\max}$ is NOT empty **do**
  Remove a curved edge $\widetilde{\mathbf{g}_a\mathbf{g}_b}$ from $H_{\max}$;
  if $\Upsilon_{\text{merge}}(\widetilde{\mathbf{g}_a\mathbf{g}_b}) > 0$ **then**
    Apply the patch merging operator on $\widetilde{\mathbf{g}_a\mathbf{g}_b}$;
    Remove the edges $\widetilde{\mathbf{g}_a\mathbf{g}_l}$, $\widetilde{\mathbf{g}_a\mathbf{g}_r}$, $\widetilde{\mathbf{g}_b\mathbf{g}_l}$ and $\widetilde{\mathbf{g}_b\mathbf{g}_r}$ from $H_{\max}$ if any of them has already been in $H_{\max}$;
  end if
end while

2.3.3.5 Discussion

The experimental results of this algorithm are shown in Figs. 2.34, 2.39 and 2.40. The distortion in cross-parameterization (measured by the L^2-stretch [65] and displayed by the color map) on the pairs of models is significantly reduced after applying the proposed domain optimization approach. The balanced boundary swapping algorithm followed by the *GreedyPatchMerging* algorithm generates results that are similar to that of the optimization which is based on the 'real' parameterization (that is, the metrics evaluated on the dense meshes). However, the speed of the balanced algorithm is about 5–15 times faster (see the statistics of times in [41]). In short, the LPBD signature introduced in [41] and the proposed balanced algorithm can efficiently improve the quality of cross-parameterization results by optimizing the shape of common base domains, which are constrained by the anchor points.

The main limitation of this approach is that, like other greedy-algorithm-based mesh optimization techniques, its results may only be local optima as it depends heavily on how good the given common base domains are. Some topological obstacles may prevent the algorithm from generating further optimized common base domains. Using the VD-based method introduced in Sect. 2.3.2.3 to generate the initial common base domains can somewhat help in this aspect.

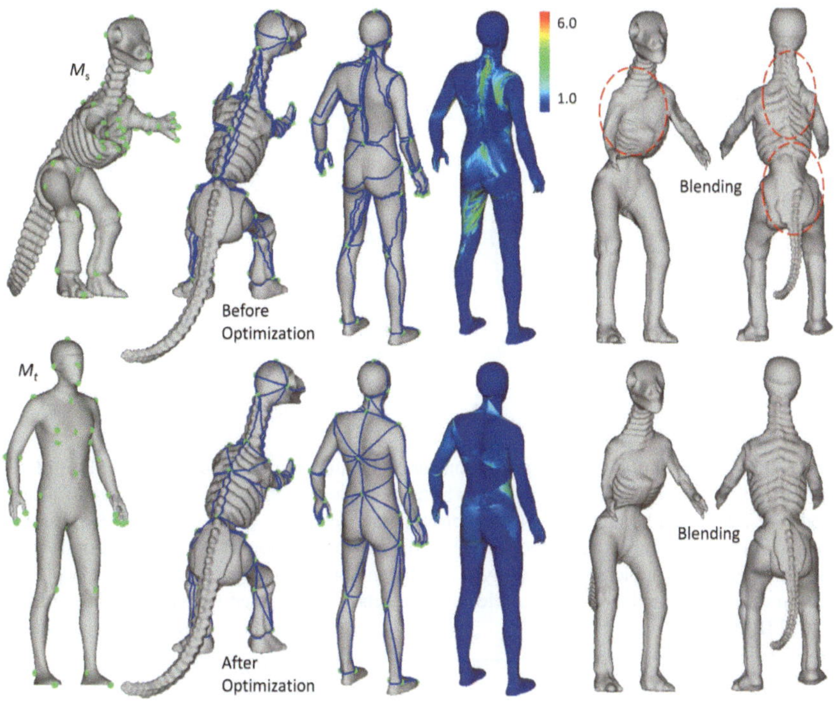

Fig. 2.39 The shape of the blended model of a dinosaur and a man can be improved by the optimized cross-parameterization. *Top row* The initial common base domains generated by [37] and the blending result with unreal distortion (*circled* by *red dash lines*) caused by the large stretch in cross-parameterization. *Bottom row* The optimized base domains lead to cross-parameterization with less distortion. The L^2-stretches are shown by the color map

2.4 Shape Space Analysis of Human Bodies

Using the cross-parameterization technique to process the mesh surface of individual human models, all the human models can be represented by mesh surfaces with a unique connectivity (that is, having the same number of vertices in the same order). Each of the individuals can be considered as an exemplar of training data. All the exemplars span the shape space of human bodies. This section presents the techniques for studying, representing and reconstructing the shape space of human bodies by the statistical model. Statistical model base on PCA [12] is first introduced for the shape space analysis, followed by a method for reconstructing a 3D human model by the processed exemplars stored in the data set.

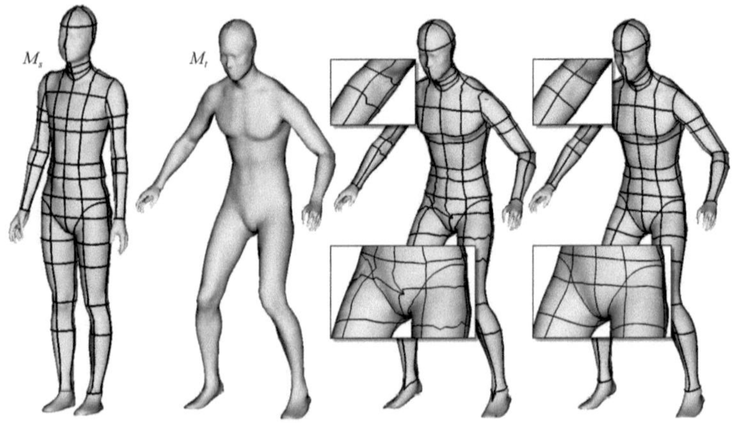

Fig. 2.40 An application of data preparation for establishing human body database. From *left* to *right*, a given source model M_s serving as a template model, an input mesh surface as a target model M_t, the result based on the original cross-parameterization [37] (with unwanted distortion on semantic feature curves), and the result based on optimized domains. The distortion in cross-parameterization is intolerable for many downstream applications (such as human space analysis and design automation of human-oriented products) (Image taken from [41] with permission ©2012 IEEE)

2.4.1 Statistical Model

PCA has been used in [2, 12, 24] to establish statistical models for analyzing human models. The main advantage is that the relationship between exemplars with low variance can be discarded after the PCA analysis; therefore, both the computational complexity and the size of data set are greatly reduced. The full data set does not need to be retained to reconstruct the original exemplars.

Assume that there are m scanned models serving as exemplars, they can be listed in a matrix

$$\mathbf{H} = [\mathbf{h}_1\ \mathbf{h}_2 \ldots \mathbf{h}_m]_{3n\times m}, \tag{2.53}$$

where $\mathbf{h}_i$ is a $3n \times 1$ vector with 3D positions of n vertices from the mesh surface of the ith model in the data set. Letting

$$\bar{\mathbf{h}} = \frac{1}{m}\sum_{i=1}^{m}\mathbf{h}_i, \tag{2.54}$$

$\mathbf{H}$ can be converted to

$$\bar{\mathbf{H}} = [(\mathbf{h}_1 - \bar{\mathbf{h}})\ (\mathbf{h}_2 - \bar{\mathbf{h}}) \ldots (\mathbf{h}_m - \bar{\mathbf{h}})]_{3n\times m}. \tag{2.55}$$

The covariance of $\bar{\mathbf{H}}$ is $\mathbf{C}_H = \bar{\mathbf{H}}(\bar{\mathbf{H}})^T$, whose dimension is $3n \times 3n$. As $m \ll 3n$, one can alternatively compute the transpose of its covariance, $\mathbf{C}_H^T = (\bar{\mathbf{H}})^T\bar{\mathbf{H}}$, to reduce the computational cost below. Applying eigenvalue decomposition on $\mathbf{C}_H^T$ as

$$\mathbf{C}_H^T\mathbf{x} = \lambda\mathbf{x}, \tag{2.56}$$

m eigenvectors, $\mathbf{x}_j$s which are $m \times 1$ vectors, can be obtained. By $\mathbf{x}_j$, the jth eigenvector of $\mathbf{C}_H$ can be determined as ([31] for reference)

$$\mathbf{y}_j = \bar{\mathbf{H}}\mathbf{x}_j, \tag{2.57}$$

where $\mathbf{y}_j$ is a $3n \times 1$ vector. The normalized eigenvectors $\hat{\mathbf{y}}_j = \mathbf{y}_j/\|\mathbf{y}_j\|$ $(j = 1, \ldots, m)$ are the principal vectors of $\bar{\mathbf{H}}$, where each is associated with an eigenvalue λ_j. In fact, each eigenvalue is proportional to the portion of the 'variance' (specifically, the sum of the squared distances between the exemplars and their multidimensional mean $\bar{\mathbf{h}}$) that is correlated with each eigenvector. The vectors are sorted so that

$$\lambda_1 \geq \lambda_2 \geq \cdots \geq \lambda_m. \tag{2.58}$$

The largest variance means that the corresponding vector $\mathbf{y}_1$ has the most dominant effect when modifying the model. The first k principal components are kept according to the percentage, r, of the total variance explained by each principal component.

$$r = \frac{\lambda_1 + \lambda_2 + \cdots + \lambda_k}{\lambda_1 + \lambda_2 + \cdots + \lambda_j + \cdots + \lambda_m} \tag{2.59}$$

The scanned and processed human models serving as a training data set can be projected (see [31] for reference) onto k-dimensional points by

$$\mathbf{b}_i = \begin{bmatrix} \hat{\mathbf{y}}_1^T \\ \hat{\mathbf{y}}_2^T \\ \vdots \\ \hat{\mathbf{y}}_k^T \end{bmatrix} (\mathbf{h}_i - \bar{\mathbf{h}}). \tag{2.60}$$

Thus, $\bar{\mathbf{H}}_{3n \times m}$ is mapped into a reduced matrix $\mathbf{B}_{k \times m} = [\mathbf{b}_i]$ $(k \ll 3n)$ spanning the linear space of exemplar human bodies, named as the *reduced exemplar matrix*. The *shape space of human models* is actually composed of the remaining principal components, $\{\hat{\mathbf{y}}_j\}$ $(j = 1, \ldots, k)$, and the reduced exemplar matrix, $\mathbf{B}$. Figures 2.41 and 2.42 show the principal components spanning the shape space of human models based on a data set with 47 female and 54 male exemplars.

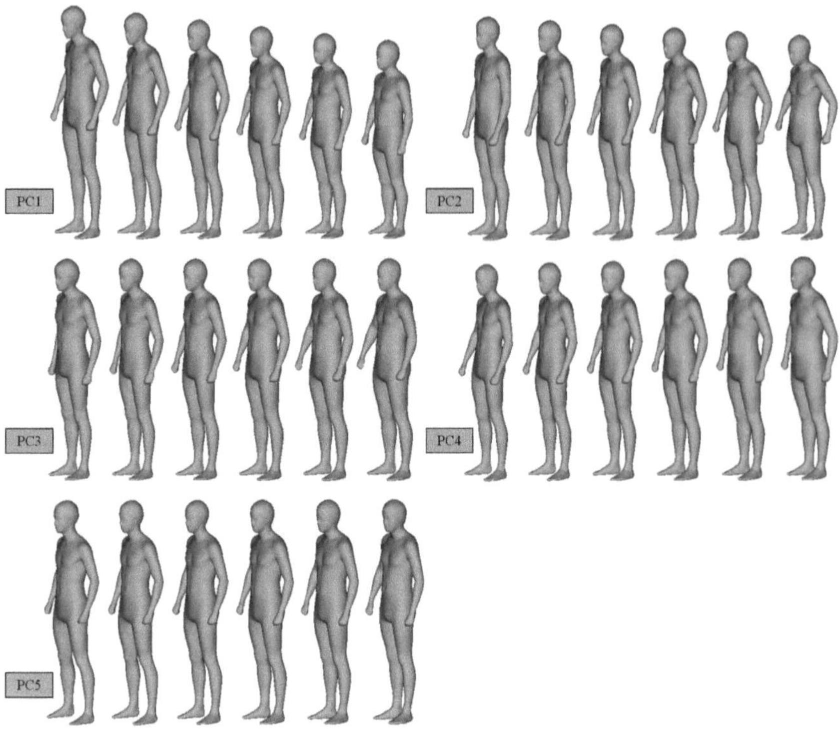

Fig. 2.41 Models spanning the first five principal components of 54 male exemplars

2.4.2 Synthesis-Based Reconstruction

A human model can be easily reconstructed in the shape space of human models represented by the above statistical model. After mapping a scanned model $\mathbf{h}_i$ into a k-dimensional point $\mathbf{b}_i$, an approximated human model $\widetilde{\mathbf{h}}_i$ can be synthesized from the k scalars in $\mathbf{b}_i$ by

$$\widetilde{\mathbf{h}}_i = [\hat{\mathbf{y}}_1 \ \hat{\mathbf{y}}_2 \ldots \hat{\mathbf{y}}_k]\mathbf{b}_i^T + \hat{\mathbf{h}}. \tag{2.61}$$

Although the value of k can be chosen by experiences, for example, let $r > 0.95$ in Eq. (2.59), it can be determined in a better way according to a user given geometric error tolerance, ε. In principle, choosing a greater value for k will make the L^2-norm,

$$L^2(\mathbf{b}_i) = \|\widetilde{\mathbf{h}}_i - \mathbf{h}_i\|^2$$

between $\widetilde{\mathbf{h}}_i$ and $\mathbf{h}_i$ smaller. Figure 2.43 shows the models reconstructed from 8, 19, 33, and 47 coefficients and the geometric error between the original models and them. The test can start from an initial guess k to see if the L^2 norms on all $\widetilde{\mathbf{h}}_i$s are smaller than ε. If not, increasing k by one and check again. Repeating this step, a minimal value of k satisfying the given geometric error tolerance can be determined.

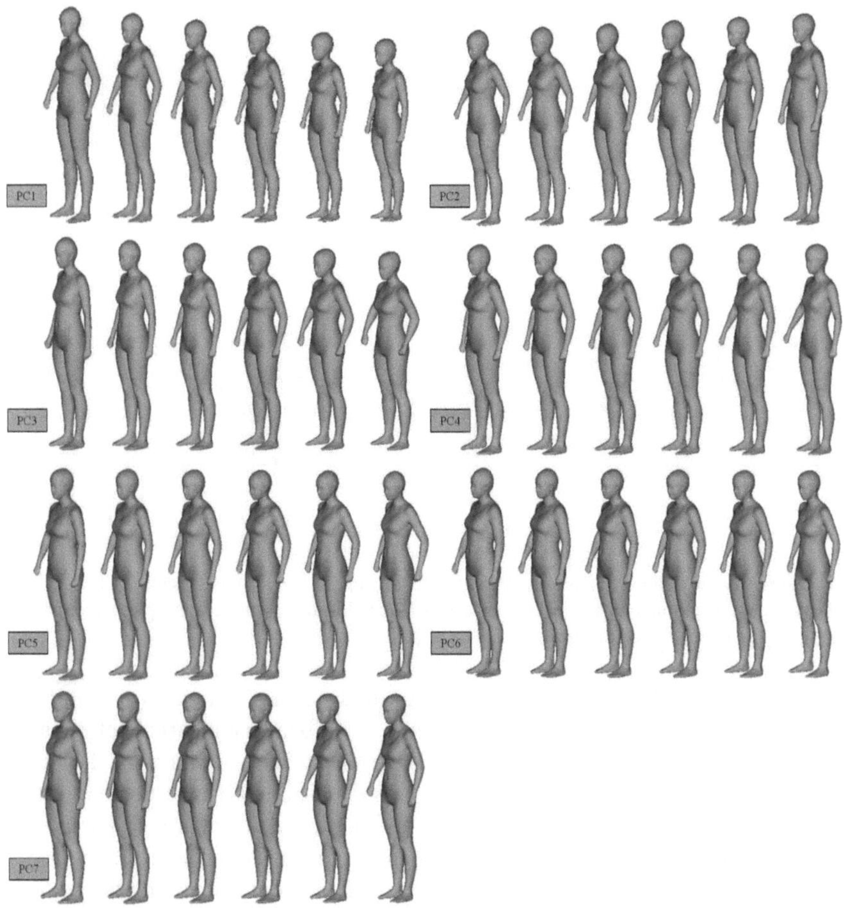

Fig. 2.42 Models spanning the first seven principal components of 47 female exemplars

2.5 Parametric Design of 3D Human Body

Automatic generation of realistic 3D human bodies is an important topic in many applications, such as biometric analysis, computer graphics, and apparel design. Several systems have been proposed, attempting to create general human models from 3D scans. However, not many small and medium-sized enterprises can afford the expensive acquisition device—3D human scanner. Therefore, there is a great demand from the market for a parametric design tool for human bodies so that the shape of a 3D human body can be generated from a set of semantic inputs (such as, height, bust girth, waist girth, hip girth, and inseam length).

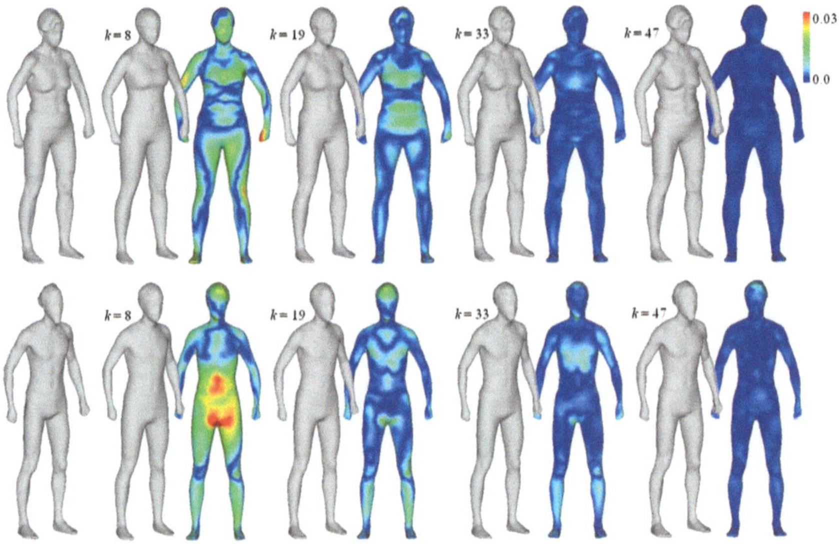

Fig. 2.43 Human bodies reconstructed using different numbers of PC coefficients (that is, different k)—using more coefficients can get better results. The color map visualizes the geometric distance between the reconstructed human model and the original model

Prior studies in literature have analyzed the body shape coefficients and their relationship with semantic parameters. Allen et al. [2] utilized parametric freeform mesh models to reconstruct human bodies from 3D scanners. Seo and Magnenat-Thalmann [68] developed a method for generating human bodies by giving a number of high level semantic constraints and evaluating the accuracy of linear-regression-based morphing functions. Linear models have been employed in other similar approaches. Scherbaum et al. [66] have concluded from their research concerning face morphing that, although nonlinear regression functions are numerically more accurate, the visual difference between them and their linear counterpart is minimal. Hasler et al. [24] adopted a similar morphing technique as a single linear equation system in the minimum norm sense for human pose modeling. They related 3D meshes and human body features (like body fat scale and weight) to generate the function of parametric design. Wang [78] adopted a nonlinear optimization approach to synthesize 3D human bodies from exemplars while satisfying the input semantic parameters; the computation, however, is very time-consuming. Some other recent related work can be found in [1, 7].

The above literature review indicates that most past studies concerning human body modeling focus on animation or human pose generation like [6] and [24]. Little work has addressed the feasibility of applying this approach to human body modeling for design and manufacturing aspects. The definition of semantic parameters for 3D human models in design and manufacturing of human-centered products is quite different from that of computer animation. Here, the feature definition given in [78] on the parameterized human bodies is employed, where the feature entities include vertices, curves and patches. Most of the semantic parameters can be measured from feature points and feature curves.

2.5.1 Correlation with Semantic Parameters

PCA helps characterize the space of human body variation, but it does not provide the correlation between the body shape and the semantic parameters. This section shows how to relate several variables by learning the relationship between the semantic parameters and the coefficients of principal components. Such relationship serves as the key to synthesizing a desired model from a set of input semantic parameters.

Suppose each scanned model has p semantic parameters, listing the parameters for all m exemplars forms a semantic parameter matrix

$$\mathbf{L} = [\mathbf{l}_1\ \mathbf{l}_2\ \cdots\ \mathbf{l}_m]_{p \times m}$$

with $\mathbf{l}_i$ being a column vector with p components. The correlation between $\mathbf{L}$ and the reduced exemplar matrix $\mathbf{B}$ can be represented by a system of linear equations. More specifically, it is assumed that

$$\mathbf{b}_i = \mathbf{R}\mathbf{l}_i + \mathbf{r} \quad (i = 1, \ldots, m) \tag{2.62}$$

where $\mathbf{R}_{k \times p}$ is the relation matrix and the vector $\mathbf{r}$ with k components is a corresponding residual. If $m > p + 1$, the value of $\mathbf{R}$ and $\mathbf{r}$ can be determined through a least-squares solution.

Given a new set of semantic parameters, $\mathbf{l}_{\text{new}}$, the corresponding coefficient in the reduced linear space $\mathbf{B}$ can be computed by

$$\mathbf{b}_{\text{new}} = \mathbf{R}\mathbf{l}_{\text{new}} + \mathbf{r}. \tag{2.63}$$

Then, the new human model can be obtained by

$$\mathbf{h}_{\text{new}} = [\hat{\mathbf{y}}_1\ \hat{\mathbf{y}}_2\ \cdots\ \hat{\mathbf{y}}_k]\mathbf{b}_{\text{new}}^T + \hat{\mathbf{h}}. \tag{2.64}$$

Notice that, if the value of p is not changed, $\mathbf{R}$ and $\mathbf{r}$ can be precomputed, and the computations in the whole regression and reconstruction procedure are just linear substitutions which can be completed in an interactive speed.

2.5.2 Feasibility Check

An example of parametric design is shown in Fig. 2.44a with the bust girth of a female model increased from 85 to 100 cm. However, the parametric design is constrained by the exemplars used. Given that most of the scanned models are within normal parameter ranges, a new model constructed with parameter values out of the ranges

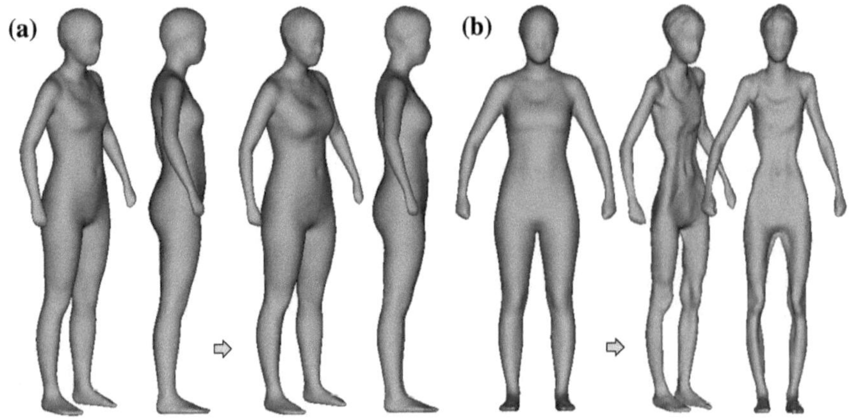

Fig. 2.44 Feasible versus infeasible changes of parameters: **a** when increasing the bust size of a model from 84 to 100 cm, a feasible human body can be reconstructed; however, **b** when decreasing the waist girth extremely from 66 to 30 cm, an infeasible model is generated. The computation employs 19 principal components

could be problematic. For a female model, it is not adequate to modify the waist size from 80 to 30 cm. Unreasonable meshes apparently occur in the output model (see Fig. 2.44b). The values out of feasible ranges may easily cause twists in the output model. Better generalization performance can be achieved by adding more extreme exemplars.

To overcome the above problem, a method was proposed in [12] to estimate the quality of the new model produced by a set of parameter values input by users. It determines whether the resultant model is of a good quality by considering the parameter ranges in the training data. Specifically, the convex hull of the m semantic parameters as sample points in the p-dimensional space, $\mathbf{l}_i$ $(i = 1, \ldots, m)$, can be used to check the feasibility of input parameters $\mathbf{l}_{\text{new}}$. For a given $\mathbf{l}_{\text{new}}$, if it is inside the convex hull, the given parameters in $\mathbf{l}_{\text{new}}$ are feasible. Otherwise, $\mathbf{l}_{\text{new}}$ is either rejected or projected onto its closest point, $\mathbf{c}(\mathbf{l}_{\text{new}})$, on the surface of the convex hull to compute the reconstructed model. An updated semantic input can be obtained.

Figure 2.45 shows an example of a gradual change of parameter values in three-dimensional space and the comparison results with the convex hull. For explanation simplification, only three semantic parameters are selected: height, bust girth, and waist girth. In this case, the height of a female model is modified from the original height of 168 cm to different heights. From Fig. 2.45, it is easy to find that only the models with a height of 160 or 165 cm are in the feasible region. The models with a height of 155 or 170 cm are outside but near the boundary of the convex hull; therefore, the reconstructed human model still looks reasonable. Other models that are far away from the convex hull look weird.

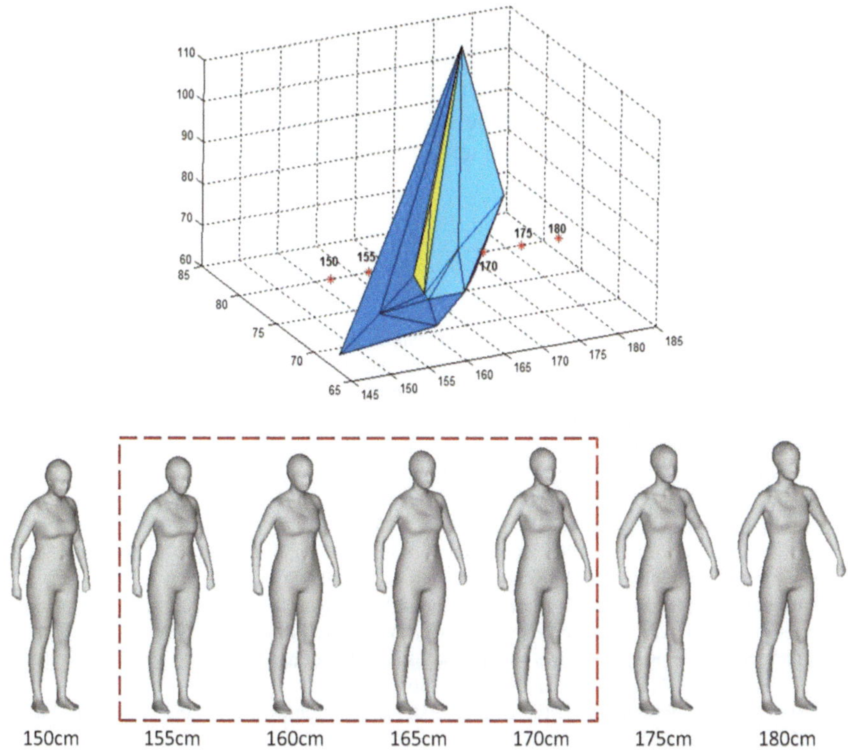

Fig. 2.45 Feasibility check on the input parameter vector by the convex hull formed by the projected semantic parameter space according to the exemplars. *Top* The convex hull of three semantic parameters: height, bust girth and waist girth. *Bottom* Changing the height of input semantic parameters, their corresponding points are shown on the *left* of the figure

2.5.3 Discussion

Figures 2.41 and 2.42 list the shape variations in the first five male and the first seven female principal components respectively from the training data set with 54 male and 47 female individuals. It is important to choose a proper threshold in the PCA step that balances the degree of data variance it produces and the complexity of the system. Enough principal components have to be kept to capture the variance in the training data. Insufficient principal components may result in inaccurate reconstructed models as analyzed in previous section. However, retaining more principal components is not always advantageous. The solution process may have additional computational load imposed by higher system complexity. Furthermore, too many components may capture unnecessary details that contain noise and thus weaken the effectiveness of the linear-regression model.

The statistical model introduced in this section has been proved to be stable even when there are only a few training data (see [12] for reference). Certainly, more training models will improve its quality and accuracy. Moreover, tests that include some noisy data as exemplars (see the study in [12]) prove that the linear-regression model introduced here is robust and tends to filter out noises.

References

1. Alexopoulos, K., Mavrikios, D., Pappas, M., Ntelis, E., Chryssolouris, G.: Multi-criteria upper body human motion adaptation. Int. J. Comput. Integr. Manuf. **20**, 57–70 (2007)
2. Allen, B., Curless, B., Popović, Z.: The space of human body shapes: reconstruction and parameterization from range scans. ACM Trans. Graph. **22**, 587–594 (2003)
3. Alliez, P., Cohen-Steiner, D., Devillers, O., Lévy, B., Desbrun, M.: Anisotropic polygonal remeshing. ACM Trans. Graph. **22**(3), 485–493 (2003)
4. Amenta, N., Bern, M., Kamvysselis, M.: A new voronoi-based surface reconstruction algorithm. In: Proceedings of SIGGRAPH'98, pp. 415–421 (1998)
5. Amenta, N., Choi, S., Kolluri, R.K.: The power crust. In: Proceedings of the Sixth ACM Symposium on Solid Modeling and Applications, pp. 249–266 (2001)
6. Anguelov, D., Srinivasan, P., Koller, D., Thrun, S., Rodgers, J., Davis, J.: Scape: shape completion and animation of people. ACM Trans. Graph. **24**, 408–416 (2005)
7. Baek, S.Y., Lee, K.: Parametric human body shape modeling framework for human-centered product design. Comput. Aided Des. **44**(1), 56–67 (2012)
8. Bernardini, F., Rushmeier, H.E.: The 3d model acquisition pipeline. Comput. Graph. Forum **21**(2), 149–172 (2002)
9. Bloomenthal, J.: Polygonization of implicit surfaces. Comput. Aided Geom. Des. **5**, 341–355 (1988)
10. Botsch, M., Kobbelt, L.: A remeshing approach to multiresolution modeling. In: SGP '04: Proceedings of the 2004 Eurographics/ACM SIGGRAPH Symposium on Geometry Processing, pp. 185–192 (2004)
11. Carr, J.C., Beatson, R.K., Cherrie, J.B., Mitchell, T.J., Fright, W.R., McCallum, B.C., Evans, T.R.: Reconstruction and representation of 3D objects with radial basis functions. In: Proceedings of SIGGRAPH 2001, pp. 67–76 (2001)
12. Chu, C.H., Tsai, Y.T., Wang, C.C.L., Kwok, T.H.: Exemplar-based statistical model for semantic parametric design of human body. Comput. Ind. **61**, 541–549 (2010)
13. Cormen, T.H., Leiserson, C.E., Rivest, R.L., Stein, C.: Introduction to Algorithms (2nd edn). MIT Press, Cambridge (2001)
14. Dey, T.K., Giesen, J., Hudson, J.: Delaunay based shape reconstruction from large data. In: Proceedings of the IEEE 2001 Symposium on Parallel and Large-Data Visualization and Graphics, pp. 19–27 (2001)
15. Dey, T.K., Goswami, S.: Tight Co Cone: a water-tight surface reconstructor. In: Proceedings of the Eighth ACM Symposium on Solid Modeling and Applications, pp. 127–134 (2003)
16. Dinh, H.Q., Turk, G., Slabaugh, G.: Reconstructing surfaces by volumetric regularization using radial basis functions. IEEE Trans. Pattern Anal. Mach. Intell. **24**, 1358–1371 (2002)
17. Eck, M., DeRose, T., Duchamp, T., Hoppe, H., Lounsbery, M., Stuetzle, W.: Multiresolution analysis of arbitrary meshes. In: SIGGRAPH '95: ACM SIGGRAPH 1995 Papers, pp. 173–182 (1995)
18. Fang, C.W., Lien, J.J.J.: Rapid image completion system using multiresolution patch-based directional and nondirectional approaches. IEEE Trans. Img. Proc. **18**, 2769–2779 (2009)
19. Floater, M.: Parametrization and smooth approximation of surface triangulations. Comput. Aided Geom. Des. **14**(3), 231–250 (1997)

20. Floater, M.: Mean value coordinates. Comput. Aided Geom. Des. **20**(1), 19–27 (2003)
21. Gatzke, T., Grimm, C., Garland, M., Zelinka, S.: Curvature maps for local shape comparison. In: SMI '05: Proceedings of the International Conference on Shape Modeling and Applications 2005, pp. 246–255. IEEE Computer Society, Washington, DC, USA (2005)
22. Guennebaud, G., Barthe, L., Paulin, M.: Interpolatory refinement for real-time processing of point-based geometry. Comput. Graph. Forum (Proc. Eurograph. 2005) **24**, 657–666 (2005)
23. Guennebaud, G., Gross, M.: Algebraic point set surfaces. ACM Trans. Graph. **26**, 23:1–23:8 (2007)
24. Hasler, N., Stoll, C., Sunkel, M., Rosenhahn, B., Seidel, H.P.: A statistical model of human pose and body shape. Comput. Graph. Forum (Proc. Eurograph. 2009) **28**, 337–346 (2009)
25. Hoffmann, C.: Geometric and Solid Modeling, Chapter 4: Robust and Error-Free Geometric Operations. http://www.cs.purdue.edu/homes/cmh/distribution/books/geo.html (2002)
26. Hoppe, H., DeRose, T., Duchamp, T., McDonald, J., Stuetzle, W.: Surface reconstruction from unorganized points. SIGGRAPH Comput. Graph. **26**, 71–78 (1992)
27. Hoppe, H., DeRose, T., Duchamp, T., McDonald, J., Stuetzle, W.: Mesh optimization. In: SIGGRAPH '93: Proceedings of the 20th Annual Conference on Computer Graphics and Interactive Techniques, pp. 19–26 (1993)
28. Hornung, A., Kobbelt, L.: Robust reconstruction of watertight 3d models from non-uniformly sampled point clouds without normal information. In: Proceedings of ACM Symposium on Geometry Processing, pp. 41–50 (2006)
29. Huang, H., Li, D., Zhang, H., Ascher, U., Cohen-Or, D.: Consolidation of unorganized point clouds for surface reconstruction. ACM Trans. Graph. **28**, 176:1–176:7 (2009)
30. Jain, V., Zhang, H.: Robust 3d shape correspondence in the spectral domain. In: SMI '06: Proceedings of the IEEE International Conference on Shape Modeling and Applications 2006, p. 19. IEEE Computer Society, Washington, DC, USA (2006). doi:http://dx.doi.org/10.1109/SMI.2006.31
31. Jolliffe, I.: Principle Component Analysis, 2nd edn. Springer, Berlin (2002)
32. Ju, T., Losasso, F., Schaefer, S., Warren, J.: Dual contouring of hermite data. ACM Trans. Graph. **21**, 339–346 (2002)
33. Karni, Z., Gotsman, C., Gortler, S.J.: Free-boundary linear parameterization of 3d meshes in the presence of constraints. In: SMI '05: Proceedings of the International Conference on Shape Modeling and Applications 2005, pp. 268–277. IEEE Computer Society (2005)
34. Kazhdan, M., Bolitho, M., Hoppe, H.: Poisson surface reconstruction. In: Proceedings of the Fourth Eurographics Symposium on Geometry Processing, pp. 61–70 (2006)
35. Khodakovsky, A., Litke, N., Schröder, P.: Globally smooth parameterizations with low distortion. In: SIGGRAPH '03: ACM SIGGRAPH 2003 Papers, pp. 350–357 (2003)
36. Kimmel, R., Sethian, J.: Computing geodesic paths on manifolds. Proc. Natl. Acad. Sci. U S A **95**, 8431–8435 (1998)
37. Kraevoy, V., Sheffer, A.: Cross-parameterization and compatible remeshing of 3d models. ACM Trans. Graph. **23**(3), 861–869 (2004)
38. Kuo, C.C., Yau, H.T.: A new combinatorial approach to surface reconstruction with sharp features. IEEE Trans. Vis. Comput. Graph. **12**, 73–82 (2006)
39. Kutulakos, K.N., Steger, E.: A theory of refractive and specular 3d shape by light-path triangulation. Int. J. Comput. Vis. **76**, 13–29 (2008)
40. Kwok, T.H., Zhang, Y., Wang, C.C.L.: Constructing common base domain by cues from Voronoi diagram. Graph. Model. **74**, 152–167 (2012)
41. Kwok, T.H., Zhang, Y., Wang, C.C.L.: Efficient optimization of common base domains for cross-parameterization. IEEE Trans. Vis. Comput. Graph. (2012)
42. Larsen, E., Gottschalk, S., Lin, M.C., Manocha, D.: Fast proximity queries with swept sphere volumes. In: Proceedings of International Conference on Robotics and Automation, pp. 3719–3726 (2000)
43. Levoy, M., Pulli, K., Curless, B., Rusinkiewicz, S., Koller, D., Pereira, L., Ginzton, M., Anderson, S., Davis, J., Ginsberg, J., Shade, J., Fulk, D.: The digital michelangelo project: 3d scanning of large statues. In: Proceedings of SIGGRAPH 2000, pp. 131–144 (2000)

44. Li, S.S.M., Wang, C.C.L., Hui, K.C.: Bending-invariant correspondence matching on 3d human bodies for feature point extraction. IEEE Trans. Autom. Sci. Eng. **8**(4), 805–814 (2011)
45. Li, W.C., Levy, B., Paul, J.C.: Mesh editing with an embedded network of curves. In: SMI '05: Proceedings of the International Conference on Shape Modeling and Applications 2005, pp. 62–71. IEEE Computer Society (2005)
46. Li, X., Bao, Y., Guo, X., Jin, M., Gu, X., Qin, H.: Globally optimal surface mapping for surfaces with arbitrary topology. IEEE Trans. Vis. Comput. Graph. **14**(4), 805–819 (2008)
47. Lipman, Y., Funkhouser, T.: Möbius voting for surface correspondence. ACM Trans. Graph. **28**, 72:1–72:12 (2009)
48. Liu, S., Chan, K.C., Wang, C.C.L.: Iterative consolidation of unorganized points. IEEE Comput. Graph. Appl. **32**, 70–83 (2012)
49. Liu, S., Wang, C.C.L.: Orienting unorganized points for surface reconstruction. Comput. Graph. **34**, 209–218 (2010)
50. Lorensen, W.E., Cline, H.E.: Marching cubes: a high resolution 3d surface construction algorithm. SIGGRAPH Comput. Graph. **21**, 163–169 (1987)
51. Mederos, B., Amenta, N., Velho, L., de Figueiredo, L.H.: Surface reconstruction from noisy point clouds. In: Proceedings of the Third Eurographics Symposium on Geometry Processing (2005)
52. Mello, V.c., Velho, L., Taubin, G.: Estimating the in/out function of a surface represented by points. In: Proceedings of the Eighth ACM Symposium on Solid Modeling and Applications, pp. 108–114 (2003)
53. Moenning, C., Dodgson, N.: Fast marching farthest point sampling. In: Proceedings of Eurographics (2003)
54. Mortenson, M.E.: Geometric Modeling. Wiley, New York (1997)
55. Mount, D., Arya, S.: ANN: a Library for Approximate Nearest Neighbor Searching. http://www.cs.umd.edu/mount/ANN/ (2006)
56. Ohtake, Y., Belyaev, A., Alexa, M., Turk, G., Seidel, H.P.: Multi-level partition of unity implicits. ACM Trans. Graph. **22**, 463–470 (2003)
57. Ohtake, Y., Belyaev, A., Seidel, H.P.: An integrating approach to meshing scattered point data. In: Proceedings of the 2005 ACM Symposium on Solid and Physical Modeling, pp. 61–69 (2005)
58. Ohtake, Y., Belyaev, A., Seidel, H.P.: 3D scattered data interpolation and approximation with multilevel compactly supported rbfs. Graph. Model. **67**, 150–165 (2005)
59. Peng, T., Gupta, S.K.: Model and algorithms for point cloud construction using digital projection patterns. J. Comput. Inf. Sci. Eng. **7**, 372–381 (2007)
60. Pietroni, N., Tarini, M., Cignoni, P.: Almost isometric mesh parameterization through abstract domains. IEEE Trans. Vis. Comput. Graph. **16**(4), 621–635 (2010)
61. Praun, E., Sweldens, W., Schröder, P.: Consistent mesh parameterizations. In: SIGGRAPH '01: Proceedings of the 28th Annual Conference on Computer Graphics and Interactive Techniques, pp. 179–184 (2001)
62. Quan, L.: Image-Based Modeling. Springer, Berlin (2010)
63. Ray, N., Li, W., Lévy, B., Sheffer, A., Alliez, P.: Periodic global parameterization. ACM Trans. Graph. **25**(4), 1460–1485 (2006)
64. Rotman, J.: An Introduction to Algebraic Geometry. Springer, Berlin (1988)
65. Sander, P., Snyder, J., Gortler, S., Hoppe, H.: Texture mapping progressive meshes. In: SIGGRAPH '01: ACM SIGGRAPH 2001, pp. 409–416 (2001)
66. Scherbaum, K., Sunkel, M., Seidel, H.P., Blanz, V.: Prediction of individual non-linear aging trajectories of faces. Comput. Graph. Forum **26**, 285–294 (2007)
67. Schreiner, J., Asirvatham, A., Praun, E., Hoppe, H.: Inter-surface mapping. ACM Trans. Graph. **23**(3), 870–877 (2004)
68. Seo, H., Magnenat-Thalmann, N.: An example-based approach to human body manipulation. Graph. Model. **66**, 1–23 (2004)
69. Shapiro, L., Brady, J.: Feature-based correspondence: an eigenvector approach. Image Vis. Comput. **10**(5), 283–288 (1992)

70. Sharf, A., Lewiner, T., Shamir, A., Kobbelt, L., Cohen-Or, D.: Competing fronts for coarse-to-fine surface reconstruction. Comput. Graph. Forum (Proc. Eurograph. 2006) **25**, 389–398 (2006)
71. Sharf, A., Lewiner, T., Shklarski, G., Toledo, S., Cohen-Or, D.: Interactive topology-aware surface reconstruction. ACM Trans. Graph. (Proc. SIGGRAPH 2007) **26**, 49:1–49:9 (2007)
72. Sheffer, A., Levy, B., Mogilnitsky, M., Bogomyakov, A.: ABF++: fast and robust angle based flattening. ACM Trans. Graph. **24**, 311–330 (2005)
73. Solinger, J.: Apparel Manufacturing Handbook: Analysis, Principles, and Practice. Bobbin Media Corp, Columbia (1988)
74. Surazhsky, V., Gotsman, C.: Explicit surface remeshing. In: SGP '03: Proceedings of the 2003 Eurographics/ACM SIGGRAPH Symposium on Geometry Processing, pp. 20–30 (2003)
75. Taylor, P.J., Shoben, M.M.: Grading for the Fashion Industry: the Theory and Practice. England Thornes, Cheltenham (1990)
76. Turk, G., O'brien, J.F.: Modelling with implicit surfaces that interpolate. ACM Trans. Graph. **21**, 855–873 (2002)
77. Wang, C.C.L.: Cyber Tape: an interactive measurement tool on polyhedral surface. Comput. Graph. **28**(5), 731–745 (2004)
78. Wang, C.C.L.: Parameterization and parametric design of mannequins. Comput. Aided Des. **37**, 83–98 (2005)
79. Wang, C.C.L.: Computing length-preserved free boundary for quasi-developable mesh segmentation. IEEE Trans. Vis. Comp. Graph. **14**(1), 25–36 (2008)
80. Wang, C.C.L., Chang, T.K.K., Yuen, M.M.F.: From laser-scanned data to feature human model: a system based on fuzzy logic concept. Comput. Aided Des. **35**, 241–253 (2003)
81. Wang, C.C.L., Hui, K.C., Tong, K.M.: Volume parameterization for design automation of customized free-form products. IEEE Trans. Autom. Sci. Eng. **4**(1), 11–21 (2007)
82. Wang, Y., Wang, C.C.L., Yuen, M.M.F.: Duplicate-skins for compatible mesh modelling. In: SPM '06: Proceedings of the 2006 ACM Symposium on Solid and Physical Modeling, pp. 207–217 (2006)
83. Yngve, G., Turk, G.: Robust creation of implicit surfaces from polygonal meshes. IEEE Trans. Vis. Comput. Graph. **8**, 346–359 (2002)

Chapter 3
Geometry of Freeform Products

3.1 Non-Manifold Data Structure for Freeform Products

In this section, a data structure that combines the boundary representation and the complex-based representation is defined. Based on this data structure, a set of basic topological operators used to manipulate the entities is created. The extended Euler operators, which perform necessary editing operations, are low-level operators. Several high-level operators, which perform commonly repeated sets of operations, are defined later in this chapter.

3.1.1 Topology

Topology is a branch of mathematics concerned with the properties of geometrical figures that remain unchanged when they are transformed by bending, stretching, shrinking, or twisting. Complex models require considerable attention to their topology. Simple elements are topologically connected using operators to form a complex model whose topology is preserved when subject to a variety of transformations. Topological definition is used to define the geometric form of the model and forms the basis of classification of object models, in particular, freeform object models. Thus, in the design of data structure of a geometric modeling system, topology forms the core of the supporting data structure. The geometric entities within a geometric model are connected by a topological network [19].

The geometric coverage of conventional *boundary representation* (B-rep) solid modelers is confined to the domain of *two-manifold* objects. For every point on the boundary of a two-manifold object, there exists a sufficiently small neighbor that is topologically the same as an open disk in $\Re^2$. If there is any point on the boundary that does not satisfy the two-manifold condition, the object is classified as *non-two-manifold*, or simply *non-manifold* (see Fig. 3.1 for an illustration). While almost all physical artifacts in the world are two-manifold objects in terms of modeling, the

C. C. L. Wang, *Geometric Modeling and Reasoning of Human-Centered Freeform Products*, DOI: 10.1007/978-1-4471-4360-4_3, © Springer-Verlag London 2013

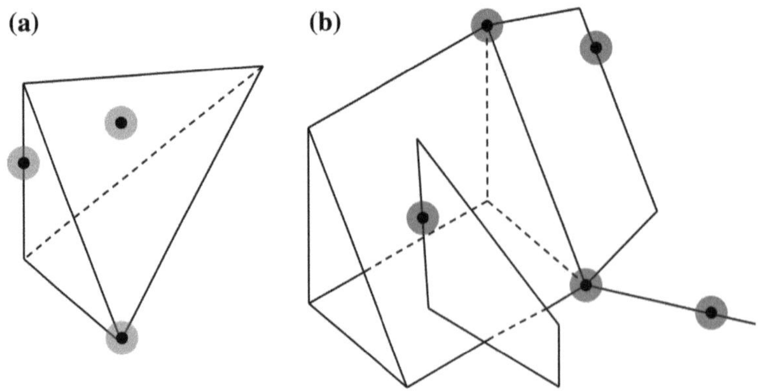

Fig. 3.1 **a** Two-manifold versus **b** non-two-manifold models, where the places to distinguish manifold and non-manifold entities are highlighted

domain of two-manifold cannot easily accommodate the entities of lower dimensionality, such as stand-alone faces and wireframe edges. Such entities are important in engineering design for representing the abstraction of a geometric shape (for example, 3D patterns in computer-aided garment design can be represented by assembled stand-alone surfaces).

One suitable mathematical definition of non-manifold geometric models is cell complexes that are defined as subsets of 3D Euclidean space [17, 22]. A geometric model represented by a cell complex is called a *complex-based non-manifold geometric model*. The concept of a cell complex is suitable for representing geometric shapes that are meaningful in engineering because its geometry domain covers a combination of wireframes, surfaces, and solid models.

Non-manifold geometric models satisfy the Euler-Poincarè formula (see [17] for reference),

$$n_v - n_e + (n_f - n_r) - (n_V - n_{V_h} + n_{V_c}) = n_C - n_{C_h} + n_{C_c} \tag{3.1}$$

where n_v, n_e, n_f, n_V, and n_C are the number of vertices, edges, faces, volumes, and complexes, respectively, while n_r, n_{V_h}, n_{V_c}, n_{C_h}, and n_{C_c} are the number of rings, holes through volumes, cavities in volumes, holes through complexes, and cavities in complexes, respectively. All non-manifold geometric models discussed in this book satisfy this formula.

3.1.2 Data Structure

There are many data structures described in the literature on modeling two-manifold objects. Most of the data structures take advantage of the fact that two faces share

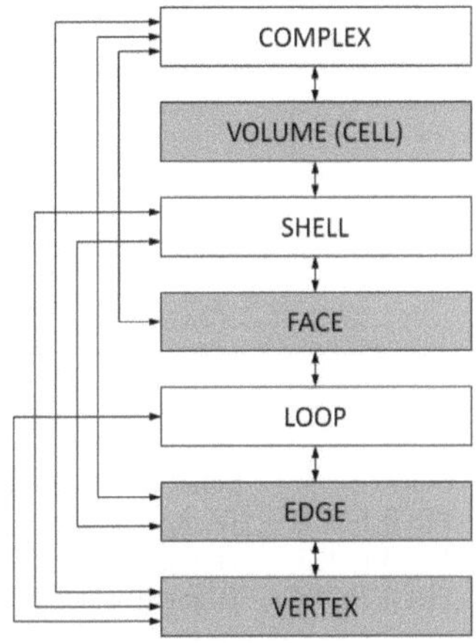

Fig. 3.2 Hierarchical structure of topological elements presented by Masuda in [17]

exactly one edge in a two-manifold model. Winged-edge data structure [1, 7] for polygonal models and half-edge data structure [2] for triangular mesh models are the most popular ones used in many geometric modeling systems. However, a more sophisticated structure is required to capture the topology of a non-manifold object. Several researchers [6, 14, 17, 31, 32] have developed different data structures for non-manifold topological modeling. A common feature of their work is the employment of the "*-use*" elements among the fundamental point set elements. A typical non-manifold topological data structure contains the following two types of elements:

- Fundamental point set elements
- Connection (or"*-use*") elements

Point set elements are the fundamental elements that make up the modeling space. Conventional names for entities of dimension *zero*, *one*, *two*, and *three* are *vertices*, *edges*, *faces*, and *volumes* (or called *cells*), respectively.

Topological elements are hierarchically interrelated. A lower dimensional element forms a part of the boundary element list of higher dimensional elements. The typical hierarchical relationship between topological elements is shown in Fig. 3.2. In this structure, a topological element can be linked to several types of higher dimensional topological elements. A face can be linked to a complex or to one or two volumes, depending on whether it is used as an adjacent face of the two volumes or as the boundary face of one or two volumes. An edge can be linked to complexes, shells, or loops. If an edge is directly linked to a complex or a shell, it is used as a wire edge in 3D space or as a volume. When an edge is linked to loops, it is used as the boundary of several faces. A vertex can be linked to edges, loops, shells, or complexes. When a vertex is linked to a loop, a shell, or a complex, it is used as an isolated vertex on a face, on a volume or in a 3D space, respectively.

Order information as well as adjacency information must be maintained to manipulate topological elements efficiently. The ordering of edges around a face and the ordering of edges around a vertex are well-known examples of ordering information

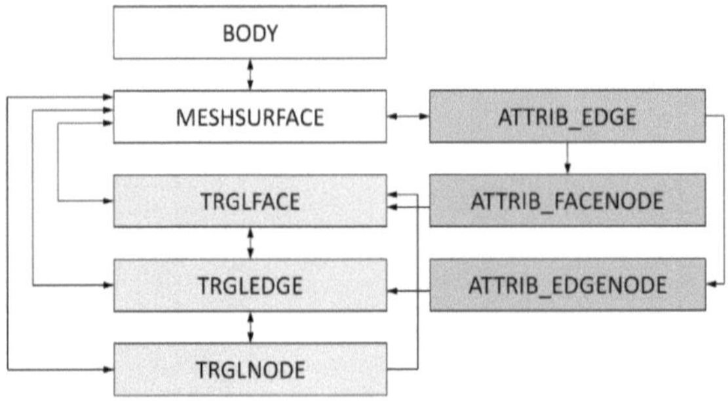

BODY	Entire model complex	MESHSURFACE	Mesh surface
TRGLFACE	Triangular face	ATTRIB_EDGE	Curves on mesh surface
TRGLEDGE	Triangular face edge	ATTRIB_FACENODE	Nodes on triangular face
TRGLNODE	Triangular face node	ATTRIB_EDGENODE	Nodes on triangular edge

Fig. 3.3 Data structure framework for modeling 3D human-centered freeform products

in solid modeling. They are also represented in the winged-edge structure or its variations. In non-manifold geometry modeling, however, it is also necessary to maintain the ordering of faces around an edge, which is termed *radial edge ordering* in the radial edge structure by Weiler [31]. Since radial edge ordering is useful for detecting closed space bounded by faces, it is explicitly maintained for effective manipulation of geometric models with multiple volumes. However, for modeling products like clothes which do not have explicitly represented volumes, such information could be neglected. For illustration purpose, a light version of non-manifold data structure is presented below, which is simplified from the comprehensive data structure presented by Wang et al. [27].

The data encapsulated in the entities must hold sufficient information to support non-manifold geometric modeling. Integrating the B-rep data structure used in ACIS [4] and the complex-based data structure by Masuda [17], the framework of the data structure can be defined as shown in Fig. 3.3.

There are several ways to view this data structure. For instance, it can be thought of as a tree with BODY as its root. A BODY has a collection of MESHSURFACEs, each of which comprises many TRGLFACEs, TRGLEDGEs, and TRGLNODEs. A TRGLFACE consists of three TRGLEDGEs. A TRGLEDGE is a line segment ended with two TRGLNODEs. Each MESHSURFACE is bounded by a set of boundary TRGLEDGEs, which has only one TRGLFACE on one side. The detailed description of each entity is shown in Table 3.1. Using this data structure, it is easy to carry out any topological and geometric manipulation on both the manifold and non-manifold triangular mesh models.

Besides topological entities, there are some objects 'attached' to the entities (such as the styling design curves in Sect. 3.3.1), which are attributes in the data structure. Four ATTRIBs are defined: ATTRIB_NODE, ATTRIB_EDGE,-ATTRIB_EDGENODE, and ATTRIB_FACENODE, where ATTRIB_EDGENODE and ATTRIB_FACENODE are derived from ATTRIB_NODE. ATTRIB_EDGE

Table 3.1 Representational entities

Entity	Comprisal	Information slots
BODY	Complex of MESHSURFACEs	(1) Bounding box of this object (2) List of MESHSURFACEs
MESHSURFACE	Complex of TRGLFACEs, TRGLEDGEs and TRGLNODEs	(1) Bounding box of this mesh surface (2) TRGLFACEs list (3) TRGLEDGEs list (4) TRGLNODEs list (5) ATTRIB_EDGEs list
TRGLFACE	Complex of three TRGLEDGEs	(1) Plane equation of this triangle (2) Direction flags of edges (3) Edges of this triangle (4) ATTRIB_FACENODES list
TRGLEDGE	Complex of two TRGLNODEs	(1) TRGLNODEs of this edge (2) Left and right TRGLFACEs (3) ATTRIB_EDGENODES list
TRGLNODE	A point	(1) Spatial position (2) Backup of spatial position (3) Planar position (4) Adjacent TRGLFACEs list (5) List of coupled TRGLNODEs

Table 3.2 Representational attributes

Attribute	Comprisal	Information slots
ATTRIB_NODE	A point	ATTRIB_EDGEs contain this node
ATTRIB_EDGE	Complex of sorted ATTRIB_NODEs	(1) ATTRIB_NODEs list (2) MESHSURFACE contains this edge
ATTRIB_EDGENODE	A point derived from ATTRIB_NODE	(1) Parametric coordinates (2) TRGLEDGE contains this edge
ATTRIB_FACENODE	A point derived from ATTRIB_NODE	(1) Barycentric coordinates (2) TRGLFACE contains this edge

NODE is the attribute node on a TRGLEDGE, and ATTRIB_FACENODE is the attribute node on a TRGLFACE. Their coordinates depend on the positions of TRGLEDGE's nodes or the positions of TRGLFACE's nodes. In detail, the coordinates of an ATTRIB_EDGENODE are represented by a parameter u related to the nodes of a TRGLEDGE while the coordinates of an ATTRIB_FACENODE are represented by (u, v, w)—the barycentric coordinates of a TRGLFACE with $u + v + w \equiv 1$ and $0 \leq u, v, w \leq 1$. An ATTRIB_EDGE is an ordered collection of ATTRIB_NODEs, which can be either ATTRIB_EDGENODEs or ATTRIB_FACENODEs. The detailed description of each attribute is shown in Table 3.2.

Table 3.3 Function description of extended Euler operators

Operator	Function of operator
make[kill]_vertex_complex	Create a single vertex complex
make[kill]_vertex_edge	Create a vertex and an edge connecting to an existing vertex (the operator can be carried out freely or inside a face)
make[kill]_edge_Chole	Connect two vertices by a new edge to form a hole in a complex
make[kill]_edge_kill[make]_complex	Connect two vertex complexes by a new edge (since the two complexes are connected, one complex should be removed)
make[kill]_face_kill[make]_Chole	Create a new face on a complex hole and remove the hole
make[kill]_vertex_ring	Create a stand-alone vertex on a face (the single vertex forms a ring)
make[kill]_edge_kill[make]_ring	Connect two stand-alone vertices by a new edge on a face (two rings are merged into one ring, thus one ring should be removed)
Split[merge]_edge	Split one edge into two edges by adding a new vertex onto the original edge

3.1.3 Topological Operators

The construction of a valid geometric model is achieved by using a proper set of topological operators. In geometric modeling, the fundamental topological operators are Euler operators [17, 32] that are consistent with the Euler-Poincarè formula. Likewise, the extended topological operators for non-manifold geometric modeling have to satisfy the same formula. Theoretically, just nine independent Euler operators and their inverse operators are sufficient to define all complex-based non-manifold geometric models [32]. When volume is not included in an object (as represented by the above data structure), only eight extended Euler operators are utilized. As illustrated in Fig. 3.4 and listed in Table 3.3, they are restricted to triangular meshes. Reverse operators are enclosed in brackets, and *Chole* denotes a hole in a complex.

When editing a model, several repeated sequences of the extended Euler operators are often used. Five of these sequences are formulated as high-level editing operators, which include *edge collapse*, *edge split*, *edge swap*, *face split*, and *face triangulation*. These high-level operators are provided to automate the performance of the extended Euler operator sequences and to increase the efficiency of topological operations. They have been frequently used in triangular mesh processing algorithms (for e.g., edge collapse, edge split, and edge swap were utilized for mesh optimization [9]; face split operator was conducted in the famous Loop Subdivision Scheme [16], Variational Subdivision Scheme [11], and Modified Butterfly Subdivision Scheme [34]; and face triangulation was applied for remeshing [28]). The detailed descriptions of these operators are listed as follows. Their sequences of extended Euler operators are shown in Table 3.4.

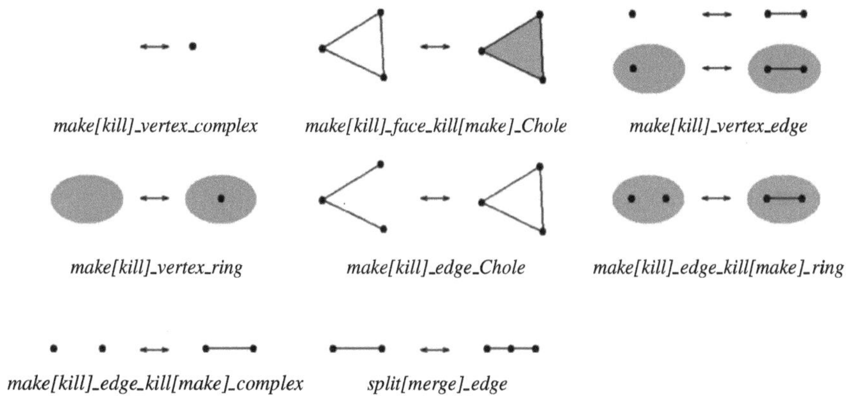

Fig. 3.4 Extended Euler operators without volume

Table 3.4 High-level operators consist of several low-level operators

High-level operators	Sequence of low-level operators
Edge split	*kill_face_make_Chole* (2 times) ⇒ *split_edge* ⇒ *make_edge_Chole* (2 times) ⇒ *make_face_kill_Chole* (4 times)
Edge collapse	*kill_face_make_Chole* (for all faces sharing $\mathbf{p}_d$) ⇒ *kill_edge_Chole* (for all edges sharing $\mathbf{p}_d$) ⇒ *kill_vertex_complex* (for $\mathbf{p}_d$) ⇒ *make_edge_Chole* (connecting to $\mathbf{p}_e$ for all original faces sharing $\mathbf{p}_d$ except f_1 and f_2) ⇒ *make_face_kill_Chole* (appropriate times)
Edge swap	*kill_face_make_Chole* (2 times) ⇒ *kill_edge_Chole* ⇒ *make_edge_Chole* ⇒ *make_face_kill_Chole* (2 times)
Face split	*kill_face_make_Chole* ⇒ *split_edge* (3 times) ⇒ *make_edge_Chole* (3 times) ⇒ *make_face_kill_Chole* (4 times)
Face triangulation	*kill_face_make_Chole* ⇒ *make_vertex_complex* (by the attribute face node), *make_vertex_edge* (by the attribute edge), *split_edge* (by the attribute edge node) ⇒ *make_edge_Chole* (appropriate times) ⇒ *make_face_kill_Chole* (appropriate times)

Edge split. For a triangular edge e, the edge split operator involves introducing an edge midpoint $\mathbf{p}_m$ that separates the edge into two new edges, e_1 and e_2. The two triangles, f_1 and f_2, sharing edge e are replaced by four new triangles, f_j ($j = 1, \ldots, 4$). One new vertex, $\mathbf{p}_m$, and four new edges, e_j ($j = 1, \ldots, 4$), are created. All topological information of entities sharing edge e and its endpoints should be altered (see Fig. 3.5a).

Edge collapse. The edge collapse operator is based on the identification of the two endpoints of edge e, thus leading to a unique point $\mathbf{p}_e$ that is either one of the original edge endpoints. This operator removes the two triangles, f_1 and f_2, their shared edge e, and two edges sharing the deleted vertex $\mathbf{p}_d$ in f_1 and f_2; and replaces the endpoints of edges sharing $\mathbf{p}_d$ to $\mathbf{p}_e$ (see Fig. 3.5a).

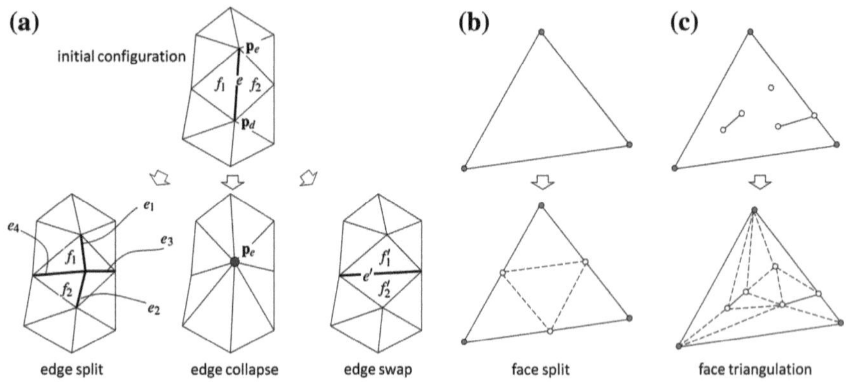

Fig. 3.5 High-level topological operators: **a** *edge collapse*, *edge split*, and *edge swap*, **b** *face split*, and **c** *face triangulation*

Edge swap. The two triangles (f_1 and f_2) sharing edge e are replaced by two new triangles (f_1' and f_2') sharing the dual edge e' of e. All topological information of entities sharing endpoints of edge e and endpoints of e' should be updated (see Fig. 3.5a).

Face split. The face split operator subdivides a triangular face into four triangular faces uniformly, where three new vertices are introduced to divide each triangular edge into two edges, and the new triangular edges and faces are constructed to link these new vertices. An illustration of the face split operator is shown in Fig. 3.5b.

Face triangulation. This operator triangulates a face with stand-alone attribute vertices and the edges on it. The vertices and edges are stored as ATTRIB_ FACENODEs, ATTRIB_EDGENODEs, and ATTRIB_EDGEs in the data structure. This operator converts ATTRIB_FACENODEs and ATTRIB_EDGENODEs to TRGLNODEs, and the TRGLEDGE with ATTRIB_EDGENODEs defined is divided into several TRGLEDGEs. The ATTRIB_EDGEs are converted to TRGLEDGEs. New TRGLEDGEs and TRGLFACEs are constructed to connect edges and nodes. The original TRGLFACE is removed. The newly constructed TRGLFACEs must be in the same orientation and not overlapped, and the newly constructed TRGLEDGEs must not intersect other triangles. One example of face triangulation is shown in Fig. 3.5c.

3.2 Constructive Design

The approach presented below extends the domain of constraint-based 3D geometric modeling from regular-shaped objects to 3D freeform objects. Every 3D freeform object M designed in this approach is related to the feature template of a human model H. The designed 3D freeform object has its own feature graph G_M, which is constructed by the constraints between G_M and H. The geometry of M determined by G_M gives only the structural geometry of M, and sketched input is used to provide the visible geometry of M by attaching some profiles Ω_M on G_M. When the geometry of G_M changes, the geometry of Ω_M changes correspondingly. The feature graph G_M of a designed freeform object M and a human model H are related by constraints. Therefore, when applying a different human model H' with the same connectivity which can be obtained by the digital human modeling method introduced in Chap. 2, the shape of M can be regenerated automatically.

3.2.1 Feature Graph

A designed 3D freeform object is constructed based on its feature graph. Each feature graph of the designed 3D freeform object is represented by a triangular mesh M^F. The notation in [9] is employed to define feature graphs here: a mesh M^F is a pair (K^F, V^F), where K^F is a simplicial complex specifying the connectivity of the mesh simplices (the adjacency of vertices, edges, and faces—i.e., the topological graph of M^F), and $V^F = \{\mathbf{v}_1, \dots, \mathbf{v}_m\}$ is the set of vertices defining the shape of the mesh in $\Re^3$. Each $\mathbf{v}_i \in V^F$ is called a *feature node* on the feature graph; each linkage between two vertices is called a *feature edge*; the faces bounded by feature edges are called *feature faces*. The feature graph is in fact a coarse mesh of the designed 3D freeform object. A refinement process is conducted to generate the detailed geometry of the designed 3D object. To better control the final shape of the designed 3D object, a new entity called *feature profile* is defined in the feature graph. A feature profile is a curve attached to a feature edge, and the feature profile is stored as a list of 3D points $\mathbf{q}_i$ $(i = 1, \dots, n)$. The first attached point $\mathbf{q}_1$ has the same position as the starting node of the edge, and the last attached point $\mathbf{q}_n$ is coincident with the end node of the edge. A parametric curve $\mathbf{c}(u)$ passing all these points is computed to describe the shape of the feature profile, where every attached point has its parameter $u_i \in [0, 1]$, $\mathbf{c}(0) = \mathbf{q}_1$, and $\mathbf{c}(1) = \mathbf{q}_n$. The parameter u_i is determined by the *chord length method* [20].

The whole feature graph consists of a node table, an edge table, and a face table. In the node table, the 3D position of each feature node is defined; in the edge table, the attached points list is the set of 3D points that defines the shape of a feature profile in $\Re^3$; and in the face table, the direction value is $+1$ when an edge is in the counterclockwise direction on a triangular face, and it is -1 when an edge is in the clockwise direction (the direction of an edge is pointing from its starting node to its end node). One example of the node table, the edge table, and the face table in a feature graph is shown in Fig. 3.6, where the bold curves are the feature profiles.

In existing constraint-based 3D geometric modeling systems [3, 10, 15], only CSG-like objects are handled. However, real applications usually require more complex geometric elements besides points, lines, and planes. In 2D approaches, the problem is usually solved by employing a two-level organization of geometric elements [21, 33]. This approach for 3D freeform object modeling also benefits from

Node Table				
1	60.0	0.0	0.0	{1, 2, 6}
2	0.0	60.0	0.0	{2, 3, 7}
3	-60.0	0.0	0.0	{3, 4, 8}
4	0.0	-60.0	0.0	{1, 4, 5}
5	0.0	0.0	60.0	{5, 6, 7, 8}

Face Table						
1	1	6	5	+1	+1	-1
2	2	7	6	+1	+1	-1
3	3	8	7	+1	+1	-1
4	4	5	8	+1	+1	-1

Edge Table				
1	4	1	5	(0, -60, 0); (22.961, -55.433, 0); ...
2	1	2	5	(60, 0, 0); (55.433, 22.961, 0); ...
3	2	3	5	(0, 60, 0); (-22.961, 55.433, 0); ...
4	3	4	5	(-60, 0, 0); (-55.433,-22.961, 0); ...
5	4	5	0	NULL
6	1	5	0	NULL
7	2	5	0	NULL
8	3	5	0	NULL

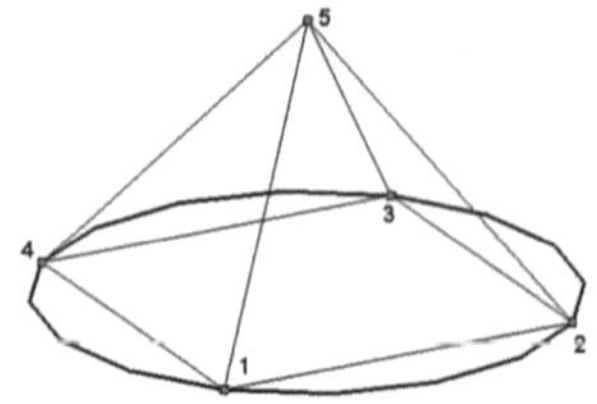

Fig. 3.6 An example of node table, edge table, and face table in a feature graph

it. Feature nodes, feature edges, and feature faces are used as the *structure geometry* entities; and feature profiles are used as the *visible geometry* entities. Since the feature profiles are attached to the feature edges, when the structural geometry is constrained, the visible geometry is also constrained.

3.2.2 Constraint Sets

After introducing the feature graph for designed 3D freeform objects, the geometric entities in the feature graph have to be constrained to the feature template of a human model. The constraints specify the relationship that must be maintained between the feature graph and the human model. In the approaches of constraint-based geometric modeling, the following classes of constraints arise naturally (see [8] for reference):

- Geometric relationship, such as concentricity or perpendicularity, and metric dimensional constraints, such as distances or angles;
- Topological relationship between entities in a model, such as incidence or connectivity;
- Semantic constraints that define validity conditions on a shape;
- Equational constraints that express the relationship between dimensional parameters, or between technological variables such as torque.

In the application of garment design (see [30] for reference), 32 different constraints are implemented, which is a 3D extension of the constraints set used in [21]. The constraints conducted in our approach can be classified into four groups: *positioning constraints*, *topological constraints*, *dimensional constraints*, and *structural constraints*. They are discussed in detail as follows. In order to apply the local propagation method (such as [23]) to solve the constraint problem, every constraint imple-

Table 3.5 Positioning constraints

Predicate	Meaning
Point($\mathbf{p}, x, y, z$)	Point $\mathbf{p}$ gets its absolute coordinates (x, y, z)
Point($\mathbf{p}, L_1, L_2$)	Point $\mathbf{p}$ is set as the intersection point of the lines L_1 and L_2
Point($\mathbf{p}, s, \mathbf{p}_1, \mathbf{p}_2$)	The position of $\mathbf{p}$ is determined by a scale factor s between $\mathbf{p}_1$ and $\mathbf{p}_2$ as $\mathbf{p} = (1 - s)\mathbf{p}_1 + s\mathbf{p}_2$
Point($\mathbf{p}, \mathbf{p}_s, \mathbf{t}, d$)	The position of $\mathbf{p}$ is determined by a unit vector $\mathbf{t}$ and a distance d from another point $\mathbf{p}_s$ as $\mathbf{p} = \mathbf{p}_s + d\mathbf{t}$
Point($\mathbf{p}, \mathbf{s}, u, v$)	On surface $\mathbf{s}(u, v)$, the position of $\mathbf{p}$ is determined by the parameters u and v
Point($\mathbf{p}, \mathbf{c}, u$)	On curve $\mathbf{c}(u)$, the position of $\mathbf{p}$ is determined by the parameter u
Line($L, \mathbf{p}_1, \mathbf{p}_2$)	Linking points $\mathbf{p}_1$ and $\mathbf{p}_2$ gets the line equation of L
Line(L, P_1, P_2)	The line equation of L is determined by the intersection of two planes P_1 and P_2
Plane($P, \mathbf{p}_1, \mathbf{p}_2, \mathbf{p}_3$)	The plane equation of P is determined by three points $\mathbf{p}_1, \mathbf{p}_2$, and $\mathbf{p}_3$
Xline(L)	The line L is along the x-axis
Yline(L)	The line L is along the y-axis
Zline(L)	The line L is along the z-axis
Profile($e, \mathbf{p}_1, \mathbf{p}_2, \mathbf{p}_3 \ldots$)	A designed profile attached to the feature edge e is approximated by a set of points $\{\mathbf{p}_1, \mathbf{p}_2, \mathbf{p}_3, \ldots\}$

Table 3.6 Topological constraints

Predicate	Meaning
On($\mathbf{p}, L$)	Point $\mathbf{p}$ lies on line L
On($\mathbf{p}, P$)	Point $\mathbf{p}$ lies on plane P
On(L, P)	Line L lies on plane P
Vertex($\mathbf{v}, \mathbf{p}$)	Point $\mathbf{p}$ is set as a feature node $\mathbf{v}$
Edge($e, \mathbf{v}_1, \mathbf{v}_2$)	Feature nodes $\mathbf{v}_1$ and $\mathbf{v}_2$ determine a feature edge e
Face($f, \mathbf{v}_1, \mathbf{v}_2, \mathbf{v}_3$)	Feature nodes $\mathbf{v}_1, \mathbf{v}_2$ and $\mathbf{v}_3$ give a feature face f in the counterclockwise direction

mented in this approach can determine the final position of a geometry element when other geometry elements in the constraint are known.

Positioning Constraints

The positioning constraints are usually defined to determine the absolute positions of geometric elements (such as point coordinates lines and plane equations). They are listed in Table 3.5.

Topological Constraints

The topological constraints are used to determine the topology of designed 3D freeform objects. This class of constraints contains a single predicate *On* to define the relationship between points, lines, and faces. It also contains other predicates to determine entities in the feature graph of a designed 3D freeform object. The topological constraints and their meanings are listed in Table 3.6.

Table 3.7 Dimensional constraints

Predicate	Meaning
$Angle(P_1, P_2, \alpha)$	The angle between planes P_1 and P_2 is α
$Distance(P_1, P_2, d)$	The distance between planes P_1 and P_2 is d when P_1 and P_2 are parallel
$Coincidence(\mathbf{p}_1, \mathbf{p}_2)$	Points $\mathbf{p}_1$ and $\mathbf{p}_2$ coincide
$Coincidence(L_1, L_2)$	Lines L_1 and L_2 coincide

Note that, since the following constraints can be changed to the representation of other constraints, they are not included in the constraints set.
$Coincidence(P_1, P_2) = Distance(P_1, P_2, 0)$—planes P_1 and P_2 are coincident. $Parallel(P_1, P_2) = Angle(P_1, P_2, 0)$—planes P_1 and P_2 are parallel. $Perpendicular(P_1, P_2) = Angle(P_1, P_2, \frac{\pi}{2})$—planes P_1 and P_2 are perpendicular

Table 3.8 Structural constraints

Predicate	Meaning
$Equal(P_1, P_2, P_3, P_4)$	$\angle(P_1, P_2) = \angle(P_3, P_4)$
$PointPos(\mathbf{p}_1, \mathbf{p}_2, x, y, z)$	The relative distances between points $\mathbf{p}_1$ and $\mathbf{p}_2$ in the coordinate directions are x, y and z
$Sum(P_1, P_2, P_3, P_4, P_5, P_6)$	$\angle(P_1, P_2) + \angle(P_3, P_4) = \angle(P_5, P_6)$
$Sum(P_1, P_2, P_3, P_4, \alpha)$	$\angle(P_1, P_2) + \angle(P_3, P_4) = \alpha$
$Symmetric(\mathbf{p}_1, \mathbf{p}_2, P)$	Points $\mathbf{p}_1$ and $\mathbf{p}_2$ are symmetric with respect to the reference plane P
$Symmetric(L_1, L_2, P)$	Lines L_1 and L_2 are symmetric with respect to the reference plane P
$Symmetric(P_1, P_2, P_r)$	Planes P_1 and P_2 are symmetric with respect to the reference plane P_r
$Difference(P_1, P_2, P_3, P_4, \alpha)$	$\angle(P_1, P_2) - \angle(P_3, P_4) = \alpha$
$Ratio(P_1, P_2, P_3, P_4, r)$	$\frac{\angle(P_1, P_2)}{\angle(P_3, P_4)} = r$

Dimensional Constraints
The dimensions between geometric elements are specified by this class of constraints (such as distance between planes), which are listed in Table 3.7.

Structural Constraints
The relationship between dimensions is constrained by the structural constraints. As mentioned by Podgorelec [21], they can also be addressed as conditional constraints. They take effect when all except one of the dimensions in the constraint equation are determined by other constraints; otherwise, they are ignored when solving the constraint problem. The structural constraints and their meanings are listed in Table 3.8.

The structural geometry of a feature graph can be determined by a local propagation-based constraint-solving algorithm [23]. Not only the geometric entities (feature nodes) but also the topological entities (feature edges and feature faces) can be constructed by computing the sequence determined by the local propagation. As mentioned by Gao and Chou [5], for a well-constrained problem, the topological

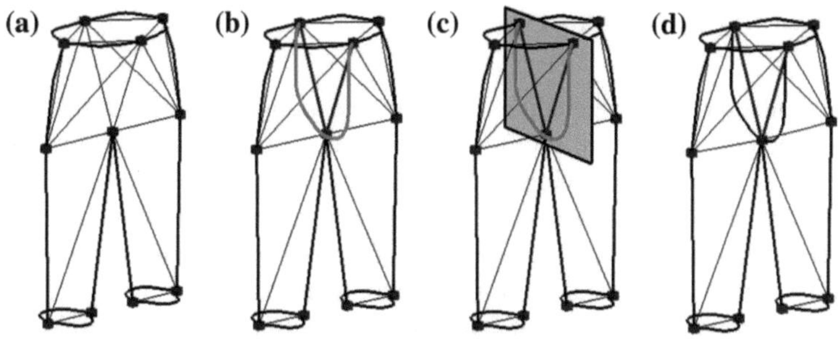

Fig. 3.7 Specifying feature profiles through a 2D stroke: **a** the feature graph before a feature profile is specified in the middle crotch, **b** an input 2D stroke, **c** the determined plane for projection, and **d** the resultant profiles attached to feature edges

completeness and the uniqueness of the designed freeform object are guaranteed by the propagation method.

3.2.3 Visible Geometry Specified by Sketch Input

After the structural geometry of the feature graph is constructed by the constraint-based approach, 2D sketches are used to specify 3D feature profiles on the feature graph. For a 2D stroke input Ψ, a plane is determined to project the 2D points onto the stroke to convert them into 3D points to be attached to feature edges. When multiple feature edges are selected, a plane that approximately passes through the selected feature edges is used. If only one feature edge is selected, the plane that bisects the dihedral angle along the chosen edge is used to project the points onto Ψ. In this way, the sketched profile faces toward the camera as much as possible. After all points on Ψ are converted into a set of points $\Psi^* \in \Re^3$, they are separated into intervals and the points in each interval are stored in the related feature edge of the interval. An example is given in Fig. 3.7.

Once the feature profile is specified, it is converted into the *Profile* $(e, \mathbf{p}_1, \mathbf{p}_2, \mathbf{p}_3, \ldots)$ constraint and stored. Therefore, next time when the same constraints are used to regenerate the designed 3D freeform object on a human model with a different shape, the positions of the attached points are shifted based on the positions of the endpoints of e. In other words, if an edge is moved from $\mathbf{v}_l\mathbf{v}_m$ to $\mathbf{v}_l^*\mathbf{v}_m^*$, the new positions of the points $\mathbf{p}_{i\ (i=0,\ldots,n-1)}$ attached to it are shifted by scaling the vector between them and $\mathbf{v}_l\mathbf{v}_m$ (more details can be found in [26]).

3.2.4 Shape Construction

This section describes the technique to construct a triangular mesh surface that interpolates specified 3D feature profiles and feature nodes on a given object feature graph. It is achieved by applying a modified *variational subdivision* scheme to the feature graph, which is defined as a coarse triangular mesh with some interpolating curves (i.e., feature profiles). The basic idea of a variational subdivision scheme is to iteratively apply a topological splitting operator to introduce new vertices in order to increase the number of degrees of freedom. The splitting operator is followed by a discrete fairing operator to increase the overall smoothness (see [11] for reference). In this approach, the scheme is applied to construct a mesh surface that interpolates not only the initial vertices but also the specified profiles.

Two operators are used in the subdivision scheme:

- The first operator is a topological splitting operator that inserts new control vertices into the mesh by the face splitting operator introduced above. The splitting operation is chosen to be uniform so that all the new vertices are regular (valence is equal to 6). The position of the inserted new vertex $\mathbf{v}^*$ that lies on the edge $\mathbf{v}_s\mathbf{v}_e$ is determined by $\mathbf{v}^* = \frac{1}{2}(\mathbf{v}_s + \mathbf{v}_e)$ if no profile is specified on $\mathbf{v}_s\mathbf{v}_e$. Otherwise, $\mathbf{v}^* = \mathbf{c}(\frac{1}{2})$ with $\mathbf{c}(u)$ being a parametric curve (such as B-Spline) interpolating all the points of the profile. $\mathbf{c}(u)$ is divided into two parts $\mathbf{c}_f(u)$ and $\mathbf{c}_b(u)$ at $\mathbf{c}(\frac{1}{2})$, and they are attached to the new edges created by splitting.
- The second operator is a smoothing operator that moves the control vertices according to weighted averages of neighboring vertices. The positions of vertices in the refined mesh are changed to achieve a global functional minimum. For example, the second-order umbrella operator in [12] can be implemented as an iterative solver for the problem. As mentioned by Kobbelt [11], since each update step only computes a linear combination of nearby vertices, the computational complexity is linear (in fact, is a constant number) if the number of umbrella iterations is bounded. In order to guarantee that the resultant fine mesh interpolates the originally given vertices, the umbrella operator must not be applied to those vertices that already belong to the initial mesh. In addition, the umbrella operator must not update the positions of the vertices that lie on the feature profiles to let the resultant fine mesh interpolate the feature profiles.

In this approach, a surface is constructed around a 3D reference model. Collisions between the constructed detail mesh surface and the human model are likely to occur during the surface construction procedure because only the structural geometry of a designed object is constrained whereas other newly introduced vertices can be moved freely. A voxel-based collision detection scheme [24] was integrated in [26] to prevent collisions between constructed surfaces and reference objects. More details can be found in [26].

3.3 Interactive Design of Tight Products

Using a feature graph to construct freeform products and constrain them to the shape of digital human models has the drawback of a relatively low flexibility of design activity. The feature graph as well as the constraints have to be updated explicitly upon every simple change. In practice, many designers wish to directly model their products in a flexible and user-friendly way. The most natural way for designers to create new fashion is to have a freehand drawing tool like drawing on a piece of paper with a pencil. This is a reason that the sketch-based interface becomes so popular in many 3D computer graphics applications. This section introduces freehand styling design tools developed by Wang et al. [29], which can be used to design products that tightly fit human bodies.

3.3.1 Styling Design by Curve Drawing

The curve creating tool of Wang's system [29] allows users to specify some points of a curve (i.e., control points). After that, a subdivision curve passing through these control points is generated by refining the line segments linking control points three times using the *Modified Butterfly Mask* [34, 35]. Here, the curve is treated as a 2D curve on the viewing plane. The control points are allowed to be moved to change the shape of the curve. After construction, the refined data points on the curve are projected onto the triangular mesh surface of a human model and linked by discrete geodesic curves. The projected styling curve is stored as a sequence of line segments in triangles (by ATTRIB_EDGEs and ATTRIB_NODEs) together with the control points for its construction.

As the viewing direction for further editing a curve may not be the same as the one creating it, the curve editing tool is developed in a different way—the styling curves are processed in 3D. First, the portion to be edited on a curve is specified. The original control points of this partial curve are employed as control points for further editing. If no control point is found (for example, the part to be edited is between two control points during construction), the curve is uniformly sampled into four to six points by its arc length and these sample points are employed as control points. Second, users can move the control points to adjust the curve. During movement, the control points and the refined data points are snapped onto the surface of the reference model. This is implemented by tracking these points to their closest points on the mesh surface in real time with the help of *Bounding Volume Hierarchy* (BVH) technique in [13]. The shape of the edited curve is again computed by the *Modified Butterfly Mask* but in 3D—so there may be interference between the curve and the mannequin, which can be eliminated by linking data points on the surface of the reference model with discrete geodesic curves. Figure 3.8 gives an illustration of this 3D curve editing tool.

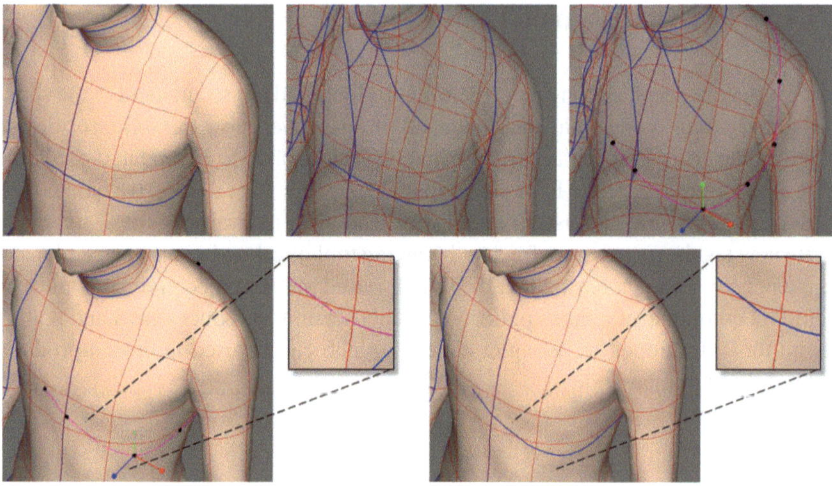

Fig. 3.8 The editing of styling curves is conducted in 3D: (*top-left*) the curve to be edited on a human model, (*top-middle*) the curve cannot be edited in a 2D manner as a part of it is at the back of the mannequin surface under the current view, (*top-right*) the curve is edited by dragging control points, (*bottom-left*) interference between the 3D curve and the mannequin occurs as only data points are snapped onto the human body (see the zoom-in view), and (*bottom-right*) the interference is eliminated by linking snapped data points with discrete geodesic curves

3.3.2 Trimming

The 3D pieces of products are generated by trimming duplicate mesh surfaces using styling curves. For the mesh surface M_H of a human model, the triangles with no styling curve are first copied into a new triangular mesh model M'_H. Second, the triangles with styling curves are re-triangulated one by one using the line segments of the styling curves. *Constrained Delaunay Triangulation* (CDT) is employed to carry out triangulation on the plane of these triangles (that is, the face triangulation operator introduced in the previous section). To make CDT robust to numerical errors, the line segments inside a triangle are processed by

1. finding intersections between line segments;
2. inserting intersections between line segments and triangle edges;
3. eliminating nearly overlapped (or partially overlapped) line segments;
4. merging nearly coincident points.

This is similar to the method presented by Mitani [18]. The resultant triangles of CDT are attached to the duplicate mesh model M'_H, and the information of whether a triangle edge is derived from a styling curve (or a feature curve) on M_H can be easily retrieved through local search using the correspondences of triangles between M'_H and M_H. Last, the triangle edges generated from the styling curves are duplicated so that the regions on M'_H surrounded by these edges are separated into pieces of mesh patches. An example can be found in Fig. 3.9.

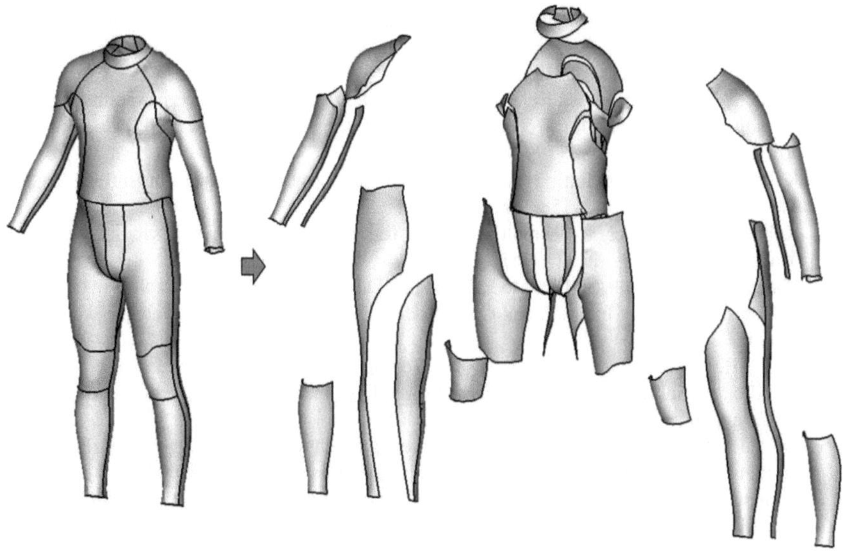

Fig. 3.9 An example of trimming the mesh surface of a human body into several pieces of mesh patches by styling curves (Image taken from [25] with permission ©2008 IEEE)

In this chapter, the constructive design tool and the freehand styling design tool have been introduced based on the geometric modeling and representation methods for freeform objects. However, the functionality of these tools is limited without the support of design transformation and the solution for optimizing manufacturability, which are covered in the following two chapters.

References

1. Baumgart, B.G.: Winged edge polyhedron representation. Technical Report, Stanford, CA, USA (1972)
2. Campagna, S., Kobbelt, L., Seidel, H.P.: Directed edges—a scalable representation for triangle meshes. ACM J. Graph. Tools **3**(4), 1–12 (1998)
3. Chen, X., Hoffmann, C.M.: On editability of feature-based design. Comput. Aid. Des. **27**, 905–914 (1995)
4. Corney, J.: 3D modeling with the ACIS kernel and toolkit. Wiley, New York (1997)
5. Gao, X.S., Chou, S.C.: Solving geometric constraint systems. i. a global propagation approach. Comput. Aided Des. **30**, 47–54 (1998)
6. Hansen, H., Christensen, N.J.: A model for n-dimensional boundary topology. In: Proceedings of the 2nd ACM Symposium on Solid Modeling and Applications, pp. 65–73 (1993)
7. Hoffmann, C.: Geometric and Solid Modeling, Chap. 4: Robust and Error-Free Geometric Operations. http://www.cs.purdue.edu/homes/cmh/distribution/books/geo.html (2002)
8. Hoffmann, C.M., Robert, J.A.: Parametric modeling. In: Handbook of computer aided geometric design, pp. 519–542. North-Holland, Amsterdam (2002)

9. Hoppe, H., DeRose, T., Duchamp, T., McDonald, J., Stuetzle, W.: Mesh optimization. In: SIGGRAPH '93: Proceedings of the 20th Annual Conference on Computer Graphics and Interactive, Techniques, pp. 19–26 (1993)
10. Hsu, C., Alt, G., Huang, Z., Beier, E., Brüderlin, B.: A constraint-based manipulator toolset for editing 3D objects. In: Proceedings of ACM 4th Symposium on Solid Modeling and Applications, pp. 168–180 (1997)
11. Kobbelt, L.: Discrete fairing and variational subdivision for freeform surface design. V. Comput. **16**(3–4), 142–158 (2000)
12. Kobbelt, L., Schröder, P.: A multiresolution framework for variational subdivision. ACM Trans. Graph. **17**, 209–237 (1998)
13. Larsen, E., Gottschalk, S., Lin, M.C., Manocha, D.: Fast proximity queries with swept sphere volumes. In: Proc. of International Conference on Robotics and Automation, pp. 3719–3726 (2000)
14. Lee, S.H., Lee, K.: Partial entity structure: a compact non-manifold boundary representation based on partial topological entities. In: Proceedings of the 6th ACM Symposium on Solid Modeling and Applications, pp. 159–170 (2001)
15. Li, Y.T., Hu, S.M., Sun, J.G.: A constructive approach to solving 3D geometric constraint systems using dependence analysis. Comput. Aided Des. **34**, 97–108 (2002)
16. Loop, C.: Smooth subdivision surfaces based on triangles. Technical, Report (1987)
17. Masuda, H.: Topological operators and boolean operations for complexbased non-manifold geometric models. Comput. Aided Des. **25**(2), 119–129 (1993)
18. Mitani, J.: A simple-to-implement method for cutting a mesh model by a hand-drawn stroke. In: Proceedings of the 2nd EUROGRAPHICS Workshop on Sketch-Based Interfaces and Modeling, pp. 35–41 (2005)
19. Mortenson, M.E.: Geometric Modeling. Wiley, New York (1997)
20. Piegl, L., Tiller, W.: The NURBS Book, 2nd edn. Springer, New York (1997)
21. Podgorelec, D.: A new constructive approach to constraint-based geometric design. Comput. Aided Des. **34**, 769–785 (2002)
22. Rossignac, J., O'Connor, M.: Sgc: A dimension-independent model for pointsets with internal structures and incomplete boundaries. In: Geometric Modeling for Product Engineering, Proceedings of the IFIP Workshop on CAD/CAM, pp. 145–180 (1989)
23. Steele, G.L., Sussman, G.L.: Constraints—a language for expressing almost-hierarchical descriptions. Artif. Intell. **14**, 1–39 (1980)
24. Vassilev, T., Spanlang, B., Chrysanthou, Y.: Fast cloth animation on walking avatars. Comput. Graph. Forum **20**, 1–8 (2001)
25. Wang, C.C.L.: A least-norm approach to flattenable mesh surface processing. In: Proceedings of IEEE International Conference on Shape Modeling and Applications 2008, pp. 131–138 (2008)
26. Wang, C.C.L., Wang, Y., Yuen, M.M.F.: Feature based 3D garment design through 2d sketches. Comput. Aided Des. **35**, 659–672 (2003)
27. Wang, C.C.L., Wang, Y., Yuen, M.M.F.: Feature-based 3D non-manifold freeform object construction. Eng. Comput. **19**(2–3), 174–190 (2003)
28. Wang, C.C.L., Yuen, M.M.F.: Freeform extrusion by sketched input. Comput. Graph. **27**(2), 255–263 (2003)
29. Wang, C.C.L., Zhang, Y., Sheung, H.: From designing products to fabricating them from planar materials. IEEE Comput. Graph. Appl. **20**, 74–85 (2010)
30. Wang, Y.: Efficient modeling methods for freeform objects. Ph.D. thesis of Hong Kong University of Science and Technology (2006)
31. Weiler, K.: The radial edge structure: a topological representation for nonmanifold geometric boundary modeling. In: M. J. Wozny, H. W. McLaughlin, and J. L. Encarnacao (eds.) Geometric modeling for CAD applications, pp. 3–36 (1986)
32. Yamaguchi, Y., Kimura, F.: Non-manifold topology based on coupling entities. IEEE Comput. Graph. Appl. **15**(1), 42–50 (1995)

33. Žalik, B.: Font design with incompletely constrained font features. In: Proceedings of the Third Pacific Conference on Computer Graphics and Applications, pp. 512–526 (1995)
34. Zorin, D., Schröder, P., Sweldens, W.: Interactive multiresolution mesh editing. In: Proceedings of SIGGRAPH '97, pp. 259–268 (1997)
35. Zorin, D., Schröder, P., DeRose, T., Kobbelt, L., Levin, A., Sweldens, W.: Siggraph 2000 course notes: Subdivision for modeling and animation. Technical Report, New York, USA (2000)

Chapter 4
Design Automation of Human-Centered Products

4.1 Transformation of Style Design

Given two human models, T and H, that are represented by piecewise linear mesh surfaces M_T and M_H with the same connectivity, a bijective mapping, $\Gamma : M_T \Rightarrow M_H$, is established by the method introduced in Chap. 2. Using Γ, the styling designs of tight products such as wetsuit coupled with a human model T can be easily transferred onto another human model H. Specifically, for every data point of styling curves on M_T, its corresponding position on M_H can be determined by this bijective mapping Γ. Linking the transferred data points by discrete geodesic curves [25] forms the curves of styling design on H, which can be used to generate cloth pieces by the mesh trimming method introduced in Chap. 3. One example of such a transformation can be found in Fig. 4.1, where the wetsuits on different human bodies with the same set of styling design curves are shown. In the right of the figure, the corresponding 2D pieces for manufacturing purpose are also shown.

The system developed by this technique is also applied to shoe design [29]. Different from the application of wetsuit design and fabrication, the styling design of shoes is conducted on a shoe last instead of the surface of a scanned foot. The surface of a shoe last for a foot is a smooth offset surface of the foot model (see Fig. 4.2 for an example). After designing the styling curves on the shoe last's surface, M_A, for the reference foot model, F_A, the curves can be transformed to the surface of another shoe last, M_B, for foot model, F_B, as long as the bijective mapping between M_A and M_B has been established. Although we can design products which are apart from the human bodies by using offset surfaces as the shoe design, the shapes of products should be embedded in the offset surfaces, which limit the type of products to be transformed. Shape warping techniques are introduced below to overcome this limitation.

C. C. L. Wang, *Geometric Modeling and Reasoning of Human-Centered Freeform Products*, DOI: 10.1007/978-1-4471-4360-4_4, © Springer-Verlag London 2013

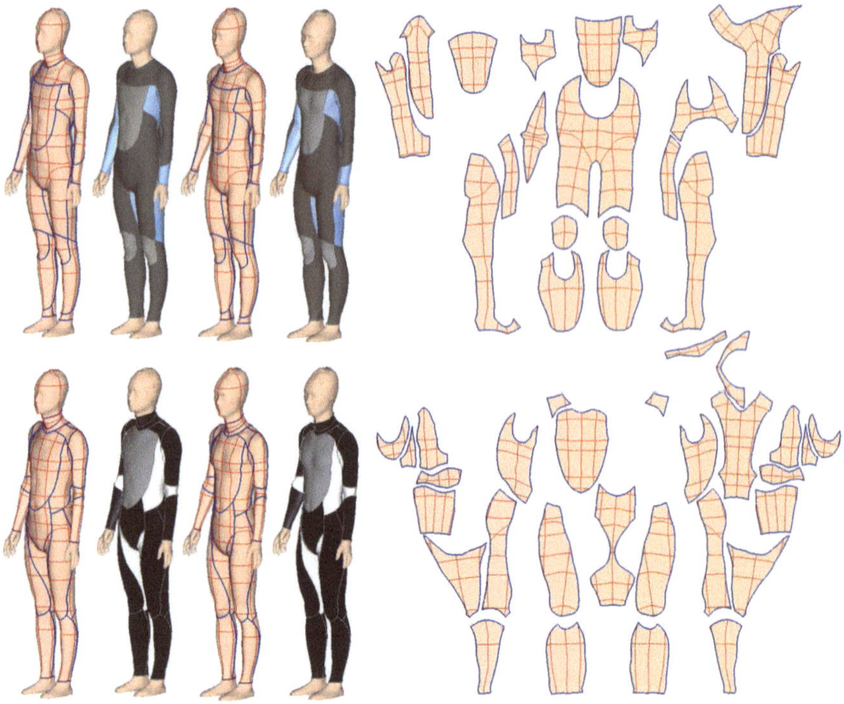

Fig. 4.1 Design automation for wetsuit: the *left* human model is the standard model used by designers and the *right* human model represents a new customer, where different styling designs can be automatically generated on the new human body with a different shape. Therefore, the 2D pieces for manufacturing can be generated automatically from the transformed wetsuit (Image taken from [29] with permission ©2010 IEEE)

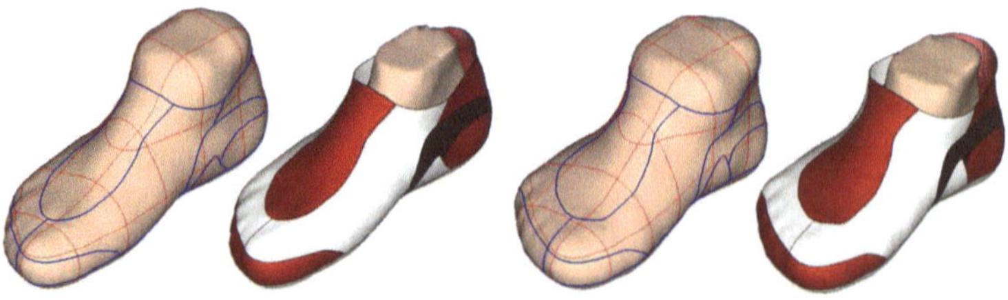

Fig. 4.2 The technique of styling curve transformation can also be used in shoe design by applying it to the offset surfaces of shoe lasts (Image taken from [29] with permission ©2010 IEEE)

4.2 Shape Warping Based on Free-Form Deformation

Using the constraint-based method introduced in Sect. 3.2 can encode the spatial relationship between the feature entities of products and the features on human bodies; therefore, when a new human body is applied, the shape of products can be reconstructed as long as the human body has the same set of predefined features. However, designers may wish to use some more flexible geometric modeling tools (such as the ones introduced in [5, 28]) to modify the shape of free-form products. In that case, the relationship between the modified product and the human model should be reconstructed. This section introduces a shape warping method based on free-form deformation for this purpose.

4.2.1 Shape Encoding onto Human Bodies

The basic problem of encoding the shape of a product onto human bodies is, in fact, how to parameterize vertices on the free-form surface M of a product by the polygons of a human model H. Similar to the above section, the human models used here are represented by piecewise linear mesh surfaces with the same connectivity, and the bijective mapping across these human models has been established by the method introduced in Chap. 2. After the parameters of each vertex on M are determined, the free-form surface M can be deformed with the shape variation of H by the mapping. This is similar to the manner of free-form deformation (FFD) [21]. Nevertheless, as mentioned by Kobayashi and Ootsubo [11], the parameterization method of FFD, which uses volumetric lattice to control a deformation, cannot be directly applied when using polygonal surfaces to control a deformation. For this reason, a method that blends linearly mapped points (similar to the blending methods in [14, 23]) is used here for the encoding/decoding of free-form products onto/from human models. This gives an efficient and effective parameterization of vertices of M on a given human model H. Each vertex in M is parameterized and weighted by a number of polygons of a human model H. These values are stored and used for mapping to the new position when the shape of the human model is changed.

To parameterize a vertex $\mathbf{q} \in M$, the following problems should be solved one by one: (1) What are the polygons on H $\mathbf{q}$ should be encoded onto? (2) What are the parameters of $\mathbf{q}$ on a polygon $P_i \in H$? (3) How to determine the weights for mapping the position of $\mathbf{q}$ by the new shape of H? The first two problems are for encoding free-form products while the third one is for decoding.

When given a point $\mathbf{q}$ on M, a fixed number of polygons on H are utilized to determine the new position of $\mathbf{q}$ when the shape of H is changed to another human body H'. These polygons are called p-polygons. Why not simply use the nearest polygon on H as the p-polygon? This is because, as proposed in [11, 23], this simple solution gives significant ramps and bumps (see [27]). The blending result of multiple p-polygons blurs the above effects. Moreover, the blending of multiple p-polygon mappings preserves symmetry. As shown in Fig. 4.3, the encoding or decoding by a single polygon mode does not maintain the vertex on the centerline of H. The vertex is centered when multiple p-polygons are used. By observation, in a single p-polygon mode, the vertex is encoded onto a polygon on the left leg of H (the green color region shows the related p-polygons). When the legs of another

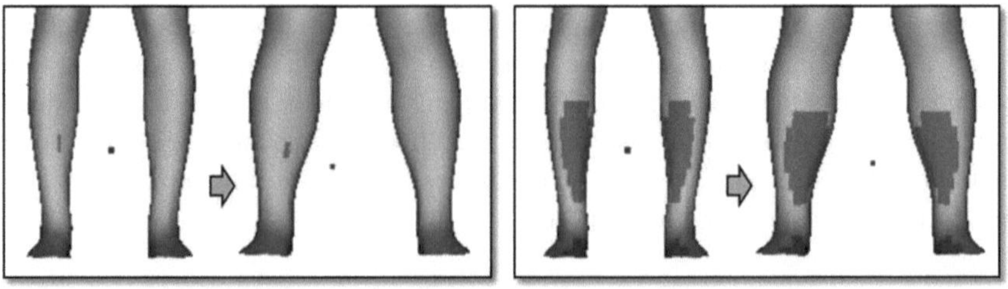

Fig. 4.3 The encoding/decoding of a vertex by a single polygon (*left*) versus by multiple p-polygons (*right*), where symmetry is preserved when multiple p-polygons are used (Image taken from [27] with permission)

human body spread out more, the vertex is dragged to the left part, which is the cause of asymmetric occurrences.

As suggested by Wang et al. [27], choosing $n_q = \lceil n_H/100 \rceil$ p-polygons usually gives good results, where n_H is the total number of polygons on H. n_q nearest polygons on H are adopted to parameterize $\mathbf{q}$. To simplify the computation, the 'nearest' here could be evaluated by the distance between $\mathbf{q}$ and the centroid of a polygon. An algorithm based on uniform space-partition can be employed to speed up the search of p-polygons. Briefly, the space around H is uniformly divided into $L \times M \times N$ boxes $\{B_{i,j,k}\}$, and each box $B_{i,j,k}$ contains a list of polygons whose centroid falls in the region of $B_{i,j,k}$. Then, Algorithm 1 is used to determine the p-polygons by searching the regions around the box containing the query point $\mathbf{q}$.

Algorithm 1 p-PolygonSearch

Input: Query point $\mathbf{q}$
Output: n_q polygons

Initialize a minimum heap Ψ and $h \Leftarrow 1$;
Determine the box B_{i_0,j_0,k_0} containing $\mathbf{q}$;
repeat
  for $si = -h$ to h **do**
    for $sj = -h$ to h **do**
      for $sk = -h$ to h **do**
        if the box B_{i_0+si,j_0+sj,k_0+sk} has NOT been visited **then**
          Insert all polygons in B_{i_0+si,j_0+sj,k_0+sk} into Ψ keyed by the distance between $\mathbf{q}$ and centroid;
        end if
      end for
    end for
  end for
  $h \Leftarrow h + 1$;
until the number of polygons in $\Psi \geq n_q$
return Top n_q polygons in Ψ;

The space-partition-based technique reduces the time for checking all polygons on H to determine the n_q nearest p-polygons. After the p-polygons are determined, the parameterization method of each polygon needs to be considered.

For a polygon $P_i = \{\mathbf{p}_1, \mathbf{p}_2, \ldots, \mathbf{p}_n\}$ on H, a local coordinate system is formed at the centroid $\mathbf{p}_c$ of P_i. Unit vectors of the local coordinate frame are defined as

$$\mathbf{x}_{P_i} = \mathbf{n}_{P_i}, \; \mathbf{y}_{P_i} = \frac{\mathbf{p}_1 - \mathbf{p}_c}{\|\mathbf{p}_1 - \mathbf{p}_c\|}, \; \mathbf{z}_{P_i} = \mathbf{x}_{P_i} \times \mathbf{y}_{P_i}, \tag{4.1}$$

where $\mathbf{n}_{P_i}$ is the normal of P_i. If P_i is degenerated into a line segment or a point, it is simply neglected during the parameterization. Therefore, the local coordinates (u_i, v_i, w_i) of a point $\mathbf{q} \in M$ on this frame is

$$\begin{bmatrix} u_i \\ v_i \\ w_i \end{bmatrix} = [\mathbf{x}_{P_i} \; \mathbf{y}_{P_i} \; \mathbf{z}_{P_i}]^T (\mathbf{q} - \mathbf{p}_c). \tag{4.2}$$

Besides (u_i, v_i, w_i), a weight ϖ_i of P_i should also be determined for the mapping process. The weight ϖ_i has a meaning of relative 'strength' of P_i against other p-polygons and is calculated by an effective function. Without loss of generality, the value of an effect function should be nonnegative and decrease monotonously according to the distance between $\mathbf{q}$ and P_i. The following effective function is suggested and used by Wang et al. [27].

$$\varpi_i = \frac{1}{\varepsilon + \|\mathbf{q} - \mathbf{p}_c\|^3} \tag{4.3}$$

where $\varepsilon = 10^{-8}$ is utilized to avoid singularity when $\|\mathbf{q} - \mathbf{p}_c\| = 0$. In summary, the encoded information of a vertex $\mathbf{q} \in M$ is n_q indices of the related p-polygons on H and the $(u_i, v_i, w_i, \varpi_i)$s of $\mathbf{q}$ according to each p-polygon.

4.2.2 Decoding for Shape Reconstruction

This section describes the method of how to warp the shape of an encoded free-form product M to a human model H so that the deformed shape $\tilde{M}$ 'fits' the shape of another human model $\tilde{H}$. On the new human model $\tilde{H}$, the geometry of each polygon P_i is changed to $\tilde{P}_i$. The new centroid $\tilde{\mathbf{p}}_c$ and axes $\tilde{\mathbf{x}}_{P_i}, \tilde{\mathbf{y}}_{P_i}, \tilde{\mathbf{z}}_{P_i}$ of $\tilde{P}_i$ are computed by the same method described in Eq. (4.1). Then, $\mathbf{q}$ is mapped to $\mathbf{q}_i$ according to $\tilde{P}_i$ as

$$\mathbf{q}_i = \tilde{\mathbf{p}}_c + u_i \tilde{\mathbf{x}}_{P_i} + v_i \tilde{\mathbf{y}}_{P_i} + w_i \tilde{\mathbf{z}}_{P_i}. \tag{4.4}$$

The new position $\mathbf{q}_j$ of $\mathbf{q}$ mapped by another polygon $\tilde{P}_j$ can be calculated in the same way. Since (u_i, v_i, w_i) represents the relative position of $\mathbf{q}$ to the polygon $\tilde{P}_i$,

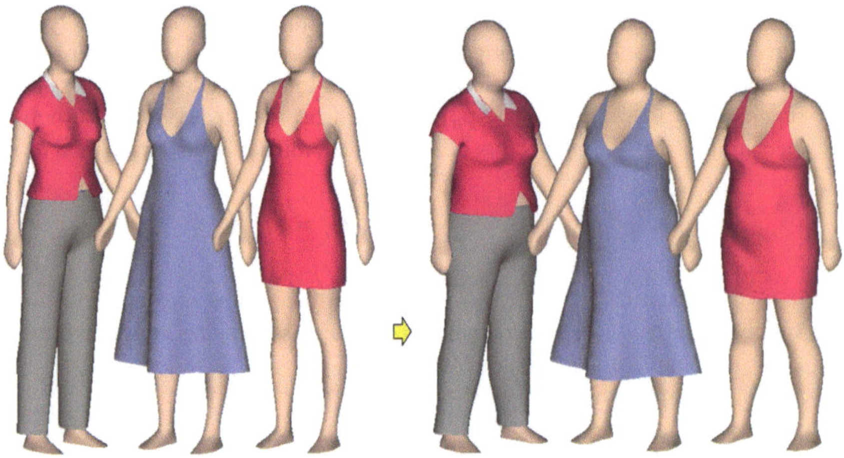

Fig. 4.4 Examples of shape warping-based design automation of apparel products: the clothes designed for a human model H_A (*left*), and the clothes automatically generated for another human model H_B (*right*)

generally, $\mathbf{q}_i$ and $\mathbf{q}_j$ are not coincident. The finally mapped point $\tilde{\mathbf{q}}$ is calculated by the following formula, which blends the points $\mathbf{q}_i$ with the weights ϖ_i.

$$\tilde{\mathbf{q}} = \frac{\sum_i \varpi_i \mathbf{q}_i}{\sum_i \varpi_i} \tag{4.5}$$

By changing the position of each $\mathbf{q} \in M$ to $\tilde{\mathbf{q}}$, the deformed mesh $\tilde{M}$ of the free-form product is determined for the human model $\tilde{H}$. Examples of using this shape warping technique in the design automation of apparel products are shown in Fig. 4.4.

4.3 Design Automation Based on Volumetric Parameterization

Considering two human bodies, H_A and H_B, which are represented by free-form polygonal meshes, a volumetric parameterization as referred to here is a forward mapping $\Psi : \Omega_A \Rightarrow \Omega_B$ from any point $\mathbf{p}$ in the space $\Omega_A \in \Re^3$ around H_A to a corresponding point $\mathbf{p}'$ in the space $\Omega_B \in \Re^3$ around H_B. Based on Ψ, a product M originally designed around H_A, which is also represented by free-form polygonal meshes, could be transferred to the shape around H_B by mapping the position of every vertex on M to a new position in Ω_B. In design automation applications, the mapping Ψ is expected to be sensitive to semantic features (which is usually described by anchor points) on the reference models. Two set of anchor points, G_A and G_B, are assigned to H_A and H_B—the anchor points could be either automatically extracted or interactively specified. The mapping is expected to provide a smooth transition between Ω_A and Ω_B (that is, the mapping $\Psi(\alpha)$ is a smooth function with the transient variable $\alpha \in [0, 1]$).

The mesh connectivities on H_A and H_B are generally inconsistent although the cross-surface parameterization method introduced in Chap. 2 can be used to produce the consistent meshes on two human models. Nevertheless, human models usually have similar features. The volumetric parameterization technique presented in this chapter makes use of this characteristic to compute Ψ on human models with different meshes.

In order to provide a smooth transition property in Ψ, the mapping is decomposed into two submappings: (1) rigid body transformation and (2) elastic mapping. The rigid body transformation is usually represented by a rotation matrix $\mathbf{R}(\theta_x, \theta_y, \theta_z)$ and a translation vector $\mathbf{t}$, which can be determined through a least-square fitting process. The procedure of computing the elastic warping component in the mapping Ψ is more complex. A three-stage approach is developed for this, which starts from computing a coarse-level warping function by anchor points. After applying the coarse-level warping function to the mesh surface of H_A, a warped human body H'_A is obtained. The shapes of H'_A and H_B are similar to each other so that a surface fitting process can be applied to H'_A to match its connectivity with that of H_B if H_A and H_B have different mesh connectivities. A bijective correspondence is therefore established between the surfaces of H_A and H_B. This correspondence can also be established by the cross-surface parameterization method introduced in Chap. 2. With the cross-surface parameterization, a volumetric paramterization can be constructed by using compactly supported radial basis function (CSRBF).

4.3.1 Rigid Body Transformation

Without loss of generality, one can assume that correspondences have been constructed between points in the spaces Ω_A and Ω_B of two reference models H_A and H_B. The simplest transient mapping $\Psi(t)$ between Ω_A and Ω_B is a linear blending: with $\Psi(\alpha) = (1-\alpha)\Omega_A + \alpha\Omega_B$. However, as demonstrated in Fig. 4.5, linear blending is not sensitive to the orientation of the reference objects where unexpected distortion may be obtained. Therefore, the mapping between Ω_A and Ω_B is decomposed into a rigid body transformation $\Gamma(\mathbf{q}) = \mathbf{R}(\theta_x, \theta_y, \theta_z)\mathbf{q} + \mathbf{t}$ and an elastic warping $E(\cdots)$ with $\Gamma(E(\Omega_A)) \equiv \Omega_B$, and hence

$$\Psi(\alpha) = \mathbf{R}(\alpha\theta_x, \alpha\theta_y, \alpha\theta_z)((1-\alpha)\Omega_A + \alpha E(\Omega_A)) + \alpha\mathbf{t}. \tag{4.6}$$

The rotation matrix $\mathbf{R}(\theta_x, \theta_y, \theta_z)$ and the translation vector $\mathbf{t}$ can be determined in a least-square sense by minimizing the following energy function defined on the anchor points (G_A and G_B prescribed on the human models)

$$J = \sum_{\mathbf{q} \in G_A} \|\mathbf{R}(\theta_x, \theta_y, \theta_z)\mathbf{q} + \mathbf{t} - \mathbf{q}'\|^2 \tag{4.7}$$

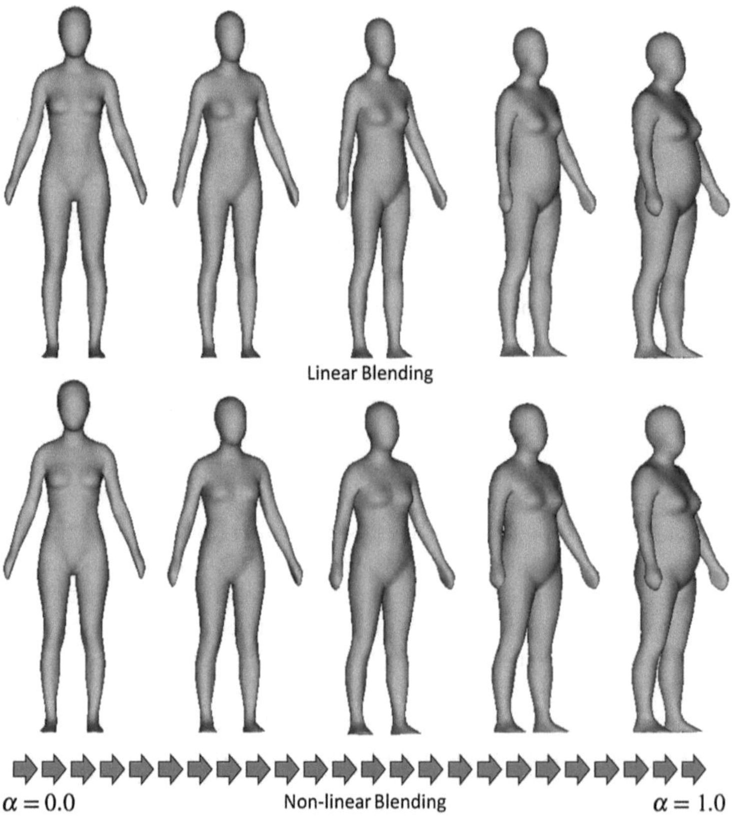

Fig. 4.5 Transient results of linear blending (*top row*) have unwanted distortion—the model becomes thin and narrow (the second and the third models in the *top row*), which is not reasonable. The nonlinear blending results obtained by decomposing the rigid body transformation from elastic warping are more reasonable (Image taken from [26] with permission ©2007 IEEE)

where $\mathbf{q}' \in G_B$ is the corresponding anchor point of $\mathbf{q} \in G_A$ and $\|\cdots\|$ is the Euclidean norm in $\Re^3$.

As mentioned by Arun et al. [2], if the solution of Eq. (4.7) is $\hat{\mathbf{R}}$ and $\hat{\mathbf{t}}$, G_B and $\hat{\mathbf{R}}G_A + \hat{\mathbf{t}}$ have the same centroid, and Eq. (4.7) can be simplified by introducing a transfer of coordinates

$$\mathbf{p}_i = \mathbf{q}_i - \mathbf{c}_A \ \ (\forall \mathbf{q}_i \in G_A) \qquad \mathbf{p}_i' = \mathbf{q}_i' - \mathbf{c}_B \ \ (\forall \mathbf{q}_i' \in G_B) \tag{4.8}$$

where $\mathbf{c}_A$ and $\mathbf{c}_B$ are the centroids of G_A and G_B, respectively, and there are n_{anc} anchor points. Thus, Eq. (4.7) can be rewritten as

$$J' = \sum_{i=1}^{n_{\mathrm{anc}}} \|\hat{\mathbf{R}}\mathbf{p}_i - \mathbf{p}_i'\|^2 \tag{4.9}$$

since $\mathbf{c}_B \equiv \hat{\mathbf{R}}\mathbf{c}_A + \hat{\mathbf{t}}$. As long as $n_{\text{anc}} > 3$, let

$$\frac{\partial J'}{\partial \hat{\mathbf{R}}} = 0, \tag{4.10}$$

the 3×3 matrix $\hat{\mathbf{R}}$ can be solved by the singular value decomposition (SVD) method [20], and the translation vector $\mathbf{t} = \hat{\mathbf{t}}$ is determined by $\mathbf{c}_B \equiv \hat{\mathbf{R}}\mathbf{c}_A + \hat{\mathbf{t}}$. Notice that $\hat{\mathbf{R}}$ determined by Eq. (4.10) is a global minimum since the objective function J' is in a quadratic form. The rotation matrix $\hat{\mathbf{R}}$ determined by SVD needs to be first converted into a quaternion $[w, (x, y, z)]$. Second, the components of the quaternion are normalized by setting $w^2 + x^2 + y^2 + z^2 = 1$. Last, the Euler angles, θ_x, θ_y, and θ_z, are separated from the normalized quaternion so that the rotation matrix $\mathbf{R}(\theta_x, \theta_y, \theta_z)$ is finalized. Details of this conversion are stated by Shoemake [22]. Once $\mathbf{R}(\theta_x, \theta_y, \theta_z)$ and $\mathbf{t}$ are determined, the rigid body transformation $\Gamma(\cdots)$ is defined.

4.3.2 RBF-Based Elastic Function

After determining the rigid body transformation $\Gamma(\cdots)$, the elastic warping $E(\cdots)$ also needs to be computed. A coarse-level warping function is expected to be a nonlinear transformation $\tilde{E} : \Re^3 \Rightarrow \Re^3$ such that

$$\tilde{E}(\mathbf{q}) = \Gamma^{-1}(\mathbf{q}) \quad (\forall \mathbf{q} \in G_A, \mathbf{q}' \in G_B) \tag{4.11}$$

with $\Gamma^{-1}(\mathbf{q}') \equiv \mathbf{R}(-\theta_x, -\theta_y, -\theta_z)(\mathbf{q}' - \mathbf{t})$. Since the anchor points in G_B have been transformed backward by $\Gamma^{-1}(\mathbf{q}')$ to have the same center and orientation as the points in G_A, the elastic function determined by Eq. (4.11) is not affected by the change in orientations of the human models. The hybrid of $\Gamma(\cdots)$ and $\tilde{E}(\cdots)$ is expected to map corresponding anchor points exactly. This is a multivariable scattered data interpolation problem, which can be solved by using radial basis functions (such as Eq. (2.22) introduced in Chap. 2).

The coefficients in RBFs can be determined by the constraints that every anchor point $\mathbf{q} \in G_A$ maps exactly to the position $\Gamma^{-1}(\mathbf{q}')$ with $\mathbf{q}' \in G_B$ (that is Eq. (4.11)). In this coarse-level elastic function, a global effect is expected, so that $g(r) = r^3$ is adopted as the radial basis function. As analyzed by Wang et al. [26], the number of anchor points in G_A and G_B should be more than three and the anchor points must not be coplanar. Therefore, the numerical system for computing the RBF-based elastic function is quite stable. The mapping determined based on anchor points is denoted by $\tilde{\Psi}$.

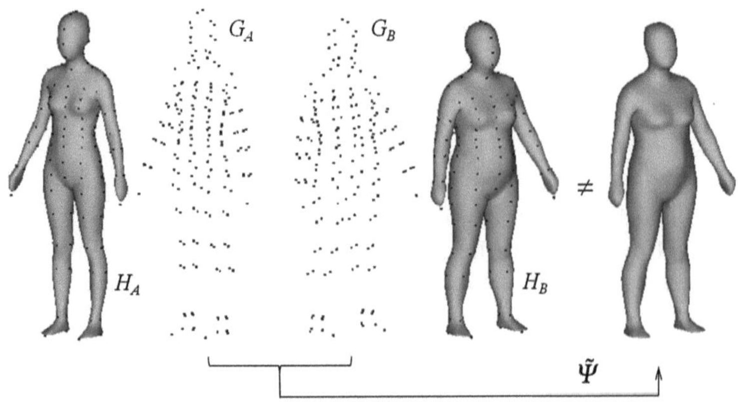

Fig. 4.6 The human model obtained by applying the transformation and warping (generated by anchor points) to the human model H_A. The resultant model is still slightly different from the human model H_B

4.3.3 Surface Fitting

The elastic warping function determined by interpolating anchor points accurately controls only the warping near anchor points. For the space between anchor points, the warping is not well- defined. For example, in Fig. 4.6, after applying the transformation and warping (generated by anchor points) on the surface of H_A, the surface obtained is still somewhat different from the surface of H_B. In order to have a more accurate mapping, the number of corresponding points needs to be increased on the surfaces of H_A and H_B. This is achieved through a procedure of surface fitting.

Surface fitting is performed by minimizing an energy function defined by the differences between the surfaces of $\tilde{E}(H_A)$ and H'_B (with $H'_B = \Gamma^{-1}(H_B)$) and by the smoothness term on the resultant surface. In fact, the surface fitting process is an evolution of $\tilde{E}(H_A)$ to obtain a deformed model H'_A, which has its shape approximates H'_B and maintains the same mesh connectivity on H_A. To accomplish the fitting, an optimization framework similar to [1] is employed. Each vertex $\mathbf{v}_i$ on the mesh surface of H_A is influenced by a transition vector $\mathbf{t}_i$. It is expected that a set of transition vectors will be found to move all vertices on H_A to a deformed surface H'_A, which matches well with H'_B.

The first objective of a good match is that H'_A should be as close as possible to the target shape H'_B. For this purpose, the objective function holds a term measuring the sum of squared distances between each vertex on H_A and the reference surface H'_B. Simply moving each vertex on H_A to its closest point on H'_B may not result in an attractive mesh since neighboring parts of H_A could be mapped to disparate parts of H'_B, and vice versa. A smoothness term is necessary during the optimization to avoid this disparity. The smoothness term defined below tries to minimize the difference between the transition vectors on neighboring vertices. In summary, the objective function is defined as:

$$E_f = \sum_{i=1}^{m} \|\mathbf{t}_i - \mathbf{t}_i^0\|^2 + \sum_{(i,j)\in \text{edges}(H_A)} \|\mathbf{t}_i - \mathbf{t}_j\|^2 \tag{4.12}$$

where $\mathbf{t}_i^0$ is the transition vector which moves the vertex $\mathbf{v}_i$ to its closest compatible point on H'_B and m is the number of vertices on H_A. As mentioned by Allen et al. [1], a point on H_A and a point on H'_B is *compatible* if the surface normal at each point is not more than $\frac{\pi}{2}$ apart so that the front-facing surfaces will not be matched with the back-facing surfaces. The $\mathbf{t}_i$s that minimizes E_f should satisfy

$$\frac{\partial E_f}{\partial \mathbf{t}_i} = \mathbf{t}_i - \mathbf{t}_i^0 + \sum_{\mathbf{v}_j \in N(\mathbf{v}_i)} (\mathbf{t}_i - \mathbf{t}_j) = 0 \tag{4.13}$$

which leads to a linear equation system

$$(|N(\mathbf{v}_i)| + 1)\mathbf{t}_i - \sum_{\mathbf{v}_j \in N(\mathbf{v}_i)} \mathbf{t}_j = \mathbf{t}_i^0 \tag{4.14}$$

where $N(\mathbf{v}_i)$ denotes the one-ring neighborhood vertices of $\mathbf{v}_i$ and $|N(\mathbf{v}_i)|$ is the number of vertices in it. Equation. (4.14) gives a diagonal-dominated linear equation system so that the convergency of using Gauss-Seidal method [20] to solve the linear system is guaranteed. Therefore, the optimized $\mathbf{t}_i$s could be determined iteratively through the update

$$\mathbf{t}_i = \frac{1}{|N(\mathbf{v}_i)| + 1} \left(\mathbf{t}_i^0 + \textstyle\sum_{\mathbf{v}_j \in N(\mathbf{v}_i)} \mathbf{t}_j \right). \tag{4.15}$$

An iterative algorithm is conducted to match vertices of H_A with the target surface H'_B.

- Step 1: The iteration starts by moving every vertex of $\mathbf{v}_i \in H_A$ to the position $\tilde{E}(\mathbf{v}_i)$.
- Step 2: $\mathbf{t}_i^0$ of every vertex $\mathbf{v}_i$ is evaluated.
- Step 3: Determine $\mathbf{t}_i$s for $\mathbf{v}_i$ which minimizes E_f by the iterative update scheme defined in Eq. (4.15).
- Step 4: Move vertex $\mathbf{v}_i$ to a new position $\mathbf{v}_i + \mathbf{t}_i$.
- Step 5: Evaluate $\mathbf{t}_i^0$s of all $\mathbf{v}_i$s, if any $\|\mathbf{t}_i^0\| > 10^{-5}$, go back to Step3.
- Step 6: Move vertex $\mathbf{v}_i$ to its closest position $\mathbf{v}_i + \mathbf{t}_i^0$ on H'_B.

In this algorithm, the most time-consuming step is to evaluate the closest compatible point of a vertex $\mathbf{v}_i$ on H'_B, from which the transient vector $\mathbf{t}_i^0$ is determined. A spatial partition method is employed to speed up the search for the closest compatible point. The bounding space of H'_B

$$[x_{\min}, x_{\max}] \times [y_{\min}, y_{\max}] \times [z_{\min}, z_{\max}]$$

is subdivided into $L \times M \times N$ subregions with a uniform width Δ, where each subregion $\Re(i, j, k) \subset \Re^3$ is defined as a subspace in the following range:

$$[x_{\min} + i\Delta, x_{\min} + (i+1)\Delta) \\ \times [y_{\min} + j\Delta, y_{\min} + (j+1)\Delta) \\ \times [z_{\min} + k\Delta, z_{\max} + (k+1)\Delta)\,.$$

A polygonal face $f \in H'_B$ is considered as contributing to a subregion $\Re(i, j, k)$ if its bounding box $B(f)$ satisfies

$$\Re(i, j, k) \cap B(f) \neq \phi.$$

Pointers to contributed triangles are held by each subregion $\Re(i, j, k)$. With this space subdivision, only the regions $i \in [l - \tau, l + \tau]$, $j \in [m - \tau, m + \tau]$, and $k \in [n - \tau, n + \tau]$ need to be searched to locate the points that are closest to a vertex in $\Re(l, m, n)$. The searching starts from $\tau = 1$; if no compatible point is found, τ is increased incrementally until a compatible point is found.

After surface fitting, the refined point correspondences between the shapes of H_A and H'_B are constructed. The correspondences can then be used to formulate a detailed elastic function $E(\cdots)$ to establish the mapping for volumetric parameterization.

4.3.4 CSRBF Approach for Volumetric Parameterization

With the help of H'_A, finer level correspondence between the points on the surfaces of H_A and H'_B is obtained. Therefore, the accurate mapping $\Psi(\alpha)$ defined in Eq. (4.6) for volumetric parameterization can be refined by computing an accurate elastic function $E(\cdots)$ based on the positions of the vertices on H_A and H'_A. The elastic function $E(\cdots)$ is evaluated in the manner of CSRBF. The coefficients of (CSRBFs) are determined by solving the linear equation system describing the position correspondences of the vertices on H_A and H'_A.

The refined elastic function $E(\cdots)$ is defined by CSRBFs in a mathematical manner so that every vertex $\mathbf{v}_i \in H_A$ is mapped to a new position $\mathbf{v}'_i \in H'_A$. This is similar to the anchor point interpolations for computing the coarse-level elastic function. Different from using globally supported RBFs in the coarse-level elastic function, the number of interpolations is significantly increased here (usually in the level of thousands or even tens of thousands). Therefore, one needs to solve a huge linear equation system which is computationally expensive. Although the fast multipole method (for example [4]) can reduce the quadratic solution time into a neatly linear time, the CSRBFs discussed below are easier to use and implement. CSRBFs with a relatively small effective distance λ make the linear equation system sparse. It can thus be solved with linear time complexity.

The refined elastic function is formulated as

$$E(\mathbf{v}) = \mathbf{a}_0 + \mathbf{A}\mathbf{v} + \sum_{i=0}^{m} \beta_i \varpi(\|\mathbf{v} - \mathbf{v}_i\|) \tag{4.16}$$

where $\mathbf{a}_0$ and $\mathbf{A}$ control the affine transformation of points, and the nonlinear term defines the warping as

$$\varpi(r) = \begin{cases} (1 - \frac{r}{\lambda})^6(\frac{35}{3}(\frac{r}{\lambda})^2 + 6\frac{r}{\lambda} + 1) & r < \lambda \\ 0 & r \geq \lambda \end{cases} \tag{4.17}$$

with λ being the support size of each radial basis function. Here, $\|\cdots\|$ denotes the Euclidean norm, $\mathbf{A} = (\mathbf{a}_1, \mathbf{a}_2, \mathbf{a}_3)^T$, and the vector coefficients $\mathbf{a}_0, \ldots, \mathbf{a}_3 \in \Re^3$ define the affine transformation of the point $\mathbf{v}$. This CSRBF was originally introduced by Buhmann [7] and has been proved to give a non-singular solution of interpolation problem. By setting

$$E(\mathbf{v}_i) = \mathbf{v}_i' \tag{4.18}$$

together with the compatibility conditions, the detailed elastic function $E(\mathbf{v})$ could be determined by the numerical solvers of sparse linear systems—for example, the bi-conjugate gradient method (see [20] for reference). In order to solve the linear equation system efficiently, Kozhekin et al. [12] sorted the vertices by the distances between them. However, in the tests conducted by Wang et al. [26], the elastic function $E(\cdots)$ could be determined at almost the same speed with or without sorting when using the bi-conjugate gradient method [20]. Using this $E(\cdots)$ and the previously obtained $\Gamma(\cdots)$, the mapping $\Psi(\alpha)$ for volumetric parameterization is explicitly defined as Eq. (4.6).

4.3.5 Discussion

The properties of the volumetric parameterization method are studied in this section. Several results are shown to help verify the properties. The advantages of this volumetric parameterization approach compared with the p-FFD-based warping introduced in Sect. 4.2 are also discussed here.

First, the mapping of spaces around reference models is established without the restriction of having the same connectivity on the reference models. The anchor points in volumetric parameterization are not necessary to be on the surfaces of the human models. More anchor points could be added in Ω_A and Ω_B to achieve a finer control of space warping—the eight anchor points at the corner of the bounding box of H_A and H_B in (Fig. 4.7) are a good example of this.

Second, the CSRBF description can be repeatedly used for the design automation of other products on the same two reference models while the p-FFD result can be determined faster but needs to be recomputed every time when a new product is considered.

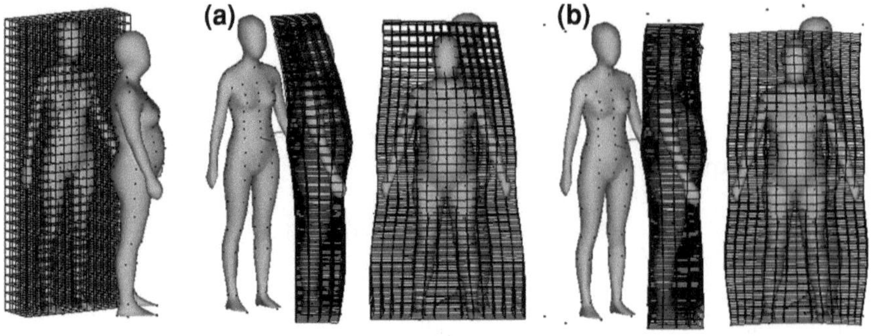

Fig. 4.7 Anchor points can be added to the space around human models. Some example space warpings **(a)** without versus **(b)** with corner anchor points are compared (Image taken from [26] with permission ©2007 IEEE)

Last, based on the separation of the rigid body transformation and the elastic warping, the mapping Ψ provides a smooth transformation between Ω_A and Ω_B, which is important for the serials 'grading'—so that unwanted distortion is avoided on the transient results.

In CSRBFs, an effective distance λ needs to be specified. For any vertex on the product M around H_A, if its distance to each of the vertices on H_A is not smaller than λ, no radial basis function will affect its position in $E(\cdots)$— in other words, its position is deformed in $E(\cdots)$ only by $\mathbf{a}_0, \ldots, \mathbf{a}_3 \in \Re^3$. Benefiting from this property, the product reconstructed by the CSRBF-based volumetric parameterization on H_B maintains its original shape at the palaces with larger distances from the surface of H_B. Figure 4.8a is an example of this where the straight profile is retained around the thighs. However, since the dress reconstructed by the p-FFD-based warping tries to maintain the distance between every vertex and the reference model, the parts near thighs show some unwanted distortion although they are far from the thighs. This difference will not be shown if the distance from every vertex on the product M to the vertices of H_A is smaller than λ because the vertices are tightly tied on the surface of H_A by the radial basis function $\varpi(\|\mathbf{v} - \mathbf{v}_i\|)$s (see (Fig. 4.8b). For the design automation of apparel products, the values of λ are chosen in the range between three and ten centimeters.

4.4 Realization of Flexible Shape Control

In the above sections, the shape optimization of free-form products to fit variations in human models is taken place in an automatic way. Using apparel products as an example, although such a method works well on tight garments such as T-shirts and jeans pants, unacceptable results with unwanted distortion on the surface regions of the garments that are not near the human body are generated. More specifically, the

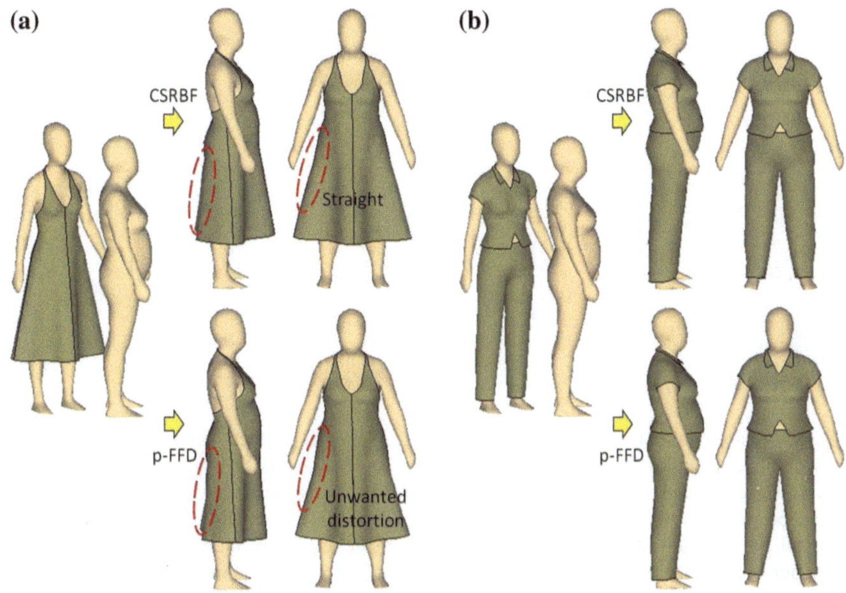

Fig. 4.8 A comparison of results obtained from CSRBF-based volumetric parameterization and from p-FFD-based shape warping: **a** product I with some parts having a long distance to the human model, and **b** product II with almost all places tightly worn on the human body (Image taken from [26] with permission ©2007 IEEE)

method relies too much on the shape of human bodies, which therefore somehow amplifies the defects of the body shape on the resized free-form products (such as the distorted shapes shown in Fig. 4.9).

This section presents a method to control the shape of apparel products during the automatic resizing procedure more flexibly. As illustrated in Fig. 4.10, the algorithm introduced by Meng et al. [17] provides a flexible shape control for automatically resizing human-oriented products in three steps. First, a product is warped from a reference human body to the space around a target human body (see (Fig. 4.10a) by the method stated in Sect. 4.2. Second, the shapes of features are optimized to match their original shapes before the warping (shown in (Fig. 4.10b). Last, the discrete surfaces of the apparel product are deformed again under an optimization framework to match their original shapes locally while interpolating the shapes of features defined in the previous step (see Fig. 4.10c). To make the approach more practical, the requirement of input free-form models for resizing is generalized. Therefore, input products to be processed in this approach are not necessary to be two-manifold mesh surfaces, which is however required by many other mesh-based deformation techniques (for example, [15, 16, 24]). Free-form products are usually represented by discrete surfaces with polygon soups, hanged edges, and even point sets. For the complex free-form products, the number of vertices on their surfaces may be huge (such as, more than one million vertices); thus, it is impractical to

Fig. 4.9 Shape control for the 3D automatic resizing of apparel products can be flexibly performed by specifying features (illustrated by the *white dashed curves*) on the apparel products. The shapes of the resized products around the features are processed to match the designed shape. This improves the shape quality of the results of automatic resizing. From *left* to *right*, the designed dress worn by a model, the side view of the designed dress, the automatic resizing result without shape control on the features using the method of [27], the result with the shapes of features preserved by [17], the automatic resizing result without shape control, and the result of [17] which gives straight profiles as the designed shape of the dress (Image taken from [17] with permission)

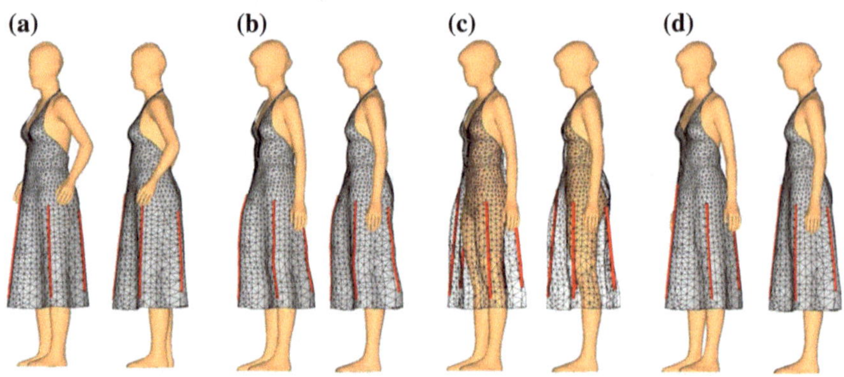

Fig. 4.10 The steps of our flexible shape control algorithm for automatic resizing: **a** the given dress with shape-preserving features shown as *red lines*, **b** the result of primary 3D resizing based on the deformation method in Sect. 4.2—the shapes of features are not preserved, **c** the shapes of features are optimized to match their shapes before resizing, and **d** the surface shape of the dress is processed to interpolate the shapes of optimized feature curves (Image taken from [17] with permission)

solve large linear systems with such sizes for surface processing [6, 30] even if the coefficient matrix is sparse. The method presented here can solve such huge linear systems in an efficient way.

The main characteristics of this method are as follows.

- A new shape matching-based method to preserve the shapes of the features defined on free-form products, where the features can be easily specified with simple interactions.
- A new shape optimization method, which can be applied to discrete surfaces containing non-manifold entities and does not need to solve large linear systems.

These result in a flexible shape control tool for the automatic resizing of free-form products.

Some non-homogeneous resizing methods have been proposed in the computer graphics area to either modify an image or deform a 3D model. A simple image operator called *seam carving* was presented by Avidan and Shamir [3] to support content-aware image resizing for both reduction and expansion. Kraevoy et al. [13] proposed a method that protects model features and structures during resizing for complex models, which is considered as a non-homogeneous resizing. Recently, Gal et al. [8] introduced a shape editing method that works by applying systematic solutions to feature curve extraction, analysis, editing, and deformation. However, this method can only be used in some mechanical objects that have explicitly sharp features. The feature preservation function provided by them [8] can hardly be applied to the products which are formed by free-form surfaces. Although these techniques can make the resized results feasible, they cannot be employed to resize human-oriented free-form products for industrial usage whereas the techniques presented in this section can.

4.4.1 Specifying Features on Product

There are many types of features used in the CAD approaches for free-form products [27], which include feature points (1D), feature curves (2D), and feature patches (3D). As flexible control tools, the features must be easy to be specified interactively. Nowadays, the sketch-based interface has become popular in many 3D applications, where sketched curves take an important role. Therefore, curves are chosen as the features in this approach to conduct a flexible shape control for the automatic resizing of free-form products. In the implementation proposed by Meng et al. [17], the feature curves are represented by B-spline curves. There are two types of feature curves. The first type is the features specified by users to control the shapes of the resized products and the second type is the boundary curves of the products. However, when the shapes of the products are represented by non-manifold entities, the boundary curves may need to be specified interactively.

As free-form products could be represented by discrete surfaces with non-manifold entities, the method used to specify feature curves should not rely on the representation of the model M to be resized. Therefore, every feature curve specified by users is processed as a set of ordered points attached to 3D free-form products. The steps are described below with the help of an example shown in Fig. 4.11.

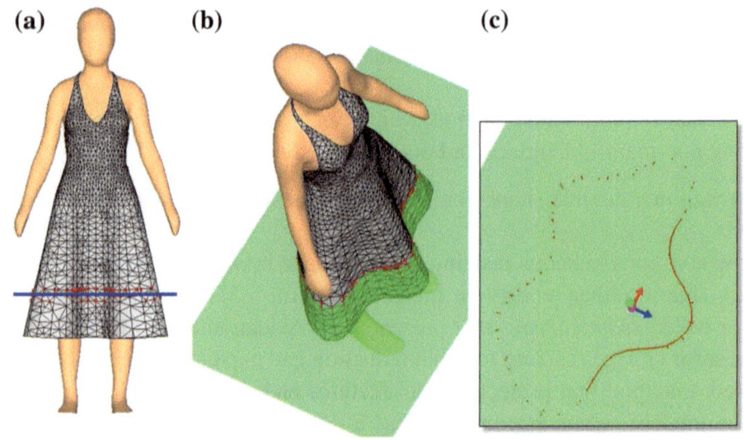

Fig. 4.11 Steps for specifying features on a product M: **a** a line segment specifies the plane on which the feature curve is defined, **b** the plane (in *green*) to hold the feature curve is rotated and the surface of M near the plane is projected onto the plane to illustrate a contour of intersections, **c** a B-spline curve (in *red*) is defined on the plane by users—the B-spline curve will be uniformly sampled into a sequence of points attached to the model M (Image taken from [17] with permission)

First, users can select a viewing direction to draw a line segment (the blue one in Fig. 4.11a). When sweeping the line segment along the viewing direction, a plane for specifying feature curves is defined (see the green plane shown in Fig. 4.11b)—it is named as *feature plane*. Second, users can draw B-Spline curves on the plane to specify the shape as well as the range of a feature curve $\mathbf{c}_f$ because an apparel product may be represented by polygon soups which cannot always generate a simple curve by intersecting a plane. Many local modification schemes (for example [19]) can be used here to control the shape of the feature curve to follow the contours of intersections (see Fig. 4.11c). Last, the feature curve is sampled into a sequence of ordered points $\bar{\mathbf{c}}_f = \{\mathbf{r}_i\}$ with $i = 1, \ldots, m$, which is considered as the discrete form of the feature curve $\mathbf{c}_f$ (see the red points in Fig. 4.11d).

All the points sampled from the feature curves are considered as being attached to the product M and are deformed together with M in the primary 3D resizing step. Each sample point $\mathbf{r}_i$ is mapped to a corresponding point $\mathbf{r}'_i \in \Re^3$ around the new human body H_B by Eq. (4.5). The positions of $\mathbf{r}'_i$s are optimized so that the feature curves preserve their shapes before resizing.

4.4.2 Shape Matching on Features

The shape of a deformed feature curve $\bar{\mathbf{c}}'_f = \{\mathbf{r}'_i\}$ must be optimized on feature plane to match its original shape defined on $\bar{\mathbf{c}}_f$. Here, the shapes of two curves are defined

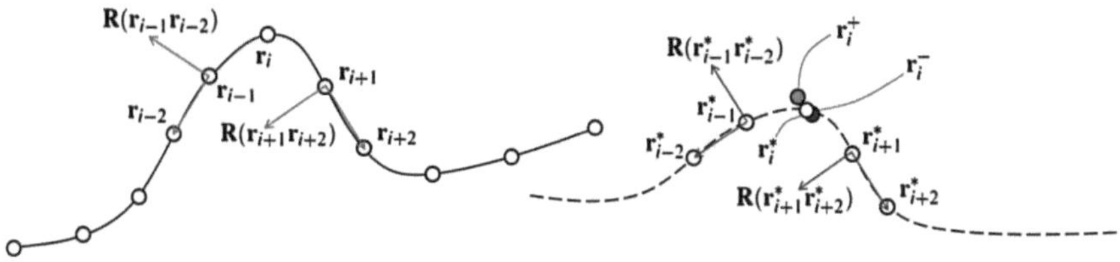

Fig. 4.12 An illustration of modeling local shape similarity by local frames on the sample points: two local frames for computing (α_i^-, β_i^-) and (α_i^+, β_i^+) at $\mathbf{r}_{i-1}$ and $\mathbf{r}_{i+1}$ respectively (*left*), and the differences between the predicted positions $\mathbf{r}_i^-$ (in *blue*) and $\mathbf{r}_i^+$ (in *yellow*) computed by the new local frames and the optimized position $\mathbf{r}_i^*$ are minimized during the optimization (*right*)

as similar when the relative distributions of the sample points on them are similar to each other. A clearer definition of the similarity is defined below and the shape of a feature curve is modified under an optimization framework.

First of all, the new feature plane holding the feature points $\{\mathbf{r}_i'\}$ must be estimated. Here, the feature plane is computed by the principal component analysis (PCA) of all the sampled points. Some auxiliary points $\{\mathbf{a}_i\}$ are randomly generated near the points $\{\mathbf{r}_i\}$ on the feature plane before the primary 3D resizing so that the feature plane can be estimated more robustly. The auxiliary points are mapped to new positions by Eq. (4.5), and the newly mapped points are then added into the PCA for computing the new feature plane.

Assume that every sample point $\mathbf{r}_i'$ is optimized to a new position $\mathbf{r}_i^*$, the local shape similarity at $\mathbf{r}_i^*$ is measured by its distance to its predicted positions defined by the local frames of the feature curve at $\mathbf{r}_{i+1}^*$ and $\mathbf{r}_{i-1}^*$. Specifically, α_i^+ is computed as the projected length of the vector $\mathbf{r}_{i+1}\mathbf{r}_i$ on the vector $\mathbf{r}_{i+1}\mathbf{r}_{i+2}$, and β_i^+ is the projected length of $\mathbf{r}_{i+1}\mathbf{r}_i$ on $\mathbf{R}(\mathbf{r}_{i+1}\mathbf{r}_{i+2})$ with $\mathbf{R}$ being an orthogonal 2D rotation matrix

$$\mathbf{R} = \begin{bmatrix} 0 & 1 \\ -1 & 0 \end{bmatrix}$$

on the feature plane. Therefore, the predicted position of $\mathbf{r}_i^*$ by the frame at $\mathbf{r}_{i+1}^*$ is defined as

$$\mathbf{r}_i^+ = \mathbf{r}_{i+1}^* + \alpha_i^+ \mathbf{r}_{i+1}^* \mathbf{r}_{i+2}^* + \beta_i^+ \mathbf{R}(\mathbf{r}_{i+1}^* \mathbf{r}_{i+2}^*). \tag{4.19}$$

Similarly, another prediction of $\mathbf{r}_i^*$ is computed by the frame at $\mathbf{r}_{i-1}^*$ as

$$\mathbf{r}_i^- = \mathbf{r}_{i-1}^* + \alpha_i^- \mathbf{r}_{i-1}^* \mathbf{r}_{i-2}^* + \beta_i^- \mathbf{R}(\mathbf{r}_{i-1}^* \mathbf{r}_{i-2}^*), \tag{4.20}$$

where α_i^- and β_i^- are the projected lengths of $\mathbf{r}_i\mathbf{r}_{i-1}$ on $\mathbf{r}_{i-1}\mathbf{r}_{i-2}$ and $\mathbf{R}(\mathbf{r}_{i-1}\mathbf{r}_{i-2})$ respectively. An illustration is given in Fig. 4.12.

When $\mathbf{r}_i^* = \mathbf{r}_i^+ = \mathbf{r}_i^-$, the local shape at $\mathbf{r}_i^*$ is defined as *similar* to the local shape at $\mathbf{r}_i$ of the feature curve $\bar{\mathbf{c}}_f$ before resizing. The global similarity energy of the optimized feature curve is then defined as

$$E_r = \sum_i \|\mathbf{r}_i^* - \mathbf{r}_i^+\|^2 + \|\mathbf{r}_i^* - \mathbf{r}_i^-\|^2, \tag{4.21}$$

where a rigid transformation and scaling of the feature curve gives zero energy value. In addition, one may also hope to keep the optimized result close to the shape before optimization. Thus, a deformation energy is defined as

$$E_d = \sum_i \|\mathbf{r}_i^* - \mathbf{r}_i'\|^2 \tag{4.22}$$

with $\mathbf{r}_i'$ being the position before optimization. The optimized positions of sample points on the feature plane can then be computed by solving the following optimization problem

$$\min_{\mathbf{r}_i^*} \varpi_r E_r + \varpi_d E_d, \tag{4.23}$$

where ϖ_r and ϖ_d are the weights to balance these two energy terms during the optimization. As the shape similarity is more important, $\varpi_r = 1.0$ and $\varpi_d = 0.1$ are used. This is a least-square nonlinear optimization problem, which can be easily solved by computing a least-square solution of an overdetermined linear equation system derived from the first derivative of Eq. (4.23).

When minimizing the global similarity energy E_r, if no point on the feature curve is fixed, any rigid transformation of the original feature curve $\bar{\mathbf{c}}_f$ on the new feature plane will lead to the minimum of E_r. The solution is not uniquely defined. When one sample point (say $\mathbf{r}_j$) on $\bar{\mathbf{c}}'_f$ is fixed during the optimization, any rotation transformation of $\bar{\mathbf{c}}_f$ around $\mathbf{r}_j$ also gives a minimal value of E_r. Therefore, at least two points should be fixed during the optimization. The selection of the two points to be fixed is based on the following heuristics:

- The points which are closer to the human body have a higher priority to be fixed since the shape of the apparel product should follow the surface of human body when near the body;
- The points close to the feature curves that have already been processed would more likely be fixed as we do not wish the optimization on these feature curves to affect the processed ones.

Therefore, two closest points to the human body with distances d_1 and d_2, respectively, are first found. Another two points nearest to the feature curves that have been processed (with distances d_3 and d_4, respectively). Among these four points, two points with the smallest values in d_k $(k = 1, \ldots, 4)$ are fixed during the optimization.

It should be noticed that, when the samples are close to the ends of the feature curves—in other words, $\mathbf{r}_i$ with $i < 3$ or $i > m - 3$, the formulae in Eqs. (4.19) and (4.20) may be invalid. Those invalid formulae are discarded from the least-square linear system to minimize the objective function in Eq. (4.23). When a feature curve is closed, the indices in Eqs. (4.19) and (4.20) are recycled to make the formulae valid.

The feature curves defined by users sometimes intersect each other. To make the optimizations conducted on two feature curves, $\bar{\mathbf{c}}_A$ and $\bar{\mathbf{c}}_B$, compatible, an intersection point $\mathbf{p}_{\text{int}}$ should be explicitly defined. First, the shape of $\bar{\mathbf{c}}_A$ is processed according to what we have proposed above. Second, when processing $\bar{\mathbf{c}}_B$, the intersection point $\mathbf{p}_{\text{int}}$ on it is fixed by $\mathbf{p}_{\text{int}}$'s position determined during the shape optimization of $\bar{\mathbf{c}}_A$. The shapes of feature curves are processed in the order of how users specify them based on the heuristic that a more important feature curve is specified earlier. Processing the intersected feature curves in different orders results in feature curves with different shapes.

The shape matching method proposed above works fine in preserving the local shapes of feature curves. For example, the zigzag shape on a given feature curve can be well preserved on the deformed feature curve. However, this method cannot preserve higher order nor global features such as circularity, orthogonality, and parallelism.

4.4.3 Sampling and Construction of Local Support

As the discrete surface of a free-form product may be represented by polygon soups which in general do not provide a correct local support, the surface M is sampled into a set of points $\bar{M}$ to make the algorithm general and robust. However, this may result in too many samples. To speed up the computation, we downsample the points of $\bar{M}$ into fewer points by

- partitioning the bounding box of $\bar{M}$ uniformly into voxels,
- and then using the sample point of $\bar{M}$ in a voxel, which is nearest to the center of the voxel, as an agent of the discrete surface $\bar{M}$.

The sampled points on feature curves are also considered as agents of $\bar{M}$. The shape optimization of the discrete surface is conducted on these agents.

For every agent point $\mathbf{a}$, the local support is established by determining its one-ring neighbors. The method of Guennebaud et al. [9] is employed here, the efficient implementation of which needs the help of a fast closest point query as [18].

1. The nearest agent point $\mathbf{a}_c$ to $\mathbf{a}$ is searched and considered as one of the one-ring neighbors of $\mathbf{a}$;
2. All the agent points $\mathbf{p}$ *on* or *above* the plane $(\mathbf{p} - \mathbf{a}_c) \cdot (\mathbf{a}_c - \mathbf{a}) = 0$ are excluded from the further neighborhood search;
3. Repeating steps 1 and 2 until no point exists in the search.

The planes determined in this search algorithm actually define a convex hull around the agent point $\mathbf{a}$, which is the only agent point inside the convex hull. All the agent points lying on the convex hull are considered as the one-ring neighbors of $\mathbf{a}$.

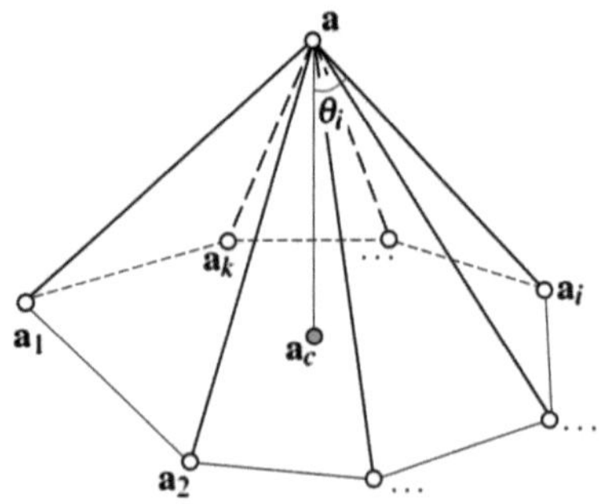

Fig. 4.13 Local shape encoding for the agents around **a**, where $\mathbf{a}_c$ is the weighted average of the one-ring neighbors of **a** as defined in Eq. (4.25)

4.4.4 Local Shape Encoding

The purpose of shape optimization of the discrete surface $\bar{M}$ is to deform the surface into a shape interpolating the feature curves (that is, the agents defined by the sample points on feature curves). The deformation taken place here wishes to maintain the local shape and the geometric details of $\bar{M}$. Therefore, a method to encode the relative spatial relationship between an agent point **a** and its one-ring neighbors is needed. Since this relationship is employed in the shape optimization of $\bar{M}$, it should have the following properties.

- The relationship between the positions of **a** and its one-ring neighbors $\mathbf{a}_i$ $(i = 1, \ldots, k)$ must be a linear combination so that the shape optimization can be solved efficiently.
- The encoding of relationship considers the nonuniformity of the distribution of $\mathbf{a}_i$ around **a**.
- As the number of agents may be large, Meng et al. [17] encode the relationship into a weakly diagonal-dominated linear equation system; therefore, the convergency of using Gauss-Seidal method[1] [20] to solve the linear system is guaranteed.

A new local coordinate encoding method is investigated below to satisfy all these requirements.

As illustrated in Fig. 4.13, the weighted average point of **a**'s one-ring neighbors, $\mathbf{a}_i$, is computed first. A weighting method according to the inverse square distance between $\mathbf{a}_i$ and **a** is adopted here as

$$\lambda_i = \frac{\|\mathbf{a}_i\mathbf{a}\|^{-2}}{\sum_{j=1}^{k}(\|\mathbf{a}_j\mathbf{a}\|^{-2})}, \tag{4.24}$$

which satisfies $\sum_{i=1}^{k} \lambda_i \equiv 1$. The weighted average

[1] When using the Gauss-Seidal method to solve a linear equation system, the solution can be obtained by updating the variables in an out-of-core manner; therefore, a huge linear equation system can still be processed easily.

$$\mathbf{a}_c = \sum_{i=1}^{k} \lambda_i \mathbf{a}_i \tag{4.25}$$

can then be obtained. The angle θ_i between the vectors $\mathbf{aa}_c$ and $\mathbf{aa}_i$ is computed for encoding the relationship between $\mathbf{a}$ and $\mathbf{a}_i$ in the local pyramid. Last, the local coordinates are encoded as

$$\mathbf{a} = \sum_{i=1}^{k} \lambda_i \mathbf{a}_i + \sum_{i=1}^{k} \lambda_i (\mathbf{a} - \mathbf{a}_i) \cos\theta_i + \mathbf{h} \tag{4.26}$$

with $\mathbf{h}$ being a compensation vector. This satisfies the convergency condition of using Gauss-Seidal method to compute the optimized positions of agent points—the linear equation system is weakly diagonal-dominated. An analysis is given in the following subsection.

4.4.5 Controlled Reconstruction

The positions of all agent points are updated, and then all vertices on the warped model of the given free-form product are moved to the blended positions by the local frames defined on the agent points.

The agent point $\mathbf{a}_j$ which does not belong to any feature curve and is far away from the human body (for example, more than $\varepsilon = 1$ cm in the application of apparel products) is updated to a new position $\mathbf{a}_j^*$. The new positions of agent points should retain their local relationship as defined in Eq. (4.26) as

$$\mathbf{a}_j^* = \sum_{i=1}^{k} \lambda_i \mathbf{a}_i^* + \sum_{i=1}^{k} \lambda_i (\mathbf{a}_j^* - \mathbf{a}_i^*) \cos\theta_i + \mathbf{h}_j. \tag{4.27}$$

When the distance between $\mathbf{a}_j$ and the human body is smaller than ε, the position of $\mathbf{a}_j$ is fixed by enforcing

$$\mathbf{a}_j^* = \mathbf{a}_j. \tag{4.28}$$

When the agent point $\mathbf{a}_j$ belongs to a feature curve, its position must be updated by

$$\mathbf{a}_j^* = \mathbf{r}_j^* \tag{4.29}$$

with $\mathbf{r}_j^*$ being its optimized position on the feature curve (by the method presented in Sect. 4.4.2). In this linear equation system, Eq. (4.27) can be updated to

$$\left(\sum_{i=1}^{k} \lambda_i (1 - \cos\theta_i)\right)\mathbf{a}_j^* - \sum_{i=1}^{k} \mathbf{a}_i^* \lambda_i (1 - \cos\theta_i) = \mathbf{h}_j,$$

which satisfies

$$|\sum_{i=1}^{k} \lambda_i (1 - \cos\theta_i)| = \sum_{i=1}^{k} |\lambda_i (1 - \cos\theta_i)|$$

by the coefficients of $\mathbf{a}_j^*$ and $\mathbf{a}_i^*$ since $\lambda_i > 0$ and $1 - \cos\theta_i \geq 0$. As long as there is an equation like Eq. (4.28) or (4.29) in the system, it is a weakly diagonal-dominated linear equation system [20] and can be solved by the Gauss-Seidal method. Specifically, the initial value of an agent that is not on the feature curves is set as $\mathbf{a}_j$. Then, the optimized positions of all such agents are computed by iteratively updating their values according to Eq. (4.27). When the L^2-norm of the update vector on all such agents is smaller than 10^{-5} or the iteration has run more than 500 times, the iteration is terminated.

To update the vertices on the warped model M' to their optimal positions, local frames are established on every agent point at its positions before and after the shape optimization. The normals of the triangles defined by $\mathbf{a}$ and its one-ring neighbors are computed, and their average normal is considered as the surface normal at $\mathbf{a}$. The orientations of normals on all agent points are corrected by using the propagation [10] based on minimal spanning tree. The normal vector at $\mathbf{a}$ is used to define a tangent plane. The local frame at $\mathbf{a}$ is established by the projection of vector $\mathbf{aa}_1$ on the tangent plane, the normal vector and their cross-product. Then, for any vertex $\mathbf{p} \in M'$, its new position can be determined by three steps.

1. Computing its local coordinates in the frames at the agents whose distance to $\mathbf{p}$ is smaller than three times of the width of the voxels used in the sampling step;
2. Applying these local coordinates to obtain new positions by the frames defined on the optimized agent points;
3. Blending the positions defined by different frames into a unique position in the manner as shown in Eq. (2.4)

By the above processes, the optimized shape M^* for a free-form product can be obtained.

4.4.6 Examples

The first example is the automatic resizing of the dresses shown in Fig. 4.9. Without applying the method proposed in this section, unwanted distortion is generated on the profiles of the dresses resized by the method of Sect. 4.2 (see the white one in Fig. 4.9). After specifying the feature curves at places where the shape control must be conducted, the profile shape of the resized model is well optimized.

Fig. 4.14 An example of a dress with multiple layers. The profile of the dress is bent (the *white* one) when applying only the primary 3D resizing. After applying the shape control technique presented in [17], the shape of the profile is well reconstructed on the resizing result (Image taken from [17] with permission)

The second example is tested on a dress with multiple layers and accessories which are represented by non-manifold surfaces. From the result shown in Fig. 4.14, it is easy to find that the approach proposed in this section is general and can be successfully applied to apparel products represented by non-manifold surfaces. More examples of apparel products can be found in [17].

References

1. Allen, B., Curless, B., Popović, Z.: The space of human body shapes: reconstruction and parameterization from range scans. ACM Trans. Graph. **22**, 587–594 (2003)
2. Arun, K.S., Huang, T.S., Blostein, S.D.: Least-squares fitting of two 3-d point sets. IEEE Trans. Pattern Anal. Mach. Intell. **9**, 698–700 (1987)
3. Avidan, S., Shamir, A.: Seam carving for content-aware image resizing. ACM Trans. Graph. **26**, 3–10 (2007)
4. Beatson, R., Greengard, L.: A short course on fast multipole methods. In: Wavelets, Multilevel Methods and Elliptic PDEs, pp. 1–37. Oxford University Press, New York (1997)
5. Botsch, M., Kobbelt, L., Pauly, M., Alliez, P., Lévy, B.: Polygonal Mesh Processing. A K Peters, Natick (2010)

6. Botsch, M., Pauly, M., Gross, M., Kobbelt, L.: Primo: coupled prisms for intuitive surface modeling. In: SGP '06: Proceedings of the Fourth Eurographics Symposium on Geometry Processing, pp. 11–20 (2006)
7. Buhmann, M.D.: A new class of radial basis functions with compact support. Math. Comput. **70**, 307–318 (2000)
8. Gal, R., Sorkine, O., Mitra, N.J., Cohen-Or, D.: Iwires: an analyze-and-edit approach to shape manipulation. ACM Trans. Graph. **28**(3), 1–10 (2009)
9. Guennebaud, G., Barthe, L., Paulin, M.: Interpolatory refinement for real-time processing of point-based geometry. Comput. Graph. Forum (Proc. Eurograph. 2005) **24**, 657–666 (2005)
10. Hoppe, H., DeRose, T., Duchamp, T., McDonald, J., Stuetzle, W.: Surface reconstruction from unorganized points. SIGGRAPH Comput. Graph. **26**, 71–78 (1992)
11. Kobayashi, K.G., Ootsubo, K.: t-ffd: free-form deformation by using triangular mesh. In: Proceedings of the Eighth ACM Symposium on Solid Modeling and Applications, pp. 226–234 (2003)
12. Kozhekin, N., Savchenko, V., Senin, M., Hagiwara, I.: An approach to surface retouching and mesh smoothing. Vis. Comput. **19**, 549–564 (2003)
13. Kraevoy, V., Sheffer, A., Shamir, A., Cohen-Or, D.: Non-homogeneous resizing of complex models. ACM Trans. Graph. **27**, 111:1–111:9 (2008)
14. Lazarus, F., Coquillart, S., Jancene, P.: Axial deformations: an intuitive deformation technique. Comput. Aided Des. **26**, 607–613 (1994)
15. Lipman, Y., Cohen-Or, D., Gal, R., Levin, D.: Volume and shape preservation via moving frame manipulation. ACM Trans. Graph. **26**(1), 5 (2007)
16. Lipman, Y., Sorkine, O., Levin, D., Cohen-Or, D.: Linear rotation-invariant coordinates for meshes. ACM Trans. Graph. **24**(3), 479–487 (2005)
17. Meng, Y., Wang, C., Jing, X.: Flexible shape control for automatic resizing of apparel products. Comput. Aided Des. **44**, 68–76 (2012)
18. Mount, D., Arya, S.: ANN: a Library for Approximate Nearest Neighbor Searching. http://www.cs.umd.edu/mount/ANN/ (2006)
19. Piegl, L., Tiller, W.: The NURBS Book (2nd edn.). Springer, Berlin (1997)
20. Press, W.H., Flannery, B.P., Teukolsky, S.A., Vetterling, W.T.: Numerical Recipes in C: the Art of Scientific Computing. Cambridge University Press, New York (1995)
21. Sederberg, T.W., Parry, S.R.: Free-form deformation of solid geometric models. SIGGRAPH Comput. Graph. **20**, 151–160 (1986)
22. Shoemake, K.: Animating rotation with quaternion curves. SIGGRAPH Comput. Graph. **19**, 245–254 (1985)
23. Singh, K., Fiume, E.: Wires: a geometric deformation technique. In: Proceedings of the 25th Annual Conference on Computer Graphics and Interactive Techniques (SIGGRAPH '98), pp. 405–414 (1998)
24. Sumner, R.W., Schmid, J., Pauly, M.: Embedded deformation for shape manipulation. ACM Trans. Graph. **26**(3), 80:1–80:7 (2007)
25. Wang, C.C.L.: Cyber Tape: an interactive measurement tool on polyhedral surface. Comput. Graph. **28**(5), 731–745 (2004)
26. Wang, C.C.L., Hui, K.C., Tong, K.M.: Volume parameterization for design automation of customized free-form products. IEEE Trans. Autom. Sci. Eng. **4**(1), 11–21 (2007)
27. Wang, C.C.L., Wang, Y., Yuen, M.M.F.: Design automation for customized apparel products. Comput. Aided Des. **37**, 675–691 (2005)
28. Wang, C.C.L., Yuen, M.M.F.: Freeform extrusion by sketched input. Comput. Graph. **27**(2), 255–263 (2003)
29. Wang, C.C.L., Zhang, Y., Sheung, H.: From designing products to fabricating them from planar materials. IEEE Comput. Graph. Appl. **20**, 74–85 (2010)
30. Yu, Y., Zhou, K., Xu, D., Shi, X., Bao, H., Guo, B., Shum, H.Y.: Mesh editing with poisson-based gradient field manipulation. ACM Trans. Graph. **23**, 644–651 (2004)

Chapter 5
Manufacturability of Products Fabricated by Planar Materials

5.1 Manufacturability Problem

The work in this research thread is motivated by the development of geometric modeling systems for freeform products fabricated by planar materials (such as metal in the ship industry, fabric in the apparel industry and toy industry, and leather in the shoe industry and furniture industry). During fabrication, 2D pieces are warped and stitched together to build final products. Ideally, the warping and the stitching should be stretch-free since a stretch will produce elastic energy in the final products which degrades fitness and creates material fatigue. Traditional design process in these industries is conducted in a trial-and-error manner. A designer drafts 2D pieces on a piece of paper and then makes a prototype to check whether the fitting is good. If the result is not satisfactory, the designer needs to modify the patterns by his experience and make another prototype. The prototyping and modification steps are repeatedly applied, which is inefficient. Another more serious problem occurs when designing in 2D instead of 3D space—the product made from the planar pieces may not approximate the desired 3D shape well. That is why 3D CAD systems are expected recently in these industries.

After designing the 3D surface patches for the shapes of products, their corresponding 2D patterns for fabrication are computed through a surface flattening process. However, the flattening is stretch-free only if a surface mesh under flattening holds the isometric mapping property (in other words, the surface is developable [5]). Current commercial CAD/CAM systems do not have the function for modeling developable freeform surfaces (for example, the piece of crumpled paper and the leather surface shown in Fig. 5.1), and existing approaches about developable surfaces in literature are weak in modeling freeform surfaces.

The research work proposed in this project relates to the study of developable surfaces in *Differential Geometry* [5], where the common forms of developable surface are generalized cylinders, conical surfaces (away from the apex), and tangent developable surfaces. The developability of a surface can be characterized by Gaussian

C. C. L. Wang, *Geometric Modeling and Reasoning of Human-Centered Freeform Products*, DOI: 10.1007/978-1-4471-4360-4_5, © Springer-Verlag London 2013

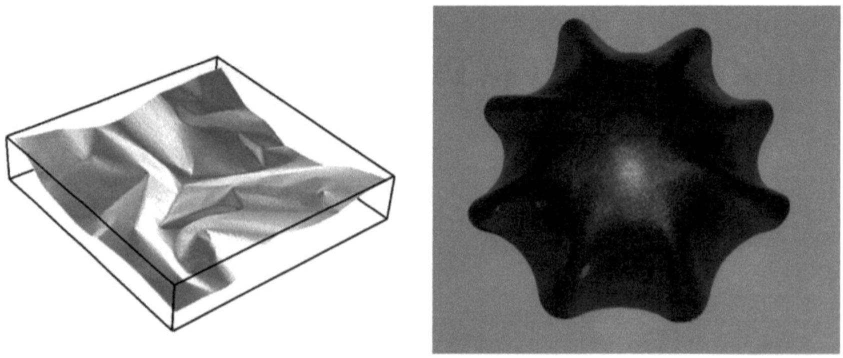

Fig. 5.1 Surfaces in reality that are hard to be modeled by continuous developable surfaces: (*left*) a *square* piece of crumpled paper and (*right*) a *spherical* leather mold (image taken from [46] with permission ©2005 ASME)

curvature which is the product of the maximum and minimum normal curvatures at a given point. In general, a surface is developable if and only if the Gaussian curvature of every point on it is zero. When computing in a discrete form (i.e., piecewise linear surfaces), such a surface is named as *flattenable mesh surface* that has zero Gaussian curvature at non-boundary vertices [55].

Freeform products automatically generated by the techniques presented in the above chapters are represented by piecewise linear surfaces, which, however, in general are not flattenable. Therefore, the aforementioned manufacturability problem exists. The approaches that try to solve this problem can be classified into two categories: (1) minimizing distortion in 3D-to-2D mapping [1, 2, 10, 13, 16, 24, 26, 28, 40, 58, 60]—named as *flattening* (or *parameterization*) methods and (2) processing 3D shapes to make them flattenable [9, 18, 27, 54, 55, 59]—named as *shape optimization* methods. Both flattening and shape optimization techniques are introduced below.

5.2 Surface Flattening Based on Energy Model

This section presents an energy-based method for flattening a 3D surface patch represented by triangular meshes into a 2D shape. A spring–mass model based on energy function is used to flatten 3D mesh surfaces into 2D patterns. The accuracy of a flattened surface can be controlled locally, which provides more flexibility for solving CAD and CAM problems.

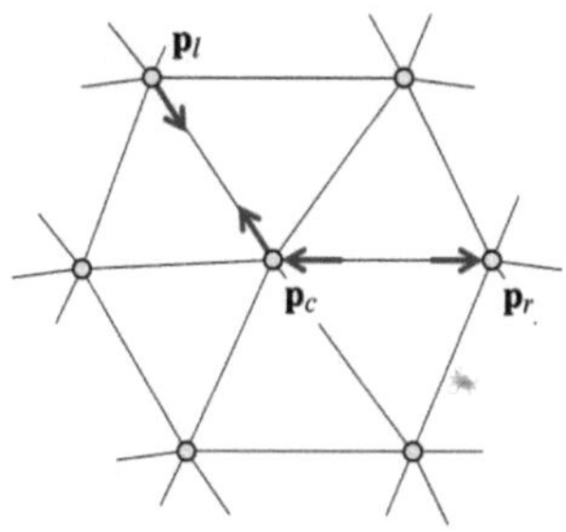

Fig. 5.2 Local configuration of a spring–mass system: nodes $\mathbf{p}_i$s are particles and the links between two particles $\mathbf{p}_i$ and $\mathbf{p}_j$ are springs. During deformation, if the distance between $\mathbf{p}_i$ and $\mathbf{p}_j$ is greater than the distance between their corresponding vertices, $\mathbf{q}_i$ and $\mathbf{q}_j$, on the 3D surface, an attraction force is applied—for example, the force between $\mathbf{p}_c$ and $\mathbf{p}_l$. Otherwise, if $\|\mathbf{p}_i\mathbf{p}_j\| < \|\mathbf{q}_i\mathbf{q}_j\|$, a repellent force is applied such as that between $\mathbf{p}_c$ and $\mathbf{p}_r$

5.2.1 Energy Function and Energy Release

A spring–mass system is established for the deformation of a planar triangular mesh F. F is represented by a pair (K, P), where K stores the connectivity of the vertices to determine the topology of the mesh, and $P = \{\mathbf{p}_1, \ldots, \mathbf{p}_m\}$ ($\mathbf{p}_i \in \Re^2$) is a set of vertex positions defining the shape of the mesh in $\Re^2$. The given mesh surface M to be flattened is represented by another pair (K, Q), where $Q = \{\mathbf{q}_1, \ldots, \mathbf{q}_m\}$, and each vertex $\mathbf{q}_i \in \Re^3$ corresponds to a planar vertex $\mathbf{p}_i$. Most of the physical parameters for the spring–mass system are derived from their corresponding geometric parameters. For example, forces, elastic deformation energy, and masses are determined by the relative displacement between mesh nodes and the area of triangles. Thus, this model is a spring–mass system in terms of geometry. The difference between the current position of the mesh and the final shape of the 2D piece can be considered as the elastic deformation energy stored in the planar spring–mass system. One example of a spring–mass system is shown in (Fig. 5.2).

5.2.1.1 Energy Function

The elastic deformation energy function and tensile force on one single particle $\mathbf{p}_i$ are defined as

$$E(\mathbf{p}_i) = \sum_{j \in N(i)} \frac{1}{2} c(\|\mathbf{p}_i\mathbf{p}_j\| - \|\mathbf{q}_i\mathbf{q}_j\|)^2, \tag{5.1}$$

$$\mathbf{f}(\mathbf{p}_i) = \sum_{j \in N(i)} c(\|\mathbf{p}_i\mathbf{p}_j\| - \|\mathbf{q}_i\mathbf{q}_j\|)\frac{\mathbf{p}_j - \mathbf{p}_i}{\|\mathbf{p}_i\mathbf{p}_j\|}, \tag{5.2}$$

where c is the spring's stiffness coefficient, $N(i)$ defines the set of one-ring neighbors of the particle $\mathbf{p}_i$, and $\mathbf{q}_i$ and $\mathbf{q}_j$ are the corresponding vertices of $\mathbf{p}_i$ and $\mathbf{p}_j$ on the given surface M.

Forces are generated by the difference between the planar triangular edges' lengths and their original lengths on M. Recall that the goal of surface flattening is to obtain a planar triangular mesh F that provides a good fit to the given 3D mesh surface. This approach is going to find a complex K and a set of vertex positions P defining a planar mesh $F = (K, P)$ that minimizes the energy function

$$E(P) = \sum_{i=1}^{m} E(\mathbf{p}_i). \tag{5.3}$$

For some applications, users may wish to control the flattening accuracy locally. Different spring constants can be defined interactively for different parts of the surface. The higher the accuracy is desired, the larger the value of the spring constants c is given.

5.2.1.2 Energy Release

Lagrange equation is usually applied to an energy model to solve dynamics problems [30]. There is no external loading applied to the spring–mass system for surface flattening. By neglecting the damping terms, the simplified Lagrange equation for releasing energy becomes

$$M\ddot{\mathbf{p}} + K\mathbf{p} = 0 \tag{5.4}$$

with M and K being the mass and stiffness matrices of the spring–mass system. The term $K\mathbf{p}$, in fact, comes from the accumulated forces on the particles.

Explicit Euler scheme is used to solve the above equation here. The acceleration of the particle $\mathbf{p}_i$ is assumed to be constant when the time step Δt is small. The balance of the whole spring–mass system is achieved once the balance for each single node is obtained. For each node $\mathbf{p}_i$, the scheme to update its position can then be derived as

$$m_i = \frac{\rho}{3} \sum_k A_k \tag{5.5}$$

$$\ddot{\mathbf{p}}(t) = \mathbf{f}_i(t)/m_i \tag{5.6}$$

$$\dot{\mathbf{p}}(t + \Delta t) = \dot{\mathbf{p}}(t) + \Delta t \ddot{\mathbf{p}}(t) \tag{5.7}$$

$$\mathbf{p}(t + \Delta t) = \mathbf{p}(t) + \Delta t \dot{\mathbf{p}}(t) \tag{5.8}$$

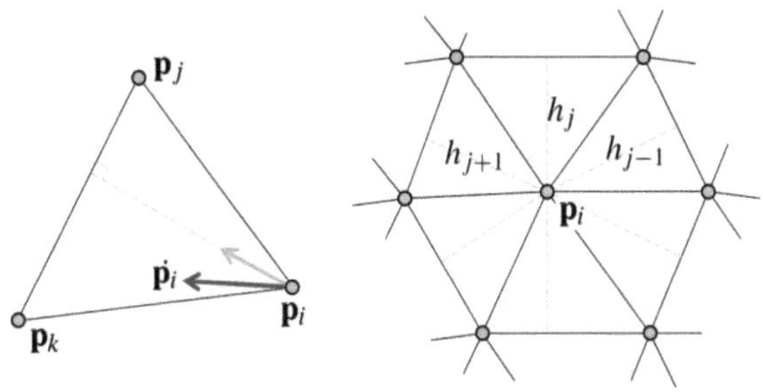

Fig. 5.3 Penalty of *triangle* overlap during energy release: (*left*) overlap happens according to the velocity direction of $\mathbf{p}_i$ and (*right*) penalty functions are added according to the distance between $\mathbf{p}_i$ and its opposite edges

where m_i is the mass of $\mathbf{p}_i$, ρ is the area density of the surface, A_k is the 3D area of the triangle k containing the particle $\mathbf{p}_i$, $\mathbf{f}_i(t)$ is the accumulated force on particle $\mathbf{p}_i$ at time t, and $\dot{\mathbf{p}}_i$ and $\ddot{\mathbf{p}}_i$ are the velocity and the acceleration of particle $\mathbf{p}_i$ respectively.

The area density, ρ, here is not the real density of the surface. In most physics-based models, ρ and C are just scale factors that make deformation more appropriate [30] (for example, $\rho = 1/\min\{m_i\}$ and $c = 0.25$). The area density, ρ, and the spring stiffness coefficient, c, can also be obtained by testing the mechanical properties. For the time step, Δt, using $\Delta t = 0.01$ can always lead to a stable result with acceptable speed.

Penalty function. During surface development, overlaps may easily occur. In (Fig. 5.3), if $\mathbf{p}_i$ moves to the right of edge $\mathbf{p}_j\mathbf{p}_k$ (where $\dot{\mathbf{p}}_i$ shows the velocity of $\mathbf{p}_i$), $\Delta\mathbf{p}_i\mathbf{p}_j\mathbf{p}_k$ overlaps other triangles.

To prevent overlaps, a penalty vector is added to move particles. The penalty vector used in Provot's method [36] is adopted here.

$$\tau(\mathbf{p}_i) = -\sum_{j=1}^{n} c_p |h_j - h_j^*| \hat{\mathbf{n}}_j, \qquad c_p = \begin{cases} 1 \ (h_j \le h_j^*) \\ 0 \ (h_j > h_j^*) \end{cases} \tag{5.9}$$

where h_j is the current distance from $\mathbf{p}_i$ to its jth opposite edge on the planar surface (see the illustration shown in Fig. 5.3), h_j^* is the Euclidean distance from $\mathbf{q}_i$ (the corresponding vertex of $\mathbf{p}_i$ on M) to its jth opposite edge on the 3D surface M, $\hat{\mathbf{n}}_j$ is a unit vector pointing from $\mathbf{p}_i$ to its opposite edge on the planar surface (see Fig. 5.3), and n is the number of opposite edges of $\mathbf{p}_i$.

Penalty function $\tau(\mathbf{p}_i)$ is a vector function. When the particle $\mathbf{p}_i$ moves to a position that is too close to its opposite edge the penalty function $\tau(\mathbf{p}_i)$ is applied to $\mathbf{p}_i$ to prevent overlapping by letting

$$\mathbf{p}_i = \mathbf{p}_i + \tau(\mathbf{p}_i). \tag{5.10}$$

Scheme of energy release. The following iterative algorithm, *EnergyRelease*(F), finds a set of vertex positions P defining a planar mesh $F = (K, P)$ that minimizes the energy function $E(F)$.

Algorithm 1 EnergyRelease

Input: A set of particles P in their initial position
Output: The final positions of P minimize $E(P)$

```
for all p_i ∈ P do
    Compute the mass of node p_i;
end for
repeat
    for all p_i ∈ P do
        Compute the accumulated force on p_i;
        Compute the new position of p_i by explicit Euler's method;
        Compute the penalty vector and apply it to p_i;
        Move p_i to the new position;
    end for
    Compute the new energy E(P);
until terminal condition is satisfied
return P;
```

The terminal condition depends on three factors: (1) the variation of energy $E(P)$ in consecutive steps is smaller than 0.05 %, (2) the difference between summed areas of triangles in 3D and 2D is small, and (3) the difference between summed edge lengths in 3D and 2D is small. If any of these conditions is satisfied, the iteration of energy release is stopped.

5.2.2 *Triangle Flattening*

This section introduces the first step of energy-based surface flattening algorithm. In this step, the triangles of the given 3D mesh are unfolded one by one to obtain the initial planar triangular mesh.
Triangle unfolding without energy release. When unfolding a triangle T on the given 3D mesh surface M, there are two possible configurations.

Unconstrained triangle unfolding. This configuration will occur when one edge $\mathbf{q}_i\mathbf{q}_j$ of the triangle T has already been unfolded to the positions $\mathbf{p}_i$ and $\mathbf{p}_j$ while the third one $\mathbf{q}_k$ is going to be located for unfolding the whole triangle T. The position of $\mathbf{q}_k$ on the 2D flattening, $\mathbf{p}_k$, is found by finding the intersection of two circles (as shown in Fig. 5.4). These circles are centered at $\mathbf{p}_i$ and $\mathbf{p}_j$ with radii $\|\mathbf{q}_i\mathbf{q}_k\|$ and $\|\mathbf{q}_j\mathbf{q}_k\|$, respectively. For a fully flattenable surface (definition is given in Sect. 5.3), the 2D position of $\mathbf{p}_k$ determined by triangle T is coincident with the 2D position of $\mathbf{p}_k$ determined by other triangles containing $\mathbf{p}_k$; but for a non-flattenable surface, they are different. Thus, constrained triangle flattening must be used.
Constrained triangle unfolding. When a triangle T shares an edge with a previously unfolded triangle T_a, a situation may arise if another edge of the triangle T is shared with another previously unfolded triangle T_d. By flattening triangle T_d, the 2D location of $\mathbf{q}_k$ is $\mathbf{p}'_k$ (as shown in Fig. 5.5). If further unfolding the triangle T by the

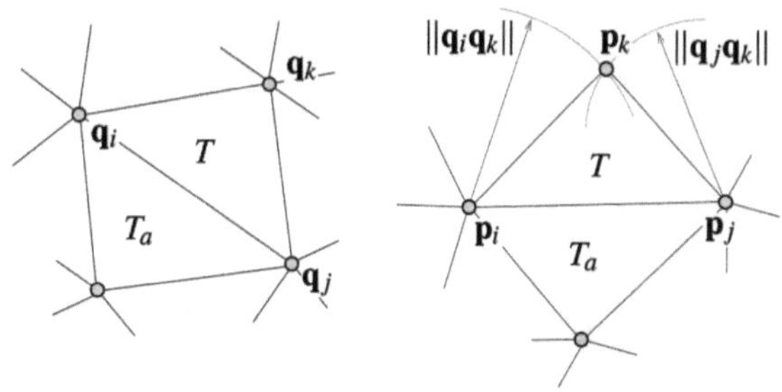

Fig. 5.4 An illustration of *unconstrained triangle unfolding*: (*left*) triangles in $\Re^3$ and (*right*) *unfolded planar triangles* and vertices

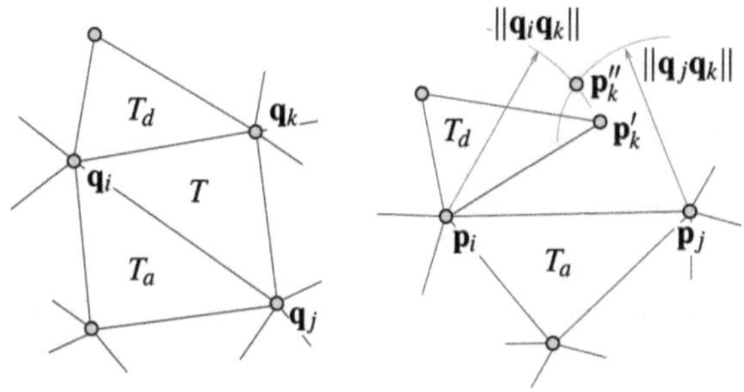

Fig. 5.5 An illustration of *constrained triangle unfolding*: (*left*) *triangles* in $\Re^3$ and (*right*) *unfolded planar triangles* and vertices

unconstrained triangle unfolding method, another location (on the 2D flattening), $\mathbf{p}''_k$, will be probably be provided for the node $\mathbf{q}_k$ (Fig. 5.5). To initially avoid this situation, an average position is used to resolve this conflict so that a unique location is produced.

The constrained triangle flattening process generates strain energy on the edges, which leads to overlap errors.

Algorithm 2 InitFlatten

Input: A given 3D triangular mesh surface M with n triangles $\{T_i\}$
Output: The initial 2D shape of P
1: Build a null spring–mass system Ψ;
2: **for** i from 1 to n **do**
3: Unfold the triangle T_i;
4: Insert the triangle T_i into Ψ to adjust the variant spring–mass system;
5: **Call** EnergyRelease(Ψ) by maximal 50 iteration steps;
6: **end for**
7: Assign the initial 2D shape of P by the positions of particles in the variant spring–mass system;
8: **return** P;

Triangle unfolding with energy release. To release the strain energy generated by the averaging process in the constrained triangle unfolding, a variant spring–mass system Ψ that contains only the unfolded triangles is constructed. Every newly

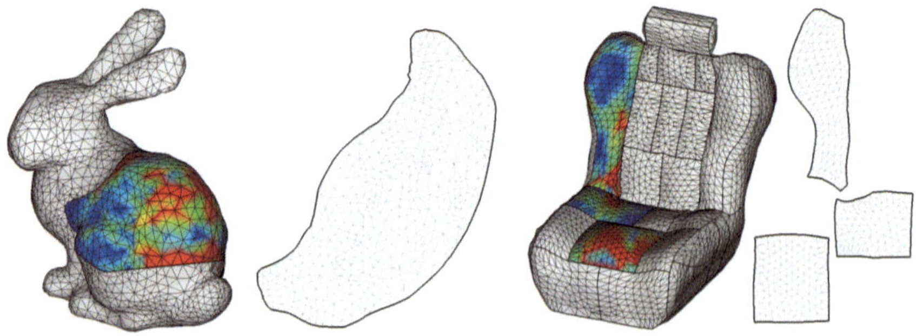

Fig. 5.6 Surface flattening results obtained by the energy-based method. The *color* maps on 3D mesh patches show the distribution of stretch energy, where *blue* denotes a small stretch and *red* represents a large stretch

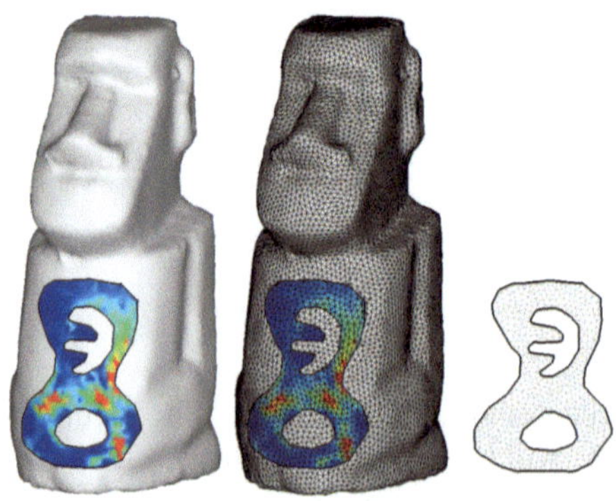

Fig. 5.7 The surface flattening result of a patch trimmed from the Moai model—patches with complex boundaries can also be successfully flattened by the energy-based approach

unfolded triangle is inserted to update this variant spring–mass system Ψ. After that, the Algorithm 1 (introduced in Sect. 5.2.1.2) is used to release the strain energy. By this way, the initial 2D piece of surface flattening can be obtained. The pseudo-code, *InitFlatten*, is given above.

5.2.3 Planar Mesh Deformation

Using a simple iterative Algorithm 1 described in Sect. 5.2.1.2 to release energy in subsequent steps, the resultant 2D piece of triangles with stretch minimized can be obtained (see Fig. 5.6 for examples).

In this section, an energy-based method is presented to determine the resultant planar shape obtained by flattening a 3D mesh surface. The advantages of this approach are its generality (for two-manifold open surface with general topology—see Fig. 5.7 for an example) and flexibility (can control the accuracy of flattening by setting different spring stiffness coefficients). This method has two major drawbacks. First,

3D shape is not updated, thus users have no idea how different the shape of a product will be when fabricating it by the flattened patches. Second, the computation is expensive—for a model with around 1 k vertices, it takes about 9 s on a desktop PC with moderate configuration.

5.3 Geometry Processing for Flattenable Mesh Surface

Different from the surface flattening approach introduced above, the techniques presented in this section focus on how to process an input model represented by piecewise linear surfaces to become surfaces that can be unfolded into plane without stretching. In short, these approaches provide tools for directly modeling flattenable freeform mesh surfaces in $\Re^3$. First, a new type of mesh surfaces—Flattenable Laplacian meshes (*FL meshes*) is presented. Second, the modeling tool for FL meshes is developed under the framework of constrained numerical optimization and the variational subdivision scheme. Furthermore, if a given polygonal mesh is almost flattenable, slight adjustments can improve its flattenability. For this reason, a local flattenable perturbation approach is also described in this section. This new local perturbation approach prevents the interference between meshes under process and their nearby objects, where the computation is based on the constrained optimization with vertex positions as variables. Last, a Least-norm solution for FL mesh processing is introduced to speed up the computation.

5.3.1 Flattenable Laplacian Meshes

The concept of *FL meshes*, which inherit advantages from both flattenable meshes and Laplacian meshes, is introduced below. The computational scheme for FL meshes is also detailed.

5.3.1.1 Flattenable Meshes

A flattenable mesh surface M is a polygonal mesh surface patch which can be flattened into a 2D pattern D without stretching any polygon on it. More specifically, the flattening involves only rotation and transformation but not deformation. However, this is only a descriptive definition. Starting with flattenable vertices, a more mathematical definition of flattenable mesh patches is given. Without loss of generality, only triangular mesh surfaces are referred in the remaining parts of this section.[1]

[1] Other types of polygonal meshes can be easily converted into triangular meshes by splitting each polygon into triangles.

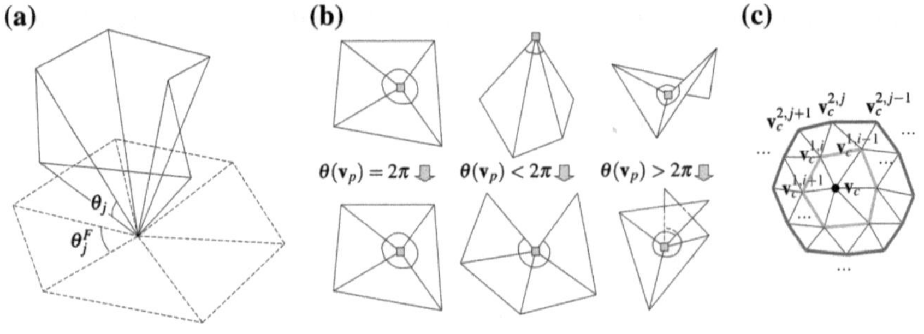

Fig. 5.8 An illustration of flattenability: **a** the inner angles *before* and *after* unfolding the triangles around a vertex, **b** the cases of $\theta(\mathbf{v}_p) = 2\pi$, $\theta(\mathbf{v}_p) < 2\pi$ and $\theta(\mathbf{v}_p) > 2\pi$, and **c** flattening on a disk-like patch can be given in a front advancing manner

Considering an interior mesh vertex and its adjacent triangular faces as shown in Fig. 5.8, the following property can be concluded.

Property 5.1 *For an inner triangular mesh vertex $\mathbf{v}_p$, if and only if the summed inner angle, $\theta(\mathbf{v}_p) = \sum_j \theta_j$, around it is identically 2π, the triangles around it can be flattened into a plane without distortion.*

This property is illustrated by Fig. 5.8. If $\theta(\mathbf{v}_p) > 2\pi$, when flattening triangles around $\mathbf{v}_p$ without stretching, the triangles will generate overlaps. If $\theta(\mathbf{v}_p) < 2\pi$, it yields gap. Therefore, the following definitions can be deduced.

Definition 5.1 An inner triangular mesh vertex $\mathbf{v}_p$ is named as *flattenable vertex* when $\theta(\mathbf{v}_p) = 2\pi$.

Definition 5.2 $\varpi(\mathbf{v}_p) = |\theta(\mathbf{v}_p) - 2\pi|$ is defined as the *flattenability* at $\mathbf{v}_p$ which is the smaller the better.

This is similar to the discrete form of Gaussian curvature [31]. We do not adopt the word *developable* (or *discrete developable*) as [14, 27, 59] is because: when discussing developable property, it is usually derived from differential geometry on regular surface points; for a sharp (or singular) vertex as shown in Fig. 5.8, which is not differentiable, it is more appropriate to name it as *flattenable/unflattenable* rather than *developable/undevelopable*.

Property 5.2 *All inner vertices of a triangular flattenable mesh patch are flattenable.*

Again, this is only a necessary condition but not sufficient. Taking a triangular mesh surface-like cylinder (with two boundary loops) as an example, even if all if its inner vertices are flattenable, it cannot be flattened into a 2D region without stretching or inserting cuts. However, the following geometry property is observed.

Proposition 5.1 *For a triangular mesh patch M in $\Re^3$ with a disk-like topology, if all its inner vertices are flattenable, it can be deformed into a patch D in $\Re^2$ without stretching any triangle.*

Proof First, if any vertex on M is not flattenable, it is known by Fig. 5.8b that the surface cannot be locally flattened into a plane without stretching.

Second, if all interior vertices on M are flattenable, is there any case that M is not flattenable? As illustrated in Fig. 5.8c, starting from an interior point $\mathbf{v}_c$, the triangles adjacent to $\mathbf{v}_c$ can be unfolded into $\Re^2$ by rotation and transformation only (without stretching) since $\theta(\mathbf{v}_c) = 2\pi$. Therefore, the shape of the pink front is formed. The vertices on the pink front can be sorted in the anticlockwise order. For a vertex $\mathbf{v}_c^{1,i}$ located on the pink front, all triangles adjacent to it can be unfolded into $\Re^2$ without stretching as that $\mathbf{v}_c^{1,i}$ is a flattenable vertex. The problem is that whether the location and orientation of planar triangles determined by the local flattening around $\mathbf{v}_c^{1,i}$ will conflict with the local flattening determined by other vertices. The answer is that if the topology of the front is disk-like, the triangles adjacent to $\mathbf{v}_c^{1,i}$ should only be adjacent to its two neighboring vertices $\mathbf{v}_c^{1,i-1}$ and $\mathbf{v}_c^{1,i+1}$, but not any other vertex on the same front. Therefore, the triangles between the pink front and the red front can be flattened without stretching.

Repeating this front advancing, as long as the front is in the disk-like topology, no stretch will be given during flattening. However, if the given mesh surface M is not in the disk-like topology, the disk-like topology cannot always be maintained on the fronts. Taking a cylinder or cone as an example, although every interior vertex on it is flattenable, some conflicts occur during the flattening of triangles in the front advancing. □

5.3.1.2 Laplacian Meshes

Let M be a given mesh patch with the graph $G = (V, E)$, where $V = 1, 2, \ldots, m$ is a set of vertices and E is a set of edges. $\mathbf{v}_i$ is used to denote the position of the vertex i in $\Re^3$, and ∂V represents the set of vertices on the boundary of M. Similar to Sorkine's work [42–45], the following equation defines the fairness or smoothness condition for $\mathbf{v}_i \in \Re^3$

$$\mathbf{v}_i - \frac{1}{|N(\mathbf{v}_i)|} \sum_{j \in N(\mathbf{v}_i)} \mathbf{v}_j = 0 \tag{5.11}$$

where $N(\mathbf{v}_i)$ is the set of one-ring neighboring vertices of $\mathbf{v}_i$ ($\mathbf{v}_i \in V \backslash \partial V$), and $|\ldots|$ denotes the number of elements in the set. The linear system can be rewritten in the matrix form

$$L\mathbf{x} = 0, \quad L\mathbf{y} = 0, \quad L\mathbf{z} = 0, \tag{5.12}$$

where $\mathbf{x}$, $\mathbf{y}$, and $\mathbf{z}$ are the $n \times 1$ vectors containing the x, y, and z coordinates of the vertices respectively. The matrix L is known in [15, 47] as the Laplacian operator below

$$L_{i,j} = \begin{cases} 1 & (i = j) \\ -\dfrac{1}{|N(\mathbf{v}_i)|} & (j \in N(\mathbf{v}_i)) \\ 0 & \text{(otherwise)} \end{cases}$$

It is known from the work of Sorkine and Cohen-Or [43] that the rank of L is $(n-k)$ where k is the number of connected components in the graph G. Therefore, for a mesh surface patch, the rank is $(n-1)$. When fixing vertices in ∂V, Eq. (5.12) gives the solution for the coordinates of inner vertices on a Laplacian mesh patch which minimize the linearized membrane energy $\int_{\Omega} \|\mathbf{v}_s\|^2 + \|\mathbf{v}_t\|^2 \mathrm{d}s\mathrm{d}t$ (Ref. [11])—the resultant mesh of Eq. (5.12) is a smooth mesh surface which is uniquely defined by G and the position of vertices in ∂V.

A Laplacian mesh surface patch in general is not a flattenable mesh (see [55] for more discussions). Therefore, surfaces like this cannot satisfy the applications in industries that need to fabricate products from planar materials. Therefore, the FL meshes defined below are requested.

Definition 5.3 A *Flattenable Laplacian mesh* is a mesh surface patch which can be flattened into 2D pieces without stretching, and meanwhile minimizes the fairness energy function defined by Laplacian operators.

5.3.1.3 Computation of FL Meshes

The computational scheme for FL meshes is presented below. Given a triangular mesh M, the positions of all inner vertices to generate an FL mesh M_{FL} are recomputed so that M_{FL} is

- a flattenable mesh surface;
- smooth (that is a Laplacian mesh);
- a good approximation of the shape of M.

The three requirements can be formulated into a constrained optimization problem

$$\arg\min_{p \in V_{\mathrm{act}}} \; w_1 J_{\mathrm{fair}} + w_2 J_{\mathrm{pos}} \quad \text{subject to} \quad \theta(\mathbf{v}_p) \equiv 2\pi \tag{5.13}$$

where $V_{\mathrm{act}} = V \setminus \partial V$, J_{fair} is the fairness term derived from the Laplacian mesh as

$$J_{\mathrm{fair}} = \frac{1}{2} \sum_{p \in V_{\mathrm{act}}} \phi(\mathbf{v}_p), \tag{5.14}$$

with $\phi(\mathbf{v}_p)$ being a piecewise function $\phi(\mathbf{v}_p) = \|L\mathbf{v}_p\|^2$ only defined on the Voronoi area of $\mathbf{v}_p$. The gradient of J_{fair} with respect to $\mathbf{v}_p$ is $\frac{\partial J_{\mathrm{fair}}}{\partial \mathbf{v}_p} = \mathbf{v}_p - \frac{1}{|N(\mathbf{v}_p)|} \sum_{k \in N(\mathbf{v}_p)} \mathbf{v}_k$, and J_{pos} is the position functional defined to minimize the difference between the new surface M_{FL} and the given surface M

$$J_{\text{pos}} = \frac{1}{2} \sum_{p \in V_{\text{act}}} \|\mathbf{v}_p - \mathbf{v}_p^0\|^2 \tag{5.15}$$

with $\mathbf{v}_p^0$ being the closest position of $\mathbf{v}_p$ on the given surface M or simply being the original position of $\mathbf{v}_p$. Two coefficients w_1 and w_2 reflect the weights of functionals in terms of fairness and position respectively. They are assigned by users. Wang's study [55] shows that choosing $w_1 = 1.0$ and $w_2 = 0.1$ gains a good tradeoff between these two terms. In the fairness term, J_{fair}, Wang [55] chooses the uniform Laplacian but not the cotangent weighted Laplacian as Desbrun et al. do [11, 31] because the uniform Laplacian acts as a regularization term to re-distribute vertices while smoothing the given surface. Furthermore, when cooperating with the variational subdivision scheme, most vertices are with regular valence; therefore, it is appropriate to employ the uniform Laplacian here. In Eq. (5.13), the position constraints are added to the objective function (but not to the constraints set)—this is based on the reason that adding these constraints will disturb the constraints of flattenability during the numerical computation. The value of w_2 must be selected as smaller than w_1; otherwise, the numerical system may become unstable.

With the Lagrange multiplier $\lambda = (\lambda_1, \lambda_2, ..., \lambda_n)$ $(n = |V_{\text{act}}|)$, the constrained optimization problem defined in Eq. (5.13) can be converted into an augmented objective function

$$J(X) = J(\mathbf{v}_1, \mathbf{v}_2, ..., \mathbf{v}_n, \lambda_1, \lambda_2, ..., \lambda_n);$$

in detail,

$$J(X) = w_1 J_{\text{fair}} + w_2 J_{\text{pos}} + \sum_{p \in V_{\text{act}}} \lambda_p (\theta(\mathbf{v}_p) - 2\pi). \tag{5.16}$$

This objective function $J(X)$ can be minimized by using the Newton's method [34] below, where a damping factor $\tau = 0.25$ is employed to increase the stability of computation. A more robust implementation is to introduce a line search mechanism in this step to determine the optimal value of τ (Ref. [35]).

Algorithm 3 Newton's Method

[t]

1: **while** $\|\delta\| > 10^{-5}$ **do**
2: Solve $\nabla^2 J(X)\delta = -\nabla J(X)$;
3: $X \leftarrow X + \tau\delta$;
4: **end while**

Using the sequential linearly constrained programming to minimize $J(X)$ by neglecting the terms that come from the second derivatives of the constraints in the Hessian matrix $\nabla^2 J(X)$, the equation $\nabla^2 J(X)\delta = -\nabla J(X)$ to be solved in each step is simplified to

$$\begin{pmatrix} H & & & \Lambda_x^T \\ & H & & \Lambda_y^T \\ & & H & \Lambda_z^T \\ \Lambda_x & \Lambda_y & \Lambda_z & 0 \end{pmatrix} \begin{pmatrix} \delta_{p_x} \\ \delta_{p_y} \\ \delta_{p_z} \\ \lambda \end{pmatrix} = \begin{pmatrix} B_{p_x} \\ B_{p_y} \\ B_{p_z} \\ B_\lambda \end{pmatrix} \tag{5.17}$$

where λ is the vector of multipliers. Notice that since λ instead of δ_λ is solved in Eq. (5.17), only the positions of inner vertices are updated by δ_p in the routine of Newton's method. Here B_{p_x}, B_{p_y} and B_{p_z} are the x-, y-, and z-components of B_p, respectively, and so are Λ_x, Λ_y and Λ_z. From Eq. (5.16), the following formulae can be obtained.

$$\begin{aligned} B_p &= \left\{-\frac{\partial}{\partial \mathbf{v}_p}(w_1 J_{\text{fair}} + w_2 J_{\text{pos}})\right\} \\ &= \left\{-w_1(\mathbf{v}_p - \tfrac{1}{|N(\mathbf{v}_p)|}\sum_{k\in N(\mathbf{v}_p)} \mathbf{v}_k) - w_2(\mathbf{v}_p - \mathbf{v}_p^0)\right\}, \end{aligned} \tag{5.18}$$

$$B_\lambda = -\left\{\frac{\partial}{\partial \lambda_p}\sum_{p\in V_{\text{act}}} \lambda_p(\theta(\mathbf{v}_p) - 2\pi)\right\} = \{2\pi - \theta(\mathbf{v}_p)\}, \tag{5.19}$$

$$H = \{h_{i,j}\},\quad h_{i,j} = \begin{cases} w_1 + w_2 & (i = j) \\ -\frac{w_1}{|N(\mathbf{v}_i)|} & (\mathbf{v}_j \in N(\mathbf{v}_i)) \\ 0 & (\text{otherwise}) \end{cases}, \tag{5.20}$$

$$\Lambda = \left\{\frac{\partial^2}{\partial \lambda_i \partial \mathbf{v}_j}\sum_{p\in V_{\text{act}}} \lambda_p(\theta(\mathbf{v}_p) - 2\pi)\right\} = \left\{\frac{\partial \theta(\mathbf{v}_i)}{\partial \mathbf{v}_j}\right\}. \tag{5.21}$$

Proposition 5.2 *The gradient of the summed inner angle $\theta(\boldsymbol{v}_p)$ at an inner vertex $\boldsymbol{v}_p \in V \setminus \partial V$ with respect to the vertex $\boldsymbol{v}_p$ itself is*

$$\frac{\partial \theta(\boldsymbol{v}_p)}{\partial \boldsymbol{v}_p} = \sum_{q\in N(\boldsymbol{v}_p)} \frac{\cot \gamma_{q,p} + \cot \xi_{q,p}}{\|\boldsymbol{v}_p \boldsymbol{v}_q\|}(\boldsymbol{v}_q - \boldsymbol{v}_p) \tag{5.22}$$

where $\xi_{q,p}$ and $\gamma_{q,p}$ are the left and right angles respectively aside the edge $\boldsymbol{v}_p\boldsymbol{v}_q$ at $\boldsymbol{v}_q$—see Fig. 5.9.

The proof of Proposition 5.2 can be found in [10].

Proposition 5.3 *The gradient of the summed inner angle $\theta(\boldsymbol{v}_q)$ at an inner vertex $\boldsymbol{v}_q \in V \setminus \partial V$ with respect to its adjacent vertex $\boldsymbol{v}_p$ is*

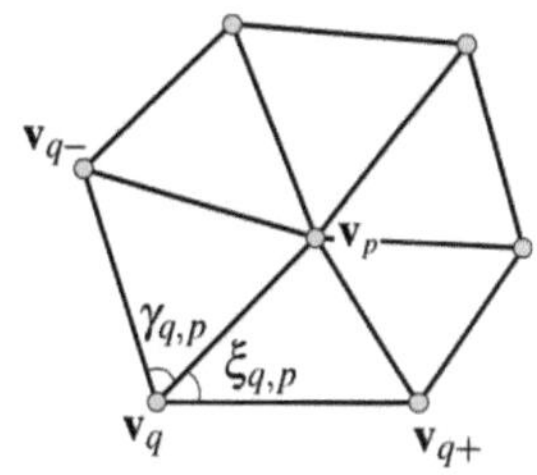

Fig. 5.9 The gradient of summed angle at a vertex with respect to the position of the vertex (or of its adjacent vertex) can be computed locally

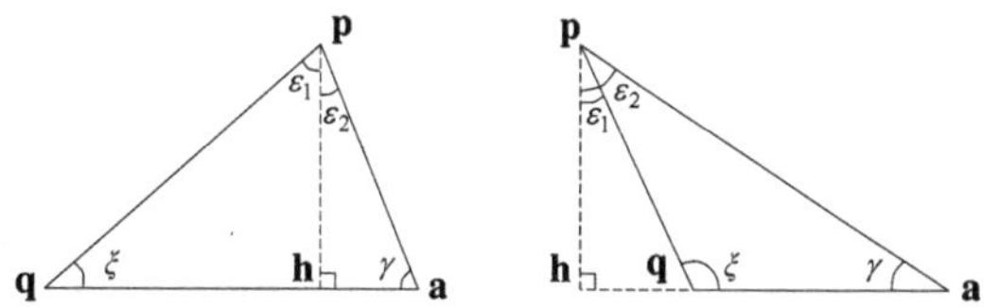

Fig. 5.10 An illustration of the proof of Proposition 5.2 and 5.3: (*left*) acute angle, and (*right*) obtuse angle

$$\begin{aligned}\frac{\partial\theta(\boldsymbol{v}_q)}{\partial\boldsymbol{v}_p} &= \frac{\cot\xi_{q,p}+\cot\gamma_{q,p}}{\|\boldsymbol{v}_p\boldsymbol{v}_q\|^2}(\boldsymbol{v}_p-\boldsymbol{v}_q)\\ &\quad-\frac{(\boldsymbol{v}_{q_+}-\boldsymbol{v}_q)}{\|\boldsymbol{v}_p\boldsymbol{v}_q\|\|\boldsymbol{v}_{q_+}\boldsymbol{v}_q\|\sin\xi_{q,p}}-\frac{(\boldsymbol{v}_{q_-}-\boldsymbol{v}_q)}{\|\boldsymbol{v}_p\boldsymbol{v}_q\|\|\boldsymbol{v}_{q_-}\boldsymbol{v}_q\|\sin\gamma_{q,p}}\end{aligned} \tag{5.23}$$

where $\mathbf{v}_{q_+}$ and $\mathbf{v}_{q_-}$ are the next and last vertices to $\mathbf{v}_q$ in $N(\mathbf{v}_p)$ anticlockwise—see the illustration in Fig. 5.9.

Proof In the summed angle $\theta(\mathbf{v}_q)$ at $\mathbf{v}_q$, only the components $\xi_{q,p}$ and $\gamma_{q,p}$ are affected by the position of $\mathbf{v}_p$. Let us consider $\xi_{q,p}$ first (see Fig. 5.10). When ξ is an acute angle, $\xi=\arccos\frac{\|\mathbf{qh}\|}{\|\mathbf{pq}\|}$ (with $\|\mathbf{qh}\|=(\mathbf{p}-\mathbf{q})\cdot\frac{\mathbf{qa}}{\|\mathbf{qa}\|}$)) yields

$$\begin{aligned}\frac{\mathrm{d}\xi}{\mathrm{d}\mathbf{p}} &= -\frac{1}{\sqrt{1-\frac{\|\mathbf{qh}\|^2}{\|\mathbf{pq}\|^2}}}\frac{\mathrm{d}}{\mathrm{d}\mathbf{p}}\left(\frac{\|\mathbf{qh}\|}{\|\mathbf{pq}\|}\right)\\ &= -\frac{\|\mathbf{pq}\|}{\|\mathbf{ph}\|\|\mathbf{pq}\|^2}\left(\|\mathbf{pq}\|\frac{\mathrm{d}\|\mathbf{qh}\|}{\mathrm{d}\mathbf{p}}-\|\mathbf{qh}\|\frac{\mathrm{d}\|\mathbf{pq}\|}{\mathrm{d}\mathbf{p}}\right)\\ &= -\frac{\|\mathbf{pq}\|}{\|\mathbf{ph}\|\|\mathbf{pq}\|^2}\left(\frac{\|\mathbf{pq}\|}{\|\mathbf{qa}\|}(\mathbf{a}-\mathbf{q})-\frac{\|\mathbf{qh}\|}{\|\mathbf{pq}\|}(\mathbf{p}-\mathbf{q})\right)\\ &= \frac{\cot\xi}{\|\mathbf{pq}\|^2}(\mathbf{p}-\mathbf{q})-\frac{1}{\|\mathbf{qa}\|\|\mathbf{pq}\|\sin\xi}(\mathbf{a}-\mathbf{q}).\end{aligned} \tag{5.24}$$

When ξ is an obtuse angle, $\xi=\arccos(-\frac{\|\mathbf{qh}\|}{\|\mathbf{pq}\|})$ (with $\|\mathbf{qh}\|=(\mathbf{q}-\mathbf{p})\cdot\frac{\mathbf{qa}}{\|\mathbf{qa}\|}$) leads to

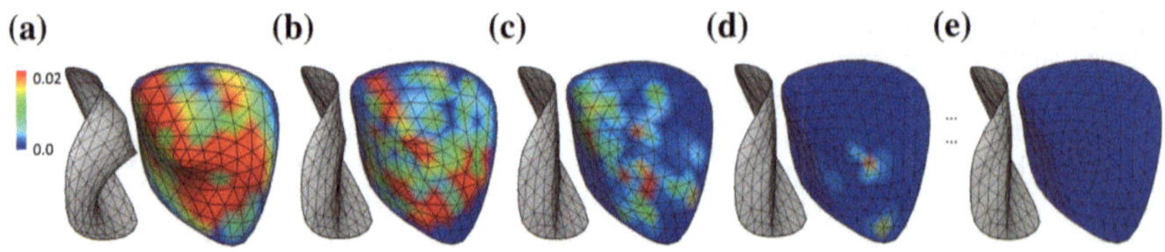

Fig. 5.11 The progressive results of computing an FL mesh: **a** the given mesh ($\varpi_{\max} = 1.92$), **b** after 2 steps ($\varpi_{\max} = 1.34 \times 10^{-1}$), **c** after 4 steps ($\varpi_{\max} = 5.46 \times 10^{-2}$), **d** after 6 steps ($\varpi_{\max} = 1.77 \times 10^{-2}$), and **e** the final FL mesh after 20 steps ($\varpi_{\max} = 3.20 \times 10^{-5}$) (Image taken from [55] with permission)

$$\begin{aligned}
\frac{\mathrm{d}\xi}{\mathrm{d}\mathbf{p}} &= -\frac{1}{\sqrt{1-\frac{\|\mathbf{qh}\|^2}{\|\mathbf{pq}\|^2}}}\frac{\mathrm{d}}{\mathrm{d}\mathbf{p}}\left(-\frac{\|\mathbf{qh}\|}{\|\mathbf{pq}\|}\right) \\
&= \frac{\|\mathbf{pq}\|}{\|\mathbf{ph}\|\,\|\mathbf{pq}\|^2}\left(\|\mathbf{pq}\|\frac{\mathrm{d}\|\mathbf{qh}\|}{\mathrm{d}\mathbf{p}} - \|\mathbf{qh}\|\frac{\mathrm{d}\|\mathbf{pq}\|}{\mathrm{d}\mathbf{p}}\right) \\
&= \frac{\|\mathbf{pq}\|}{\|\mathbf{ph}\|\,\|\mathbf{pq}\|^2}\left(\frac{\|\mathbf{pq}\|}{\|\mathbf{qa}\|}(\mathbf{q}-\mathbf{a}) - \frac{\|\mathbf{qh}\|}{\|\mathbf{pq}\|}(\mathbf{p}-\mathbf{q})\right) \\
&= \frac{\cot\xi}{\|\mathbf{pq}\|^2}(\mathbf{p}-\mathbf{q}) - \frac{1}{\|\mathbf{qa}\|\,\|\mathbf{pq}\|\sin\xi}(\mathbf{a}-\mathbf{q}).
\end{aligned} \tag{5.25}$$

The part corresponding to $\gamma_{q,p}$ can be derived in a similar way. □

Based on Eqs. (5.24) and (5.25), Proposition 5.2 can also be proved as follows.

$$\frac{d\varepsilon_1}{d\mathbf{p}} = \frac{d}{d\mathbf{p}}\left(\frac{\pi}{2}-\xi\right) = -\frac{d\xi}{d\mathbf{p}} = \frac{1}{\|\mathbf{qa}\|\,\|\mathbf{pq}\|\sin\xi}(\mathbf{a}-\mathbf{q}) + \frac{\cot\xi}{\|\mathbf{pq}\|^2}(\mathbf{q}-\mathbf{p})$$

and

$$\frac{d\varepsilon_2}{d\mathbf{p}} = \frac{d}{d\mathbf{p}}\left(\frac{\pi}{2}-\gamma\right) = -\frac{d\gamma}{d\mathbf{p}} = \frac{1}{\|\mathbf{qa}\|\,\|\mathbf{pa}\|\sin\gamma}(\mathbf{q}-\mathbf{a}) + \frac{\cot\gamma}{\|\mathbf{pq}\|^2}(\mathbf{a}-\mathbf{p}).$$

As $\|\mathbf{pq}\|\sin\xi = \|\mathbf{pa}\|\sin\gamma = \|\mathbf{ph}\|$, the first terms are eliminated when summing all apex angles around $\mathbf{p}$.

By this formulation, the FL mesh patch M_{FL} can be modeled from a given polygonal mesh M using the sequential linearly constrained programming. Figure 5.11 gives the progressive results of how an FL mesh is obtained from the given mesh surface patch in Fig. 5.11a.

Defining the maximal vertex flattenability $\varpi_{\max} = \max\{\varpi(\mathbf{v}_p),\ \forall \mathbf{v}_p \in V_{\mathrm{act}}\}$, after six steps of iteration, the mesh patch is similar to the final FL mesh, and its $\varpi_{\max}$ is also close to $\varpi_{\max}$ on the final FL mesh. In conclusion, the computation of FL meshes converges very fast. However, the computation may fail on some meshes with topological obstructions. To overcome the topological obstructions, if a high value

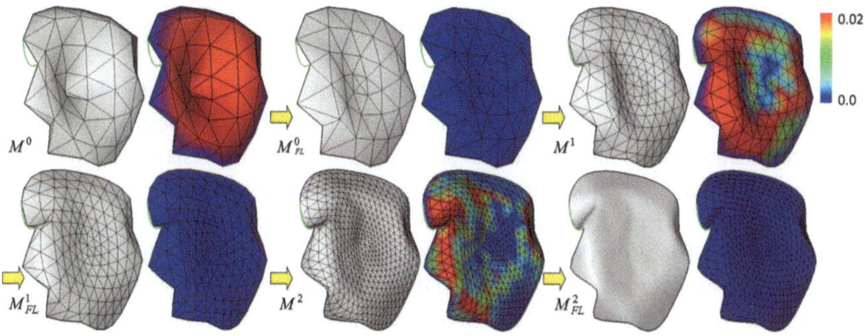

Fig. 5.12 Variational subdivision with FL meshes (Image taken from [55] with permission)

of $\varpi(\mathbf{v}_p)$ keeps showing on the vertex $\mathbf{v}_p$ during the computation of FL meshes, the triangles around $\mathbf{v}_p$ should be refined by a strategy similar to the $\sqrt{3}$-subdivision [20] to add more degrees of freedom to the mesh under processing.

5.3.2 Variational Subdivision of FL Meshes

The FL mesh introduced above can be integrated with subdivision schemes to generate a high quality FL mesh patch from a coarse control mesh. Unlike [27], the subdivision schemes with fixed masks (such as the Loop, the Doo-Sabin, and the Catmull–Clark scheme) are not adopted here since the vertices will be adjusted later to make the surface flattenable—the fixed linear combination rules are, in fact, not followed. Instead, a modified variational subdivision algorithm akin to [21] is adopted, which consists of three steps:

Step 1. A topological splitting operator is conducted to introduce new vertices to increase the number of degrees of freedom (in other words, M^{i+1} is obtained from M^i).
Step 2. Discrete fairing operators are applied to move the newly created vertices to increase the overall smoothness and interpolate the user-specified boundary curves.
Step 3. The FL mesh processing method is applied to M^{i+1} to compute an FL mesh surface M^{i+1}_{FL}.

Iterating these three steps, a hierarchical sequence of FL meshes is generated (for example, the sequence shown in Fig. 5.12).

In the first step, instead of uniformly applying the one-to-four triangle subdivision, triangles are split based on the lengths of their edges. More specifically, the average edge length $\overline{L}$ on M^i is first computed, and only the edges whose length is greater than $0.5\overline{L}$ are split. Therefore, each triangle on M^i is adaptively converted into one (with no edge split), two (with one edge split), three (with two edges split), or four triangles (with all edges split) on M^{i+1}.

The second step of this modified variational subdivision scheme moves the newly created vertices in the first step to either increase the overall smoothness or interpolate the user-specified boundary curve. As shown in Fig. 5.12 this scheme allows users to specify some interpolation curves on the boundary edges (the green curves on the control mesh M^0 in Fig. 5.12). A refined mesh should interpolate these curves during the subdivision. Therefore, when a boundary edge with an interpolation curve is split, the newly created vertex should be moved to the middle of the curve; meanwhile, the curve is split into two curves which are attached to the two newly created edges. Some of the boundary edges have no interpolation curve specified; on these edges, the splitting vertex is expected to be moved to a place that makes the boundary curve smooth. The mask for boundary vertices in the modified Butterfly subdivision scheme [64] is used to determine their new positions by

$$\mathbf{v}_{\text{new}} = \frac{9}{16}(\mathbf{v}_i + \mathbf{v}_{i+1}) - \frac{1}{16}(\mathbf{v}_{i-1} + \mathbf{v}_{i+2}), \tag{5.26}$$

where $\mathbf{v}_i\mathbf{v}_{i+1}$ is the boundary edge to be split by the newly inserted vertex $\mathbf{v}_{\text{new}}$, and $\mathbf{v}_{i-1}\mathbf{v}_i$ and $\mathbf{v}_{i+1}\mathbf{v}_{i+2}$ are the neighboring edges of $\mathbf{v}_i\mathbf{v}_{i+1}$ on the boundary of the surface patch. The positions of the newly created inner vertices are determined by iteratively applying the first-order umbrella operator and the second-order umbrella operator in succession. In detail, after applying the first-order umbrella operator to all vertices for 10 runs, the vertices are moved by the second-order umbrella operator for 100 runs. The two operators from Kobbelt and Schröder [21] are listed below.

$$\mathbf{u}(\mathbf{v}_i) = \frac{1}{|N(\mathbf{v}_i)|} \sum_{j \in N(\mathbf{v}_i)} \mathbf{v}_j - \mathbf{v}_i, \tag{5.27}$$

$$\mathbf{u}^2(\mathbf{v}_i) = \frac{1}{|N(\mathbf{v}_i)|} \sum_{j \in N(\mathbf{v}_i)} \mathbf{u}(\mathbf{v}_j) - \mathbf{u}(\mathbf{v}_i), \tag{5.28}$$

thus the first-order umbrella operator is defined as

$$\mathbf{v}_i \leftarrow \mathbf{v}_i + \mathbf{u}(\mathbf{v}_i), \tag{5.29}$$

and the second-order umbrella operator is

$$\mathbf{v}_i \leftarrow \mathbf{v}_i - \frac{1}{\nu}\mathbf{u}^2(\mathbf{v}_i) \tag{5.30}$$

with $\nu = 1 + |N(\mathbf{v}_i)|^{-1} \sum_{j \in N(\mathbf{v}_i)} |N(\mathbf{v}_j)|^{-1}$.

This subdivision algorithm introduces a useful method for modeling flattenable mesh surfaces. Surfaces generated by this method are not the same as the results

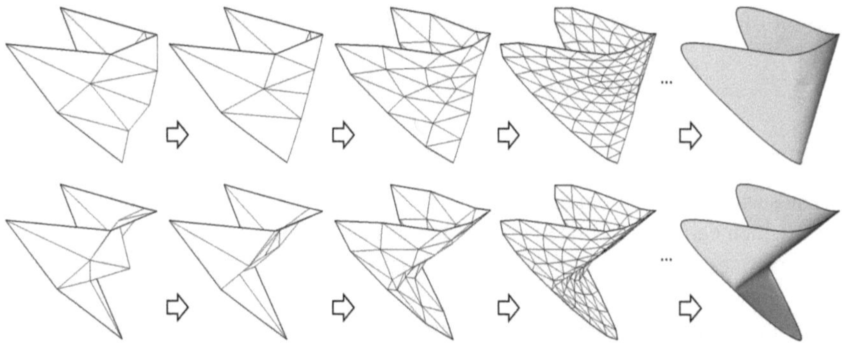

Fig. 5.13 With the same boundary and the same connectivity on a given control mesh, different FL meshes are generated if different initial shapes are given (Image taken from [55] with permission)

from other developable meshes modeling approaches, which are derived from the theorems of continuous differential geometry. Both smooth surfaces and surfaces with crumpled effect (see Fig. 5.12) can be modeled by this approach. Two other examples of subdividing FL meshes, which have the effect of paper craft, are shown in Fig. 5.13.

5.3.3 Local Flattenable Perturbation

Mesh surfaces generated in some applications may be almost flattenable (i.e., with only a few non-flattenable vertices). For those surfaces, a local perturbation approach is developed to increase their flattenability with slight position adjustment. The approach is based on local optimization, so it can be finished in a relatively short time compared with the computation of FL meshes.

In the local perturbation, only the vertices falling in the disk region of a central vertex $\mathbf{v}_c$ are moved in each step. Suppose $N_r(\mathbf{v}_c)$ denotes the r-rings of neighbors of $\mathbf{v}_c$ and $\partial N_r(\mathbf{v}_c)$ represents the vertices on the rth-ring of $\mathbf{v}_c$, the relationship between the rings is defined as

$$N_{r+1}(\mathbf{v}_c) = N_r(\mathbf{v}_c) + \partial N_{r+1}(\mathbf{v}_c).$$

The set of vertices that are moved in the local perturbation around $\mathbf{v}_c$ is defined as

$$V_{\text{act}} = N_r(\mathbf{v}_c) \bigcap (V \backslash \partial V). \tag{5.31}$$

The tests conducted by Wang [55] choose $r = 3$. Choosing a larger support size certainly makes the computation more robust to overcome a stuck point—local optimum, but it has a longer computing time. When moving vertices in V_{act}, not only vertices in V_{act} but also the vertices adjacent to ∂V_{act} have their flattenability changed.

The set of vertices whose flattenability is affected by the movement of vertices in V_{act} is

$$V_{\text{dev}} = N_{r+1}(\mathbf{v}_c) \bigcap (V \backslash \partial V). \tag{5.32}$$

When moving vertices in V_{act}, the following properties are usually demanded.

- The surface shape after perturbation still approximates the original surface around $\mathbf{v}_c$, which can be formulated as the same functional J_{pos} in Eq. (5.13);
- All vertices in V_{dev} are as flattenable as possible (i.e., $\forall \mathbf{v}_p \in V_{\text{dev}}$ letting $\varpi(\mathbf{v}_p) \simeq 0$);
- The interference between the given surface and a reference object H is prevented.

The last requirement is important for human-oriented product design (such as garments and shoes) because the products represented by a collection of flattenable mesh patches will be worn by human beings. All these factors are formulated into a constrained optimization problem

$$\begin{array}{ll} \arg\min_{p \in V_{\text{act}}} J_{\text{pos}} & \\ \quad \text{subject to} \;\; \theta(\mathbf{v}_q) \equiv 2\pi & (q \in V_{\text{dev}}) \\ \qquad\qquad\quad (\mathbf{v}_r - \mathbf{h}_r) \cdot \mathbf{n}_{\mathbf{h}_r} \geq \epsilon & (r \in V_{\text{collid}}) \end{array} \tag{5.33}$$

where $\mathbf{h}_r$ is the current closest point of $\mathbf{v}_r$ on the reference body H, $\mathbf{n}_{\mathbf{h}_r}$ is the unit normal vector at $\mathbf{h}_r \in H$ facing outwards, and $V_{\text{collid}} \subset V_{\text{act}}$ is the set of vertices whose distance to the human body is smaller than the user-specified collision tolerance ϵ.

Again, the optimization problem is solved by using the sequential linearly constrained programming, where the Lagrangian function below is adopted.

$$\begin{aligned} J_{\text{local}}(X) = & \sum_{p \in V_{\text{dev}}} \frac{1}{2} \|\mathbf{v}_p - \mathbf{v}_p^0\|^2 \\ & + \sum_{q \in V_{\text{dev}}} \lambda_q (\theta(\mathbf{v}_q) - 2\pi) \\ & + \sum_{r \in V_{\text{collid}}} \lambda_r ((\mathbf{v}_r - \mathbf{h}_r) \cdot \mathbf{n}_{\mathbf{h}_r} - \epsilon) \end{aligned} \tag{5.34}$$

The linear equation system solved at each iteration is

$$\begin{pmatrix} I & & & \Lambda_x^T \\ & I & & \Lambda_y^T \\ & & I & \Lambda_z^T \\ \Lambda_x & \Lambda_y & \Lambda_z & 0 \end{pmatrix} \begin{pmatrix} \delta_{p_x} \\ \delta_{p_y} \\ \delta_{p_z} \\ \lambda \end{pmatrix} = \begin{pmatrix} B_{p_x} \\ B_{p_y} \\ B_{p_z} \\ B_\lambda \end{pmatrix} \tag{5.35}$$

where

$$B_p = \{\mathbf{v}_p^0 - \mathbf{v}_p\}, \, B_\lambda = \begin{pmatrix} 2\pi - \theta(\mathbf{v}_q) \\ -((\mathbf{v}_r - \mathbf{h}_r) \cdot \mathbf{n}_{\mathbf{h}_r} - \epsilon) \end{pmatrix},$$

$$\Lambda = \begin{pmatrix} \alpha_{q,p} \\ \beta_{r,p} \end{pmatrix} = \begin{pmatrix} \partial\theta(\mathbf{v}_q)/\partial\mathbf{v}_p \\ \delta_{r,p}\mathbf{n}_{\mathbf{h}_r} \end{pmatrix}$$

with the Kronecker delta $\delta_{i,j} = \begin{cases} 1 \; (i = j) \\ 0 \; (i \neq j) \end{cases}$.

Not surprisingly, numerical computation for the problem defined in Eq. (5.34) is not as stable as the one in Eq. (5.13) because the fairness term defined in Eq. (5.13) takes the role of regularization to the mesh system which prevents singularity. To avoid singular results, Wang [55] employs Singular value decomposition (SVD) [35] to solve the linear equation system in Eq. (5.34). First, λ in

$$(\Lambda_x\Lambda_x^T + \Lambda_y\Lambda_y^T + \Lambda_z\Lambda_z^T)\lambda = \Lambda_x B_{p_x} + \Lambda_y B_{p_y} + \Lambda_z B_{p_z} - B_\lambda \qquad (5.36)$$

is solved by SVD. δ_{p_x}, δ_{p_y} and δ_{p_z} are subsequently computed by

$$\delta_{p_x} = B_{p_x} - \Lambda_x^T\lambda, \qquad \delta_{p_y} = B_{p_y} - \Lambda_y^T\lambda, \qquad \delta_{p_z} = B_{p_z} - \Lambda_z^T\lambda. \qquad (5.37)$$

Based on these formulae (Eqs. 5.35– 5.37), Algorithm 4 can be used to improve the flattenability of a given mesh patch by local perturbation (the pseudo-code has been listed). The basic idea is that by storing all vertices into a maximum heap keyed by the flattenability on vertices, the regions around k-most non-flattenable vertices are iteratively perturbed while preventing the interference with the reference body H. Notice that the value of k is controlled by the number of vertices whose flattenability is greater than $0.5\varpi_{\max}$.

Algorithm 4 LocalFlattenablePerturbation

For a given mesh patch M, compute the flattenability on all vertices $\mathbf{v}_p \in V \backslash \partial V$, and record the maximal flattenability value $\varpi_{\max}$;
Save the current positions of all vertices as the current optimal record;
Search the closest tracking point of all vertices on the reference body H;
Insert all vertices into a maximum heap Υ;
repeat
  while Υ is not empty **AND** the flattenability of its top node is greater than $0.5\varpi_{\max}$ **do**
    Remove the top node $\mathbf{v}_t$ from Υ;
    Determine V_{act}, V_{dev}, and V_{collid} by a user-specified ring number r;
    repeat
      Compute $(\delta_{p_x}, \delta_{p_y}, \delta_{p_z})$ for all $\mathbf{v}_p \in V_{\text{act}}$ using Eqs. (5.35– 5.37);
      $\mathbf{v}_p \leftarrow \mathbf{v}_p + \tau(\delta_{p_x}, \delta_{p_y}, \delta_{p_z})$;
    until the process has been iterated for more than 5 times;
    Update the closest tracking point of all vertices in V_{act};
  end while
  if the current maximal flattenability is smaller than $\varpi_{\max}$ **then**
    Update $\varpi_{\max}$ and save the current positions of all vertices as an optimal result;
  end if
  Add all the removed nodes back into Υ, and update the position of all moved vertices in Υ by their new flattenability;
until the terminal condition is satisfied.

A soft update strategy for Newton's method is employed here; for example, $\tau = 0.1$ is usually chosen to stabilize the computation. In Algorithm 4, there are two time-consuming space searching steps—Steps 3 and 13. In these steps, the publicly available PQP query library [22] can be employed to determine the closest point of a vertex $\mathbf{v}_t \in V_{act}$ on the human model H.

There is no theoretical guarantee of the convergency of the computation conducted in Algorithm 4. However, Steps 15–17 ensure that this algorithm always outputs results that are not worse than the inputs by keeping the optimal results during the computation. The terminal condition of the outer iteration loop (Step 19) consists of three parts. The iteration is stopped if (1) $\varpi_{\max} < \varepsilon$ ($\varepsilon = 0.001$ is chosen in [55]), (2) the iteration has been repeated 10 times without updating $\varpi_{\max}$, or (3) the iteration has been repeated more than 50 times.

As the computation of this local computation scheme can be completed in a short time, it can be further developed into an interactive tool to let users select a small region to process and improve the flattenability in the region immediately after the selection. The surface can, therefore, process in an interactive manner.

5.3.3.1 Limitations

This approach for computing flattenable mesh surfaces has the following limitations.

- The current method only preserves G^0 continuity on the boundary of an FL mesh surface, which is acceptable for sheet manufacturing of soft materials (say, the apparel industry). For those industries dealing with stiffer materials (say, the ship industry), G^1 continuity may be expected. Although a method to approximate G^1 continuity is proposed by Wang [53], it is still an open problem to achieve an exact G^1 continuity across the boundary of FL meshes.
- The FL mesh is not a connectivity invariant representation (i.e., for an input with the same geometry shape but with different mesh connectivity, the resultant FL could be different). This limitation can be somewhat overcome if a remeshing step like [4] is conducted to remesh the input surface into a semi-regular mesh with almost uniform triangle size.
- The collision response method, which is currently implemented in the local flattenable perturbation, is relatively primitive. Ideally, the collision handling algorithm should provide a response that changes the direction of colliding vertices in order to simulate the slippage of vertices on the reference bodies. A geometric distribution correction method such as [50] can help on this issue.

5.3.4 Least-Norm Solution for FL Mesh Processing

FL mesh processing is formulated as a constrained optimization problem in Eq. (5.13). Another way to compute an FL mesh from an input piecewise linear surface M is introduced below, in which the minimization problem

$$\min_{p \in V_{\text{act}}} J_{\text{pos}} \quad \text{s.t.} \quad \theta(\mathbf{v}_p) \equiv 2\pi \tag{5.38}$$

is reformulated as a minimum-norm problem in terms of estimation error. Similar to Eq. (5.13), V_{act} is the collection of interior vertices, and J_{pos} is defined as Eq. (5.15) with $\mathbf{v}_p^0$ being the closest position of $\mathbf{v}_p$ on the given surface M (or simply being the original position of $\mathbf{v}_p$).

5.3.4.1 Reformulation

After linearizing the constraints in Eq. (5.38), the flattenable mesh modeling problem can be converted into a minimum-norm problem and solved by a least-norm solution. As discussed in [62], carefully selecting alternative variables can make the linearization more accurate so that the computation converges faster than the Newton steps. In this formulation, the update vectors of interior vertices' positions are chosen as variables.

First of all, the updated position $\mathbf{v}_p$ of a vertex is assumed to be an ideal position satisfying the flattenable constraint (i.e., the 2π constraint on its summed inner angles). The relationship between $\mathbf{v}_p$ and $\mathbf{v}_p^0$ can be represented by

$$\mathbf{v}_p = \mathbf{v}_p^0 + \mathbf{d}_p \tag{5.39}$$

The current position $\mathbf{v}_p^0$ can be considered as an initial guess, and $\mathbf{d}_p$ represents the estimation error. Therefore, the problem in Eq. (5.38) can be rephrased as to determine a set of minimal update vectors $\mathbf{d}_p$ on interior vertices so that the flattenable constraint is satisfied on every interior vertex. Let n_{act} be the number of vertices in V_{act}, there are totally $3n_{\text{act}}$ variables to be computed as each $\mathbf{d}_p$ has three components. However, only n_{act} flattenable constraints are given. Therefore, this is an under-determined system of equations. Among the infinitely many solutions, a reasonable choice is to obtain the solution with minimal-norm of estimation error.

The flattenable constraint defined on every interior vertex $\mathbf{v}_p$ is

$$\theta(\mathbf{v}_p) = 2\pi \tag{5.40}$$

where the value of $\theta(\mathbf{v}_p)$ is in terms of positions of $\mathbf{v}_p$ and $\mathbf{v}_q$ ($\forall \mathbf{v}_q \in N(\mathbf{v}_p)$). Here, $N(\mathbf{v}_p)$ denotes the one-ring neighboring vertices of $\mathbf{v}_p$. To linearize the nonlinear expression in Eq. (5.40), the Taylor expansion for multi-variables is applied to $\theta(\mathbf{v}_p)$ as

$$\theta(\mathbf{v}_p) = \theta(\mathbf{v}_p^0) + \nabla\theta(\mathbf{v}_p^0)^T(\mathbf{v}_p - \mathbf{v}_p^0, \mathbf{v}_q - \mathbf{v}_q^0)^T + \ldots \tag{5.41}$$

by neglecting the high-order terms. More specifically, the nonlinear flattenable constraint in Eq. (5.40) can be rewritten as

$$\theta_p(\mathbf{v}_p^0) \cdot \mathbf{d}_p + \sum_{q \in N(p)} \theta_q(\mathbf{v}_p^0) \cdot \mathbf{d}_q \approx 2\pi - \theta(\mathbf{v}_p^0) \tag{5.42}$$

where $\theta_p(\mathbf{v}_p^0)$ and $\theta_q(\mathbf{v}_p^0)$ are the derivatives of $\theta(\mathbf{v}_p)$ in terms of $\mathbf{v}_p$ and $\mathbf{v}_q$ at current positions. Formulae for the derivatives, θ_p and θ_q, have been given in Eqs. (5.22) and (5.23), which can efficiently be calculated. The error introduced by this approximation is quadratic to the estimation errors $\mathbf{d}_p$ and $\mathbf{d}_q$, thus the computation converges fast.

5.3.4.2 Least-Norm Computation

By linearizing the flattenable constraint into Eq. (5.42), the set of flattenable constraints is converted into n_{act} linear equations with only $3n_{\text{act}}$ unknown variables—an under-determined linear equation system. Letting $\mathbf{d}$ denote a vector with $3n_{\text{act}}$ components formed by the n_{act} update vectors $\mathbf{d}_p$, together with Eq. (5.42), the flattenable mesh processing problem becomes

$$\min \ \|\mathbf{d}\|^2 \quad \text{s.t.} \ \mathbf{A}\mathbf{d} = \mathbf{b} \tag{5.43}$$

where $\mathbf{A}\mathbf{d} = \mathbf{b}$ is from Eq. (5.42). Clearly, it now becomes a least-norm problem. The matrix $\mathbf{A}$ has full rank as the flattenable constraints are independent. From the literature of mathematics, it is known that for a full rank coefficient matrix $\mathbf{A}$, the above least-norm problem has a unique solution (cf. [49])

$$\mathbf{d} = \mathbf{A}^T(\mathbf{A}\mathbf{A}^T)^{-1}\mathbf{b}. \tag{5.44}$$

The value of $\mathbf{d}$ can be solved by finding a solution to the normal equation

$$(\mathbf{A}\mathbf{A}^T)\mathbf{x} = \mathbf{b} \tag{5.45}$$

followed by a substitution that $\mathbf{d} = \mathbf{A}^T\mathbf{x}$. The matrix $\mathbf{A}$ is sparse, so the solution in Eq. (5.45) can be efficiently solved. For example, the *SuperLU* solver [25] packaged by the *OpenNL* interface [23] can be employed.

The squared norm $\|\mathbf{d}\|^2$ of the vector $\mathbf{d}$ is exactly the shape approximation term J_{pos} in Eq. (5.38). Therefore, when the current positions $\mathbf{v}_p^0$ of vertices are close to the optimal positions $\mathbf{v}_p$ (i.e., the approximation error introduced in Eq. (5.42) is small), the optimal solution can directly be obtained. Although it is not easy to find a good initial guess, the estimation error becomes smaller and smaller if the positions of vertices are repeatedly updated after computing the least-norm solution of $\mathbf{d}_p$. The newly updated positions are adopted as the current position in the next round of evaluation. The outline of the flattenable mesh processing method is listed in Algorithm 5. The terminal condition consists of two parts. The iteration is stopped

Algorithm 5 Least-NormFLMeshProcessing

1: **repeat**
2: Setup the linearized constraints of Eq. (5.42) in $\mathbf{A}^T$;
3: Compute $\mathbf{x}$ by Eq. (5.45);
4: Compute the update vectors $\mathbf{d} = \mathbf{A}^T\mathbf{x}$;
5: Update the position of each vertex $\mathbf{v}_p \in V_{\text{act}}$;
6: **until** the terminal condition is satisfied.

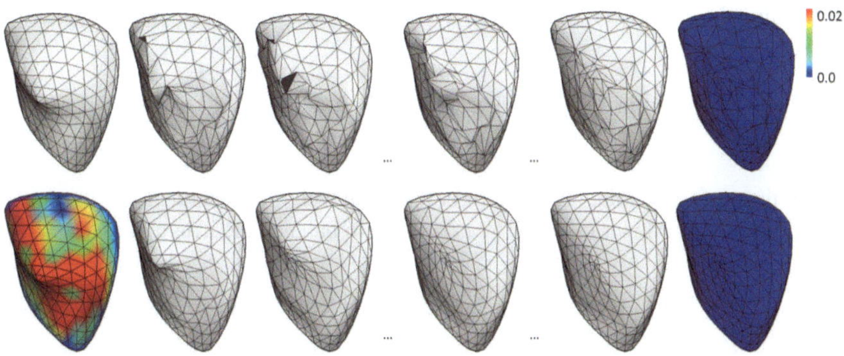

Fig. 5.14 An example of flattenable mesh processing: (*top row*) the progressive results by the direct update of vertex positions—the computation even diverges in some examples, and (*bottom row*) the results by the least-square update. The *colors* represent the difference between each vertex's summed inner angle and 2π (Image taken from [54] with permission ©2008 IEEE)

if (1) $\|\mathbf{d}\|^2 < \varepsilon$ or (2) the iteration has been repeated more than 50 times. $\varepsilon = \max\{n_{\text{act}} \times 10^{-8}, 10^{-5}\}$ is chosen in Wang's implementation [54].

5.3.4.3 Least-Square Update

The above outline of flattenable mesh surface processing works well when the given surface M is nearly flattenable. However, if M is far from flattenable, the movement of vertices during the above flattenable mesh processing routine may break the regularity of vertex distribution on the surface. One example is shown in Fig. 5.14a. There are two causes of this instability: (1) the formulation of flattenable mesh processing does not consider the distribution of vertices, and (2) the linearized flattenable constraints in Eq. (5.42) are only in the first-order of accuracy. As no consideration is given to the distribution of vertices, they are freely moved during the mesh processing. This gives more degrees of freedom to process meshes; however, when M is far from flattenable, it easily leads to instability and the first-order approximation of flattenable constraints may drive the vertices some inaccurate places.

In order to improve the robustness of FL mesh processing based on least-norm solution, the shape of mesh surface can be updated through a least-square solution, which is stimulated by the Laplacian mesh editing technique [42]. Let $\mathbf{v}_p^* = \mathbf{v}_p^0 + \mathbf{d}_p$, where $\mathbf{d}_p$ is the update vector determined in Eq. (5.44), the updated positions of vertices are computed by

$$\begin{pmatrix} \mathbf{L} \\ \mathbf{I} \end{pmatrix} \mathbf{v} = \begin{pmatrix} \mathbf{g} \\ \mathbf{v}^* \end{pmatrix} \tag{5.46}$$

where $\mathbf{L}$ denotes the discrete graph Laplacian operator, $\mathbf{v}$ is a vector consisting of all vertices in V_{act}, $\mathbf{v}^*$ is a collection of their corresponding $\mathbf{v}_p^*$, and $\mathbf{g}$ is a vector containing those static boundary vertices. Solving the above least-square problem is in fact

$$\min \left\{ \sum_{p \in V_{\text{act}}} \|\mathbf{L}(\mathbf{v}_p)\|^2 + \sum_{p \in V_{\text{act}}} \|\mathbf{v}_p - \mathbf{v}_p^*\|^2 \right\}. \tag{5.47}$$

Either the uniform Laplacian operator or the cotan weighted discrete Laplacian operator proposed by Meyer et al. [31] can be used to balance the irregular meshes as

$$\mathbf{L}(\mathbf{v}_p) = \frac{1}{2A(\mathbf{v}_p)} \sum_{q \in N(p)} (\cot \gamma_{pq} + \cot \xi_{pq})(\mathbf{v}_p - \mathbf{v}_q) \tag{5.48}$$

with γ_{pq} and ξ_{pq} being the opposite angles in two triangles adjacent to edge $\mathbf{v}_p\mathbf{v}_q$ (see the illustration shown in Fig. 5.9). The static boundary vertices in $\mathbf{L}(\mathbf{v}_p)$ are moved to the vector $\mathbf{g}$ on the right. Notice that $\bar{A}(\mathbf{v}_p)$ here is a normalized mixed area at $\mathbf{v}_p$ as

$$\bar{A}(\mathbf{v}_p) = A_m(\mathbf{v}_p) / \left(\frac{1}{n_{\text{act}}} \sum_{p \in V_{\text{act}}} A_m(\mathbf{v}_p) \right). \tag{5.49}$$

The formula for $A_m(\mathbf{v}_p)$ has been given by Meyer et al. [31]. Without this normalization, the value of coefficients in $\mathbf{L}(\mathbf{v}_p)$ is much greater if the dimension of the mesh surface is small, and vice versa. If the coefficients are large, which means the weight on the Laplacian term in Eq. (5.48) is great, $\mathbf{v}_p$ hardly converges to the position $\mathbf{v}_p^*$. However, if the coefficients are small, the Laplacian term in Eq. (5.48) becomes useless.

This least-square position update is adopted in step 5 of Algorithm 5 with a fixed Laplacian matrix $\mathbf{L}$ computed at the beginning of the iteration. However, the positions of vertices computed by Eq. (5.46) will never be equal to $\mathbf{v}_p^*$. Therefore, when the mesh surface under processing becomes nearly flattenable, the positions are updated by moving $\mathbf{v}_p$ to $\mathbf{v}_p + \mathbf{d}_p$ directly. In Wang's implementation [54], a hybrid condition is employed to change least-square update to direct update—either 1) $\theta_{\text{err}}^i < 0.001$ or 2) $(\theta_{\text{err}}^{i-1} - \theta_{\text{err}}^i) < 0.01$ after 20 iteration steps.

$$\theta_{\text{err}}^i = \max\{|\theta(\mathbf{v}_p) - 2\pi|\} \tag{5.50}$$

defines the flattenable error after the ith iteration. An example of using this scheme to update positions is shown in Fig. 5.14b.

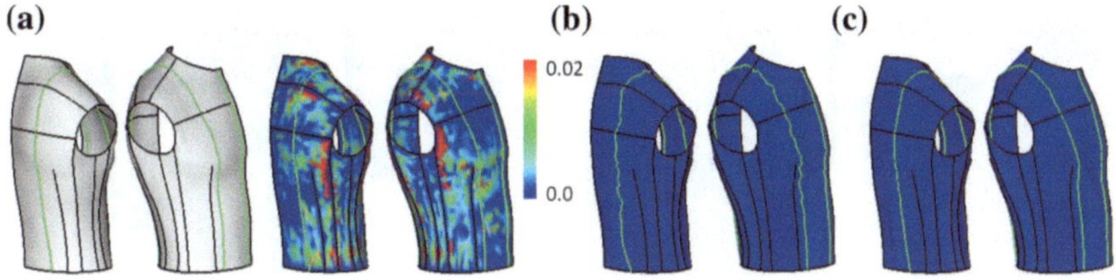

Fig. 5.15 An example of using weights on vertices to preserve the shape of semantic feature curves: **a** the input surface for wetsuit with feature curves in *green*, **b** the flattenable mesh processing result without using weight on vertices ($\theta_{\text{err}} = 1.10 \times 10^{-5}$), **c** the result of adding higher weight to the vertices of the feature curves ($\theta_{\text{err}} = 1.15 \times 10^{-5}$). Notice that the uniform Laplacian is used in this test, and the distortion of features in **b** will not be such significant if the cotan weighted Laplacian is employed (Image taken from [54] with permission ©2008 IEEE)

5.3.4.4 Weights on Vertices

The formulae of least-norm solution assume that all vertices are equally important during the flattenable mesh processing. However, different weights need to be assigned to vertices in practice. For example, the vertices adjacent to the boundary of a given surface are expected to have smaller movement than other interior vertices so that the tangential continuity across the boundary can be approximated. In garment design applications, semantic features are defined on some vertices. These vertices are often required to have much smaller movement than others to preserve the shape of feature curves (see Fig. 5.15). All these can be implemented by introducing different weights to the vertices.

Without loss of generality, let w_p denote the weights we wish to add to the vertex $\mathbf{v}_p$, where a greater w_p will lead to a smaller movement $\mathbf{d}_p$. By a simple change of variables

$$\mathbf{r}_p = w_p \mathbf{d}_p \tag{5.51}$$

the least-norm solution of $\mathbf{r}_p$ can then be computed as

$$\min \ \|\mathbf{r}\|^2 \quad \text{s.t.} \ \mathbf{AWr} = \mathbf{b} \tag{5.52}$$

where $\mathbf{r}$ is the vector containing all $\mathbf{r}_p$ and $\mathbf{W}$ is a diagonal matrix with $1/w_p$ as the diagonal element. After solving $\mathbf{r}$ by

$$\mathbf{r} = (\mathbf{AW})^T \left((\mathbf{AW})(\mathbf{AW})^T \right)^{-1} \mathbf{b} \tag{5.53}$$

the value of $\mathbf{d}$ can be determined by $\mathbf{d} = \mathbf{Wr}$. Similarly, the least-square update is changed to

$$\begin{pmatrix} \mathbf{L} \\ \mathbf{W}^{-1} \end{pmatrix} \mathbf{v} = \begin{pmatrix} \mathbf{g} \\ \mathbf{W}^{-1} \mathbf{v}^* \end{pmatrix} \tag{5.54}$$

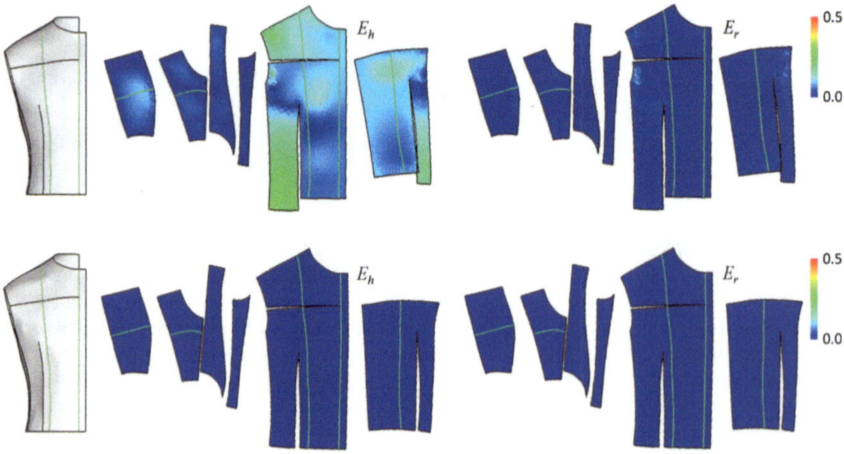

Fig. 5.16 The comparison of planar pieces flattened from (*top row*) the original meshes and (*bottom row*) the processed flattenable meshes. The least-square conformable mapping [24] is adopted to generate the planar pieces, and the *color* maps illustrate the values of the homogeneity of distortion E_h and the aspect ratio E_r (see [3] for reference) (Image taken from [54] with permission ©2008 IEEE)

to make the vertices with a larger weight to be less free.

The weight w_p should be carefully selected. A too small value will have no effect on the computational results whereas a too high weight can cause numerical problems. Wang's tests [54] choose $w_p = 10$ for the vertices to be constrained and $w_p = 1$ for other vertices. Figure 5.15 shows an example from the garment industry for keeping the semantic feature curves by using different weights on vertices.

After the flattenable mesh processing, all processed surface patches can be flattened into planes with almost no distortion. The comparison of the example shown in Fig. 5.15 before versus after processing is given in Fig. 5.16. Here, the least-square conformal mapping method [24] is used to compute planar meshes. To better illustrate the distortion introduced during flattening, the color maps for the homogeneity of distortion E_h and the aspect ratio E_r are also included. The computation formulae of E_h and E_r [3] are described below.

To measure the distortion between two triangles $\Delta\mathbf{apq}$ and $\Delta\mathbf{a}^p\mathbf{p}^p\mathbf{q}^p$, they are placed in the same planar coordinate frame as shown in Fig. 5.17 with planar coordinates

$$\mathbf{a} = (0,0),\ \mathbf{p} = (\mathbf{p}_x, 0),\ \mathbf{q} = (\mathbf{q}_x, \mathbf{q}_y),$$
$$\mathbf{a}^p = (0,0),\ \mathbf{p}^p = (\mathbf{p}_x^p, \mathbf{p}_y^p),\ \mathbf{q}^p = (\mathbf{q}_x^p, \mathbf{q}_y^p).$$

Then, a distortion tensor $\mathbf{f}$ is defined as

$$\mathbf{f} = \frac{1}{\mathbf{p}_x^p\mathbf{q}_y^p - \mathbf{p}_y^p\mathbf{q}_x^p}\begin{bmatrix} (\mathbf{p}_x\mathbf{q}_y^p - \mathbf{q}_x\mathbf{p}_y^p) & (\mathbf{p}_x^p\mathbf{q}_x - \mathbf{p}_x\mathbf{q}_x^p) \\ (-\mathbf{q}_y\mathbf{p}_y^p) & (\mathbf{q}_y\mathbf{p}_x^p) \end{bmatrix}. \tag{5.55}$$

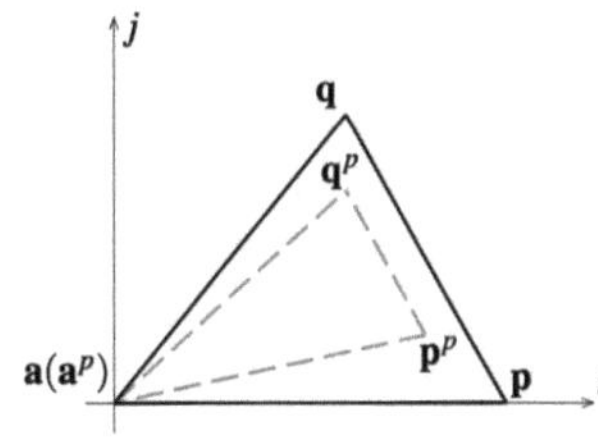

Fig. 5.17 Two corresponding *triangles* are placed in the same planar coordinate system for computing the homogeneity of distortion E_h and the aspect ratio E_r

Computing two coefficients

$$d_p = \sqrt{x_c(\omega)^2 + y_c(\omega)^2} \quad \text{and} \quad d_q = \sqrt{x_c(\omega + \tfrac{\pi}{2})^2 + y_c(\omega + \tfrac{\pi}{2})^2}$$

with

$$x_c(\omega) = f_{11} \cos\omega + f_{11} \sin\omega$$
$$y_c(\omega) = f_{21} \cos\omega + f_{22} \sin\omega$$

at the principal direction

$$\omega = \tfrac{1}{2} \tan^{-1}\left(\frac{2(f_{11}f_{12}+f_{21}f_{22})}{f_{11}^2+f_{21}^2-f_{12}^2-f_{22}^2}\right),$$

the *homogeneity of distortion* (E_h) on these two triangles is defined as

$$E_h(\mathbf{f}) = |d_p d_q - 1|, \quad E_h(\mathbf{f}) \in [0, +\infty), \tag{5.56}$$

and the *aspect ratio* (E_r) can be measured by

$$E_r(\mathbf{f}) = |\frac{d_p}{d_q} - 1|, \quad E_r(\mathbf{f}) \in [0, +\infty). \tag{5.57}$$

Both E_h and E_r should be equal to zero on all triangles for an isometric mapping—in other words, if a flattening result gives $E_h = 0$ and $E_r = 0$, it is stretch-free.

5.3.4.5 Limitations

The least-norm approach for flattenable mesh processing presented in this section has the following limitations.

- Although the least-square update scheme improves the stability of numerical computation, it does not really converge to the solution for a flattenable mesh. This is why the direct update strategy is adopted in the last few steps of computation. The current method presented by Wang [54] is based on experimental tests. A better condition for changing the update scheme needs to be studied.
- Many engineering applications wish to preserve the G^1 continuity across the boundary of assembled patches. In the current setting of the flattenable mesh processing, the constraint can only be added softly by using higher weights on the

vertices adjacent to boundaries. There will be numerical problem if making the G^1 continuity constraint hard by fixing these vertices. Moreover, numerical problem occurs if the weights chosen on these vertices are too great. The G^1 continuity preservation is still a problem to be solved.

- With the Laplacian operator employed in the least-square update scheme, the processed mesh surfaces always shrink in some sense. This is fine for applications where the 2D material pieces used to fabricate the final product can be slightly stretched. However, it becomes a problem in applications where no stretch is allowed. A control of shrinking or inflating effect (see [33]) needs to be considered.

5.4 Segmentation for Production

All techniques introduced above focus on flattening and processing a surface patch with a disk-like topology. A more difficult problem is how to segment a given model into nearly developable atlases so that a nearly stretch-free flattening can be computed. The criterion for segmentation is needed to evaluate the possibility of flattening a given surface patch, which should be fast computed. In this section, a method is introduced to compute the length-preserved free boundary (LPFB) of a mesh patch which speeds up the mesh parameterization. The distortion on parameterization can then be employed as the criterion in a trial-and-error algorithm for segmenting a given model into nearly developable atlases. The computation of LPFB is formulated as a numerical optimization problem in the angle space, where the angle excess on the boundary is minimized while the constraints derived from the closed-path theorem and the lengths of edges are preserved.

From differential geometry [5], it is known that only a fully developable surface, which is a small class among all freeform surfaces, can be flattened to give an isometric mapping. Therefore, to be practical, a method is needed to segment an input mesh model P into nearly developable (also called *quasi-developable*) patches instead of fully developable ones. An application for such patches is toy fabrication (as shown in Fig. 5.18), where a small number of patches are desired since more patches yield more stitching work in the fabrication. Most recent attempts for this are [14] and [38], where the authors partition given models into several-conical patches. However, conical surface is not the only type of developable surfaces. A more general criterion for surface flattening is needed. This is Wang's motivation of computing LPFB [52] to speed up parameterization so that the distortion on the resultant planar mesh can serve as the criterion for quasi-developable mesh segmentation.

The most straightforward criterion for developability is the integral, $\sigma = \int |\kappa_G| dA$, of Gaussian curvature κ_G or any discrete form of its variants [31, 39, 61] over the given surface patch P. However, although discrete Gaussian curvature works well on dart insertion to reduce the stretch of flattening where local developability is counted, it is weak in distinguishing the degree of global developability on nearly developable surfaces. For the examples shown in Fig. 5.19, both have few non-developable vertices which give non-zero Gaussian curvature, and the value of

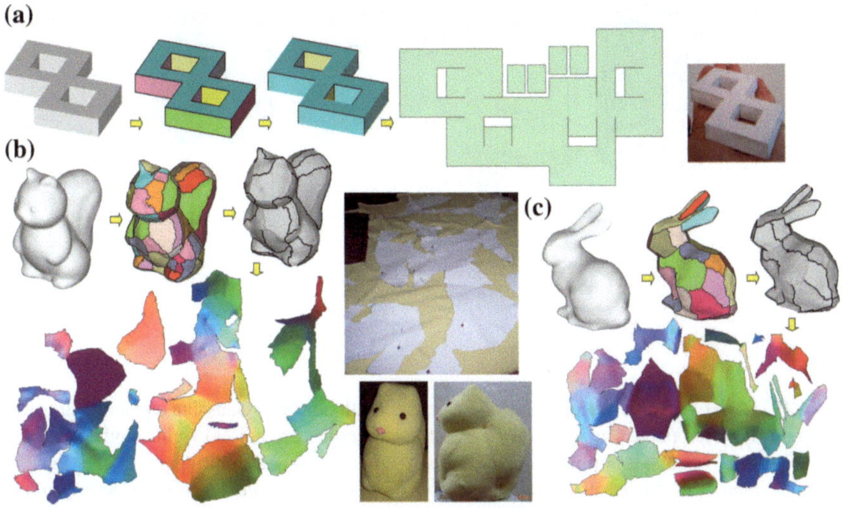

Fig. 5.18 Testing the mesh segmentation algorithm on three models: **a** a CSG-like model, **b** a squirrel model, and **c** a bunny rabbit. The results are given in three steps. The *colorful segmentations* of the given models are the results from error-controlled VSA, and then the results after LPFB-based patch merging are given in *gray*. The flattening layout is shown in colors which represent the normal vectors on the original 3D models. Last, the photographs of the physical models made by these patterns are shown (Image taken from [52] with permission ©2008 IEEE)

Gaussian curvature integral on the cylinder is greater than that on the cube. However, the stretch on the cube is even more significant, which can be seen from the texture checkboard. Furthermore, using σ as a criterion cannot prevent the generation of non-flattenable patches led by topological obstructions (for example, a cylinder cannot be flattenable without inserting a cut to link its two openings). Notice that when working on models with polygonal meshes, the definition of developability is generalized—the less stretch is given on a flattened mesh surface, the higher developability the surface has.

From observation, Wang [52] finds that, for a fully developable surface P, both the area and the edge lengths on its flattening D are coherent with P. Denoting the boundaries of P and D as ∂P and ∂D, $\|\partial P\| = \|\partial D\|$ is preserved on fully developable surfaces. For a general surface P, by computing an optimal LPFB which minimizes the angle distortion between ∂P and ∂D while preserving $\|\partial P\| = \|\partial D\|$, the distortion on the flattening of P with ∂D fixed can serve as a good indicator for the developability—the less distortion, the more developable patch P is. Experimental tests show that fixing LPFB acts as an amplifier for detecting the developability of a surface by the distortion on D. More specifically, for quasi-developable surfaces (like the cylinder in Fig. 5.19), fixing LPFB reduces unexpected distortion from noisy points but increases distortion on non-developable surfaces (as the cube in Fig. 5.19). Moreover, fixing ∂D speeds up the computation D since the computation of x- and y-coordinates can be decoupled to reduce the dimension of the linear system.

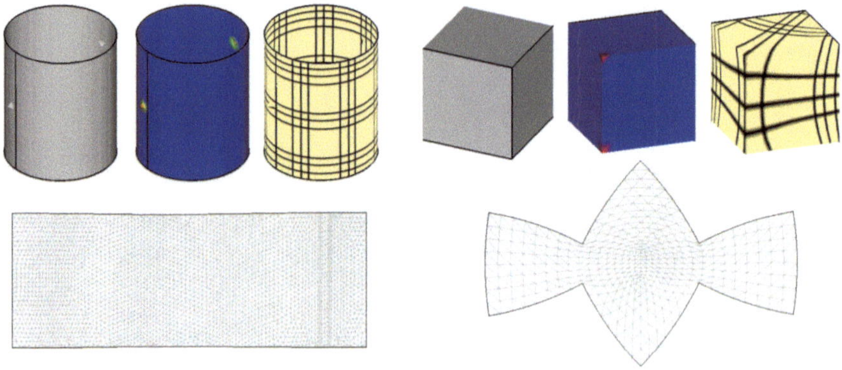

Fig. 5.19 Both of the two examples have few vertices with non-zero Gaussian curvature; however, one is nearly developable (the noisy *cylinder*—the sharp points are noises) while another (the *cube* with the given five cuts) is non-developable, where the *black curves* are cutting paths. The Gaussian curvature map is displayed with the help of *colors*, where *blue color* represents zero Gaussian curvature and *red* denotes the maximal value—the two corner vertices on the *back* of the cube are with high Gaussian curvature since no cut passes them. The *bottom row* gives the LPFB and their corresponding surface flattening. The texture checkboard determined from the flattening is also used to illustrate the stretch—the smaller distortion is shown, the higher developability is with the surface. The integral of discrete Gaussian curvature (which is computed by the approach of Meyer et al. [31]) over area of the cylinder example is 8.930, which is even greater than the integral result of the cube example that is 3.142—but the cylinder can obviously be flattened with a smaller stretch (Image taken from [52] with permission ©2008 IEEE)

The approach presented in this section computes the LPFB and employs it to accelerate the intrinsic parameterization (IP) of interior mesh regions, where the distortion on resultant 2D meshes is used in a trial-and-error approach as the criterion for segmenting a given 3D mesh into quasi-developable atlases.

5.4.1 LPFB Computation

5.4.1.1 Basic Method

The basic methodology for computing a LPFB is presented here. The problem of determining the planar coordinates of boundary vertices on ∂D is formulated as a constrained optimization problem, where the objective functional is derived from the following two energy terms for the differences between ∂D and ∂P.

Boundary length energy. With the length-preservation requirement between ∂D and ∂P, the first energy term is defined by the edge length on boundaries. For a polygonal edge e on ∂P, letting l_e^0 and l_e represent its corresponding lengths on ∂P and ∂D, the boundary length energy is defined as

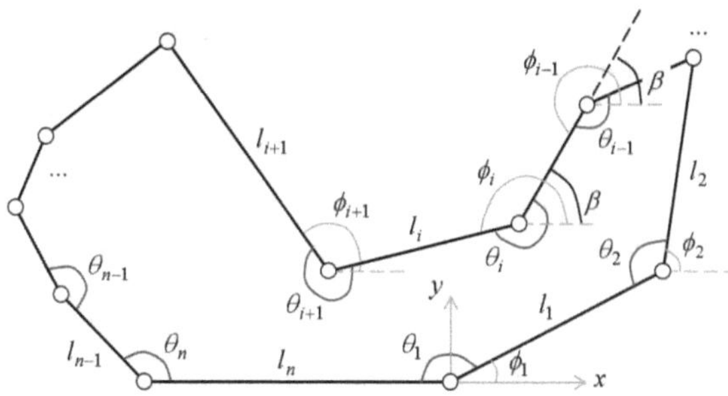

Fig. 5.20 Closed-path constraints on a planar boundary

$$\Pi_L = \frac{1}{n_e} \sum_{e \in \partial P} |l_e - l_e^0|, \tag{5.58}$$

where n_e represents the number of edges on ∂P. Introducing the boundary length energy is straightforward. However, since $\partial D \in \Re^2$ and $\partial P \in \Re^3$, it is not easy to measure the shape similarity between ∂D and ∂P by Π_L. Therefore, the second energy term is introduced.

Boundary morphological energy. Angles are employed to measure the morphological difference between ∂D and ∂P. For a vertex v on ∂P, suppose that its surface inner angle on P is α_v and its inner turning angle on D is θ_v (see the illustration in Fig. 5.20), the shape similarity between ∂D and ∂P is evaluated with the help of the following boundary angle error together with the boundary length error:

$$\Pi_\theta = \frac{1}{n_e} \sum_{v \in \partial P} |\alpha_v - \theta_v|. \tag{5.59}$$

In Π_θ, n_e is the number of vertices on ∂P which is, in fact, the same as the number of edges on ∂P, and the surface inner angle α_v is the sum of all vertex angles at v on P.

Based on the above boundary energy terms, one can compute an optimal LPFB of P by the functional

$$\arg\min_{v \in \partial D} \Pi_\theta \quad \text{s.t. } \Pi_L = 0. \tag{5.60}$$

However, as both Π_θ and Π_L are complex in terms of the planar position of boundary vertices, directly solving the problem defined in Eq. (5.60) is quite slow and thus impractical. The optimization functional needs to be reformulated. Wang [52] solves the above optimization problem in the angle space, which greatly simplifies the formulae. The morphological term is converted into

$$J_E = \sum_i \frac{1}{2}(\theta_i - \alpha_i)^2, \tag{5.61}$$

where θ_i is the inner turning angle of a boundary vertex $v_i \in \partial P$, α_i is the surface inner angle at v_i on P, and the index i of boundary vertex is given in an anticlockwise direction.

From the closed-path theorem (see [32]), it is known that, if a simple non-self-intersection planar closed path is in anticlockwise direction, the total turning is 2π. As shown in Fig. 5.20, its total turning by accumulating vertex turning angles is

$$\sum_{i=1}^{n_e}(\pi - \theta_i),$$

so the constraint below must be satisfied in the angle space:

$$n_e\pi - \sum_{i=1}^{n_e} \theta_i \equiv 2\pi. \tag{5.62}$$

Besides J_E and the constraint comes from the closed-path theorem, position coincident constraints need to be added to boundary vertices. After determining the inner turning angles θ_is and placing v_1 at the origin, the planar coordinates (x_i, y_i) of a boundary vertex v_i becomes

$$x_i = \sum_{k=1}^{i-1} l_k \cos\phi_k, \quad y_i = \sum_{k=1}^{i-1} l_k \sin\phi_k. \tag{5.63}$$

As illustrated in Fig. 5.20, $\theta_i = 2\pi - (\phi_i - \beta)$ is given at the vertex v_i with $\beta = \phi_{i-1} - \pi$, which leads to $\phi_i = \pi - \theta_i + \phi_{i-1}$. Together with $\phi_1 = \pi - \theta_1$, the general formula for ϕ_i can be derived in terms of θ_i as

$$\phi_i = i\pi - \sum_{b=1}^{i} \theta_b. \tag{5.64}$$

In order to ensure that ∂D is closed, (x_{n_e+1}, y_{n_e+1}) must coincide with the origin, which leads to

$$\sum_{i=1}^{n_e} l_i \cos\phi_i \equiv 0, \quad \sum_{i=1}^{n_e} l_i \sin\phi_i \equiv 0. \tag{5.65}$$

Therefore, the LPFB of a given patch P can be determined by the following constrained optimization problem defined in the angle space:

$$\begin{aligned} &\arg\min_{\theta_i} \sum_i \tfrac{1}{2}(\theta_i - \alpha_i)^2 \\ &\text{s.t. } n_e\pi - \sum_{i=1}^{n_e}\theta_i \equiv 2\pi, \ \sum_{i=1}^{n_e} l_i\cos\phi_i \equiv 0, \ \sum_{i=1}^{n_e} l_i \sin\phi_i \equiv 0. \end{aligned} \tag{5.66}$$

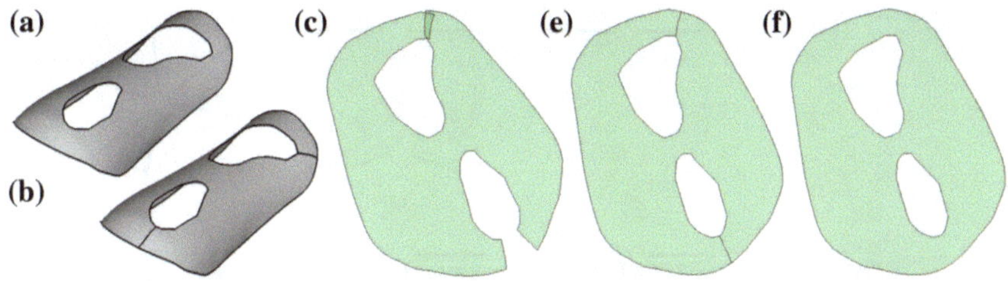

Fig. 5.21 Virtual cutting scheme for a surface patch with multiple boundary loops: **a** the given mesh patch P, **b** a duplicate patch P_d of P is generated and cut by the shortest path from *inner* loops to the *outer* boundary, **c** LPFB of P_d determined by applying the basic method, **d** LPFB obtained by adding position constraints of the vertices to the cutting path, and **e** the resultant D computed by the virtual cutting scheme

The efficient numerical implementation for solving Eq. (5.66) is given below. After obtaining the optimal θ_is, the planar coordinates of boundary vertices can be easily computed by Eqs. (5.63) and (5.64).

5.4.1.2 Virtual Cutting Scheme

The basic method stated above works well for mesh surfaces with a disk-like topology; however, it shows problems when computing LPFB of surfaces with multiple boundary loops. Therefore, a virtual cutting scheme is developed below to solve this problem. The algorithm consists of six steps (illustrated with the example shown in Fig. 5.21):

Step 1. Construct a duplicate patch P_d of P, where each vertex $v_i^d \in P_d$ has a corresponding vertex $v_i \in P$.
Step 2. For every inner boundary loop on P_d, the shortest path from it to the outer loop is determined by the Dijkstra's algorithm [8] with multiple sources.
Step 3. Cut P_d along the shortest paths (see Fig. 5.21b).
Step 4. Apply the basic method of LPFB to P_d (the result is shown in Fig. 5.21c).
Step 5. Add the coincident constraints of the vertices on cutting paths to the optimization framework so that an improved LPFB is obtained (as Fig. 5.21d).
Step 6. By the correspondences between $v_i^d \in P_d$ and $v_i \in P$, the planar coordinates for all boundary vertices on P can be determined (see Fig. 5.21e).

The technical details of this scheme are described as follows.

Steps 1 and 2 are trivial. In Step 3, duplicate edges are iteratively introduced to the edges that belong to the cutting paths by three operators (see Fig. 5.22), which are:

- *Hole open*. For a given edge e on a cutting path, when neither of the two vertices are on a boundary, this operator is applied to construct a hole by converting e into

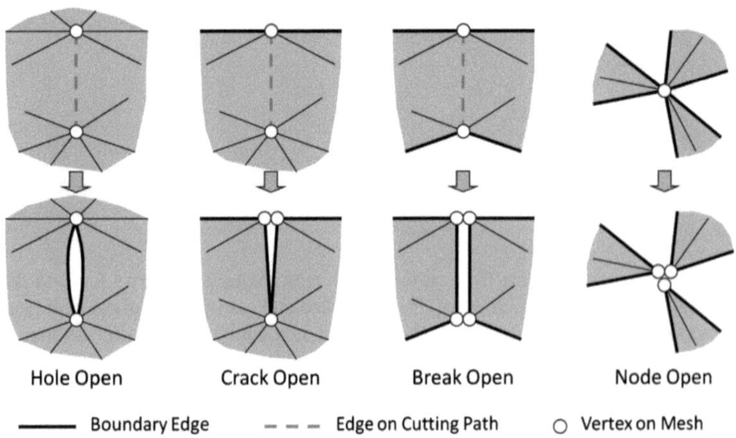

Fig. 5.22 A mesh surface can be cut along a path by iteratively applying four operators: **a** hole open, **b** crack open, **c** break open, and **d** node open

a boundary edge and adding another boundary edge coincident with e. Then, both vertices have become boundary vertices.

Crack open. When a vertex v_s on e is on the boundary, this operator is applied to create a crack along e by duplicating a vertex v_{new} to v_s and a new edge e_{new} for e, where e_{new} links v_{new} and v_d—another vertex on e.

Break open. When both v_s and v_d on e are on the boundary, we fully separate the left and the right parts of e by this break open operator (see Fig. 5.22).

After turning the edges on a cutting path into boundary edges, the following *Node Open* operator is applied to the boundary vertices finally.

- *Node open.* For a boundary vertex v_s linked with n $(n > 2)$ boundary edges, $\big((n-2)/2\big)$ new vertices coinciding with v_s are constructed, and the edges and faces linking to v_s are separated so that every vertex is linked with only two boundary edges (see Fig. 5.22).

On the resultant patch P_d with a single loop, the coincident pairs are stored as the information will be used in Step 5.

Step 4 uses the basic method for computing LPFB—an example result after this step is shown in Fig. 5.21c, where the newly created cutting vertices do not coincide with their parents.

After getting the result from the basic method, Step 5 further deforms the boundary to make the vertices stored in the coincident pairs become coincident. For this purpose, the constraints defined in Eq. (5.66) needs to be adjusted. Suppose that $\varphi(p)$ and $\chi(p)$ represent the indices of the vertices in the pth coincident pair, without loss of generality, letting $\varphi(p) < \chi(p)$, the coincident constraint of $v_{\varphi(p)}$ and $v_{\chi(p)}$ yields

$$\sum\nolimits_{k=\varphi(p)}^{\chi(p)-1} l_k \cos\phi_k \equiv 0, \qquad \sum\nolimits_{k=\varphi(p)}^{\chi(p)-1} l_k \sin\phi_k \equiv 0$$

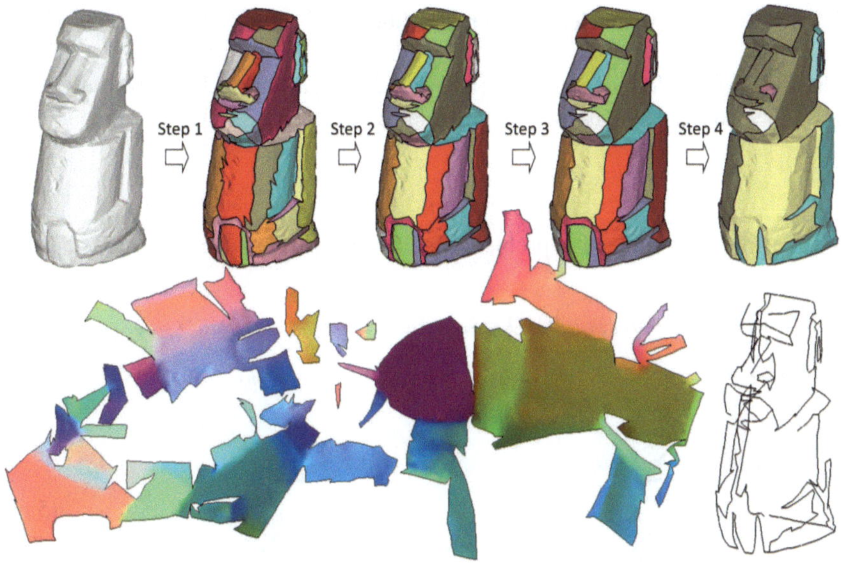

Fig. 5.23 An illustration of the quasi-developable mesh segmentation algorithm: (*leftmost*) the given model, (*middle left*) after error-controlled VSA (with 68 patches), (*middle*) after minimum-cut-based boundary refinement (still with 68 patches), (*middle right*) after boundary denoizing, (*rightmost*) the result from LPFB-based patch merging (with seven patches), and (*bottom row*) the surface flattening results where the colors represent normal vectors on the original 3D model (Image taken from [52] with permission ©2008 IEEE)

by Eq. (5.63). Together with $\varphi(0) = 1$ and $\chi(0) = n_e + 1$, the constrained optimization problem in Eq. (5.66) is reformulated to

$$\begin{gathered} \arg\min_{\theta_i} \sum_i \tfrac{1}{2}(\theta_i - \alpha_i)^2 \\ \text{s.t. } n_e\pi - \sum_{i=1}^{n_e} \theta_i \equiv 2\pi,\ \sum_{k=\varphi(p)}^{\chi(p)-1} l_k \cos\phi_k \equiv 0,\ \sum_{k=\varphi(p)}^{\chi(p)-1} l_k \sin\phi_k \equiv 0, \cdots, \end{gathered} \tag{5.67}$$

where $0 \le p \le m$ with m pairs of coincident vertices that were constructed in Step 3.

After solving the optimization problem in Eq. (5.67), an LPFB is determined with all the child-and-parent vertices on cutting paths coincided (see Fig. 5.21d for an example). The planar coordinates of vertices on ∂P are then computed from the corresponding vertices on ∂P_d in the last step.

5.4.1.3 Efficient Numerical Implementation

The only problem left for computing LPFB on a given mesh patch is how to efficiently solve the constrained optimization problem defined above. Using the Lagrange multiplier

$$(\lambda_\theta, \lambda_{0x}, \lambda_{0y}, \dots, \lambda_{px}, \lambda_{py}, \dots, \lambda_{mx}, \lambda_{my})$$

the constrained optimization problem can be converted into an augmented objective function,

$$J(X) = J(\theta_1, \dots, \theta_{n_e}, \lambda_\theta, \lambda_{0x}, \lambda_{0y}, \dots, \lambda_{px}, \lambda_{py}, \dots, \lambda_{mx}, \lambda_{my})$$

with $X = (\theta_1, \dots, \theta_{n_e}, \lambda_\theta, \lambda_{0x}, \lambda_{0y}, \dots, \lambda_{px}, \lambda_{py}, \dots, \lambda_{mx}, \lambda_{my})$. In detail,

$$\begin{aligned} J(X) = {} & \sum_{i=1}^{n_e} \frac{1}{2}(\theta_i - \alpha_i)^2 + \lambda_\theta \Big((n_e - 2)\pi - \sum_{k=1}^{n_e} \theta_k\Big) \\ & + \sum_{p=0}^{m} \lambda_{px} \sum_{k=\varphi(p)}^{\chi(p)-1} l_k \cos\phi_k + \sum_{p=0}^{m} \lambda_{py} \sum_{k=\varphi(p)}^{\chi(p)-1} l_k \sin\phi_k, \end{aligned} \tag{5.68}$$

which can be minimized by Newton's method [34]. Similar to the FL mesh computation presented in Sect. 5.4, the sequential linearly constrained programming is used to minimize $J(X)$ by neglecting the terms derived from the second derivatives of the constraints in the Hessian matrix $\nabla^2 J(X)$. The equation

$$\nabla^2 J(X)\delta = -\nabla J(X)$$

solved at each step is simplified to

$$\begin{bmatrix} I & \Lambda^T \\ \Lambda & 0 \end{bmatrix} \begin{bmatrix} \delta_\theta \\ \delta_\lambda \end{bmatrix} = \begin{bmatrix} B_\theta \\ B_\lambda \end{bmatrix}. \tag{5.69}$$

In this equation, Λ, B_θ, and B_λ can all be efficiently evaluated.

Remark 5.1 B_λ is computed by

$$B_\lambda = \left(-\frac{\partial J}{\partial \lambda_\theta}, -\frac{\partial J}{\partial \lambda_{0x}}, -\frac{\partial J}{\partial \lambda_{0y}}, \dots, -\frac{\partial J}{\partial \lambda_{mx}}, -\frac{\partial J}{\partial \lambda_{my}}\right), \tag{5.70}$$

where

$$-\frac{\partial J}{\partial \lambda_\theta} = -(n_e - 2)\pi + \sum_{k=1}^{n_e} \theta_k, \tag{5.71}$$

$$-\frac{\partial J}{\partial \lambda_{px}} = -\sum_{k=\varphi(p)}^{\chi(p)-1} l_k \cos\phi_k, \qquad -\frac{\partial J}{\partial \lambda_{px}} = -\sum_{k=\varphi(p)}^{\chi(p)-1} l_k \sin\phi_k. \tag{5.72}$$

Remark 5.2 $B_\theta = \{-\partial J/\partial\theta_i\}$ can be efficiently evaluated by the following recursion formulae

$$\begin{aligned} b_{\theta_1} &= -(\theta_1 - \alpha_1) + \lambda_\theta + \sum_{p=0}^{m} \sum_{k=\varphi(p)}^{\chi(p)-1} \left(-\lambda_{px} \sin\phi_k + \lambda_{py} \cos\phi_k\right) l_k, \\ b_{\theta_{i+1}} &= b_{\theta_i} + (\theta_i - \alpha_i) - (\theta_{i+1} - \alpha_{i+1}) + \sum_{p=0}^{m} A(p,i) \end{aligned} \tag{5.73}$$

with

$$A(p,i) = \begin{cases} (\lambda_{px} \sin\phi_i - \lambda_{py} \cos\phi_i) l_i, & \varphi(p) \le i < \chi(p) \\ 0, & \text{otherwise.} \end{cases} \tag{5.74}$$

Proof From $B_\theta = \{-\partial J/\partial\theta_i\}$, it is easy to obtain

$$b_{\theta_i} = -(\theta_i - \alpha_i) + \lambda_\theta + \sum_{p=0}^{m} \sum_{k=\varphi(p)}^{\chi(p)-1} \left(-\lambda_{px} \sin\phi_k + \lambda_{py} \cos\phi_k\right) l_k \frac{\partial\phi_k}{\partial\theta_i}. \tag{5.75}$$

Defining

$$\Gamma(p,i) \equiv \sum_{k=\varphi(p)}^{\chi(p)-1} \left(-\lambda_{px} \sin\phi_k + \lambda_{py} \cos\phi_k\right) l_k \frac{\partial\phi_k}{\partial\theta_i},$$

the following formula can be derived

$$\Gamma(p,i) \equiv \begin{cases} \sum_{k=\varphi(p)}^{\chi(p)-1} (-\lambda_{px} \sin\phi_k + \lambda_{py} \cos\phi_k) l_k, & i \le \varphi(p) < \chi(p) \\ \sum_{k=i}^{\chi(p)-1} (-\lambda_{px} \sin\phi_k + \lambda_{py} \cos\phi_k) l_k, & \varphi(p) < i < \chi(p) \\ 0, & \varphi(p) < \chi(p) \le i \end{cases} \tag{5.76}$$

with the fact that

$$\frac{\partial\phi_k}{\partial\theta_i} = \begin{cases} 0, & k < i \\ -1, & k \ge i \end{cases}.$$

This can be further simplified as follows.

Case 1 When $i \ge \chi(p)$, $i+1 > \chi(p)$, it can have

$$\Gamma(p,i+1) = \Gamma(p,i) = 0,$$

which leads to $A(p,i) = 0$.

Case 2 For $i = \chi(p) - 1$, $\Gamma(p,i+1) = 0$,

$$\Gamma(p,i) = (-\lambda_{px} \sin\phi_k + \lambda_{py} \cos\phi_k) l_i,$$

thus

$$A(p,i) = (\lambda_{px}\sin\phi_i - \lambda_{py}\cos\phi_i)l_i.$$

Case 3 When $\varphi(p) \le i < \chi(p) - 1$,

$$\Gamma(p,i+1) = \sum_{k=i+1}^{\chi(p)-1}(-\lambda_{px}\sin\phi_k + \lambda_{py}\cos\phi_k)l_k,$$

$$\Gamma(p,i) = \sum_{k=i}^{\chi(p)-1}(-\lambda_{px}\sin\phi_k + \lambda_{py}\cos\phi_k)l_k,$$

so it can be concluded that

$$A(p,i) = \Gamma(p,i+1) - \Gamma(p,i) = (\lambda_{px}\sin\phi_i - \lambda_{py}\cos\phi_i)l_i.$$

Case 4 $i = \varphi(p) - 1$, that is $i + 1 = \varphi(p)$, which leads to

$$\Gamma(p,i+1) = \Gamma(p,i) = \sum_{k=\varphi(p)}^{\chi(p)-1}(-\lambda_{px}\sin\phi_k + \lambda_{py}\cos\phi_k)l_k,$$

thus $A(p,i) = 0$.

Case 5 $i < \varphi(p) - 1$, that is $i + 1 < \varphi(p)$; for the same reason as the above case, $A(p,i) = 0$ can be obtained.
By summarizing all these five cases, the formula of $A(p,i)$ can be obtained.

Q.E.D.

Remark 5.3 The dimension of Λ is $(2m+3) \times n_e$, and its ith column vector is

$$\Lambda_i = \frac{\partial^2 J}{\partial\lambda\partial\theta_i} = \begin{pmatrix} \frac{\partial}{\partial\theta_i}((n-2)\pi - \sum_{k=1}^{n_e}\theta_k) \\ \frac{\partial}{\partial\theta_i}\sum_{k=\varphi(p)}^{\chi(p)-1} l_k\cos\phi_k \\ \frac{\partial}{\partial\theta_i}\sum_{k=\varphi(p)}^{\chi(p)-1} l_k\sin\phi_k \\ \cdots \end{pmatrix}_{(2m+3)\times 1} \tag{5.77}$$

with $p = 0, 1, \cdots, m$. Every element of Λ_i can be evaluated by

$$\begin{aligned} \Lambda_{1,i} &= -1, \\ \Lambda_{2p+2,i+1} &= \Lambda_{2p+2,i} + B(p,i), \\ \Lambda_{2p+3,i+1} &= \Lambda_{2p+3,i} + D(p,i) \end{aligned} \tag{5.78}$$

with

$$B(p,i) = \begin{cases} -l_i\sin\phi_i, & \varphi(p) \le i < \chi(p) \\ 0, & \text{otherwise,} \end{cases} \tag{5.79}$$

$$D(p,i)=\begin{cases} l_i\cos\phi_i, & \varphi(p)\le i<\chi(p)\\ 0, & \text{otherwise.}\end{cases} \tag{5.80}$$

Proof By $\Lambda_i=\partial^2 J/\partial\lambda\partial\theta_i$, it is easy to obtain

$$\begin{aligned}\Lambda_{1,i}&=-1,\\ \Lambda_{2p+2,i+1}&=-\sum_{k=\varphi(p)}^{\chi(p)-1} l_k\sin\phi_k\frac{\partial\phi_k}{\partial\theta_i},\\ \Lambda_{2p+3,i+1}&=\sum_{k=\varphi(p)}^{\chi(p)-1} l_k\cos\phi_k\frac{\partial\phi_k}{\partial\theta_i}.\end{aligned}$$

The following proof is ignored as it is similar to the proof of Remark 5.2.

Q.E.D.

With these remarks, Eq. (5.69) can then be solved by

$$\Lambda\Lambda^T\delta_\lambda=\Lambda B_\theta-B_\lambda, \tag{5.81}$$

$$\delta_\theta=B_\theta-\Lambda^T\delta_\lambda, \tag{5.82}$$

where $\Lambda\Lambda^T$ is $(2m+3)\times(2m+3)$ as Λ is $(2m+3)\times n_e$. Since m is usually small ($m\ll n_e$), Eq. (5.81) can be efficiently solved using *Gaussian elimination* as $rank(\Lambda\Lambda^T)=2m+3$, or it can be solved by SVD if not full rank. By δ_λ determined in Eq. (5.81), δ_θ can be easily computed from Eq. (5.82). Same as all other nonlinear problems, a good initial value can speed up the above computation. By Wang's tests [52], when choosing $\theta_i=\alpha_i$ and $\lambda_\theta=\lambda_{px}=\lambda_{py}=1$, the computation converges in several tens of steps.

5.4.2 Quasi-Developable Mesh Segmentation

Given a mesh model H, it is segmented into nearly developable (or flattenable) patches by a trial-and-error algorithm. The basic idea is that: H is segmented into nearly planar charts where are then incrementally merged back into larger quasi-developable surface patches through a trial-and-error procedure where the merging criteria are efficiently evaluated with the help of LPFB. As illustrated in Fig. 5.23, the segmentation algorithm consists of four steps.

Step 1: Error-controlled variational shape approximation. First, a given model H is segmented into nearly planar patches by the variational shape approximation (VSA) algorithm [7]. In this step, the shape approximation error instead of the proxy number is controlled. Starting from one seed, more seeds are incrementally added into the k-proxy clustering algorithm until the maximal approximation error max $L^{2,1}$ shown on all charts is smaller than a given tolerance. The error-controlled VSA results in a number of small patches on a complex model.

Step 2: Boundary refinement based on minimum-cut. In this step, the boundaries of patches are refined by the minimum-cut method akin to [17] so that the zigzag effect is improved. Briefly, a fuzzy area is determined around the boundaries between different charts. After converting the faces in the fuzzy area to the nodes of a weighted graph, the re-partition of triangles is found by a minimal-cut algorithm on the graph weighted by dihedral angles (see [8] for reference).

Step 3: (Optional) LSE-based boundary denoizing. In order to reduce the effect from noise, the vertices located on the refined boundaries are further adjusted. This is an optional step—those noise-free models do not need this. For the faces f on a chart around a boundary vertex, the normal vectors $\mathbf{n}_f$ are assigned as the chart normal vector, which is the normalized vector by summing up the area-weighted normals of triangles on the chart (see [7] for details). Then, the position of every boundary vertex $\mathbf{v}_i$ is updated by minimizing the following least-square error (LSE) defined on the faces:

$$E(\mathbf{v}_i) = \sum_{j \in N_v(i)} \sum_{f \in F_{ij}} (\mathbf{n}_f \cdot (\mathbf{v}_i - \mathbf{v}_j))^2,$$

where $N_v(i)$ denotes the one-ring neighboring vertices of $\mathbf{v}_i$ and F_{ij} represents the two faces that are adjacent to the edge $\mathbf{v}_i\mathbf{v}_j$. $E(\mathbf{v}_i)$ can be iteratively minimized by the following update given by Taubin [48]:

$$\mathbf{v}_i' \leftarrow \mathbf{v}_i + \tau \sum_{j \in N_v(i)} \sum_{f \in F_{ij}} \mathbf{n}_f \mathbf{n}_f^T (\mathbf{v}_j - \mathbf{v}_i)$$

with $\tau = 0.25$ being the relaxation factor.

Step 4: LPFB-based patch merging. The neighboring small charts are incrementally merged into a larger patch if the new patch P is still quasi-developable. The LPFB of P is computed first. Then, the IP [10] is employed to determine the flattening result D while fixing the boundary vertices. The distortion on D is employed to evaluate the developability. The detailed LPFB-based patch merging is presented below.

5.4.2.1 Measurement for Classification

The measurement employed in this segmentation algorithm relies on the flattening Ω of P in D with a fixed LPFB. For this reason, the following linear equation system from the IP method [10] is employed by Wang [52] to calculate the planar coordinates of interior vertices:

$$MU = V \tag{5.83}$$

where U is the vector of planar coordinates, V is the vector for boundary condition that

$$V = \begin{cases} \psi(i), & i \in \partial P \\ 0, & \text{otherwise} \end{cases} \tag{5.84}$$

with $\psi(i)$ representing the planar coordinates of vertex $\mathbf{v}_i$ on the LPFB of P, and M is a sparse matrix whose coefficients $m_{i,j}$ are given by

$$\forall i \in (P - \partial P), \quad m_{i,j} = \begin{cases} \cot \gamma_{i,j} + \cot \xi_{i,j}, & j \in N_v(i) \\ -\sum\limits_{k \in N_v(i)} m_{i,k} & i = j \\ 0 & \text{otherwise.} \end{cases} \tag{5.85}$$

$$\forall i \in \partial P, \quad m_{i,j} = \begin{cases} 1, & i = j \\ 0, & i \neq j. \end{cases} \tag{5.86}$$

In Eq. (5.85), $N_v(i)$ denotes the one-ring neighbors of vertex $\mathbf{v}_i$, and $\gamma_{i,j}$ and $\xi_{i,j}$ represent the opposite angles in the two neighboring triangles of the edge $\mathbf{v}_i\mathbf{v}_j$ on P respectively. Why is this computation faster than directly applying the IP? The reason is that for a surface P with n vertices, the dimension of the above linear system is $n \times n$ since the computation of x- and y-coordinates can be separated. However, for the method of Desbrun et al. [10], when a free boundary is needed, both x- and y-coordinates should be computed together in a linear system. The lower bound of the computing time for solving sparse linear system is linear—therefore at least half of the computing time is reduced. However, in general, the time of computation depends on the pattern of the matrix M. With the conjugate gradient solver, the upper bound is in quadratic complexity. Wang's tests [52] show that by reducing the dimension of M by half, the speedup is usually more than double.

By Ω determined from LPFB of P, a hybrid classification using three types of measurement is conducted. The three types of measurement include the area distortion $A(\Omega)$, the $L^2(\Omega)$ norm of texture stretch, and the global self-overlaps $S(\Omega)$. The area distortion computes the percentage of area change in D and P by

$$A(\Omega) = \frac{|\sum\limits_T A_T - \sum\limits_T A_T^0|}{\sum\limits_T A_T^0} \times 100\,\%, \tag{5.87}$$

where A_T and A_T^0 are the areas of a triangle in 2D and 3D respectively. Although a small $A(\Omega)$ is the necessary condition for P being quasi-developable, it is possible to have extreme cases in which a small $A(\Omega)$ is given on a non-developable surface. Therefore, the L^2 norm of texture stretches [37] is conducted to exclude such extreme cases. It measures the surface distortion on Ω, where 1.0 is the lower bound on any parameterization. However, above two types of measurement cannot distinguish the global self-overlapping on Ω (for example, the self-overlapping in Fig. 5.21c). Numerically detecting the self-overlaps is usually time-consuming. To speed up, Wang [52] seeks help from graphics hardware. After drawing all triangles in the blending mode with a transparency value of 0.25, the pixels drawn more than once (which lead to overlaps) can be easily detected by their pixel color value. When an overlap is found, the measurement $S(\Omega)$ gives 1; otherwise, $S(\Omega) = 0$ is returned. As long as the resolution of discretization is high enough, self-intersections can be effectively detected in a given tolerance.

5.4.2.2 LPFB-Based Patch Merging

The method presented by Wang [52] incrementally merges neighboring small patches into a larger patch if the new patch can still be flattened without stretching. Therefore, the measurement for classification needs to be repeatedly evaluated—the LPFB-based technique shows its advantage in the speed here.

The following boundary-based merging algorithm progressively merges small nearly planar patches into larger nearly developable patches.

1. Every boundary curve between different patches is inserted into a maximum heap Ψ which sorts the area of two patches aside the boundary curve.
2. If Ψ is empty, stop the algorithm.
3. Remove the top curve c_t from Ψ.
4. Duplicate a patch P simulating the merged patch by eliminating c_t.
5. Compute the LPFB and then the parameterization Ω of P from the fixed LPFB.
6. Evaluate $A(\Omega)$; if $A(\Omega) > 10\,\%$, go back to step 2.
7. Evaluate $L^2(\Omega)$; if $L^2(\Omega) > 2.0$, go back to step 2.
8. Evaluate $S(\Omega)$; if $S(\Omega) = 1$, go back to step 2.
9. Merge the patches aside c_t into a new one, P_{new}, by removing c_t.
10. For the remaining boundary curves on P_{new}, update their positions in the heap Ψ.
11. Go back to Step 2.

This algorithm can efficiently merge small planar patches into large nearly developable patches. The two thresholds, 10 % and 2.0, were determined from a supervised learning process [12] with a set of training mesh surfaces. Figure 5.23d, e shows an example for the LPFB-based patch merging, and more examples can be found in Fig. 5.18.

5.4.3 Discussion on Limitations

Serving as a model to compute the distortion criterion for developable mesh segmentation, the resultant mesh from LPFB + IP needs to be robustly computed. The computation method presented in this paper relies on the surface inner angles on boundary vertices since they are the reference angles to reach in the numerical optimization framework. If the boundary of a segmented patch contains vertices with a high curvature, the result of LPFB may be weird. However, this is naturally avoided by VSA—the first step of this segmentation algorithm. Moreover, in practice, there is no guarantee that noise will not be shown on the boundary of a surface patch. Therefore, the first limitation of this approach is that the method for computing LPFB is not robust enough. This is also why boundary denoizing (Step 3) is needed in the segmentation algorithm. Wang's tests [52] show that the method works well if the boundary denoizing step is applied together.

The second limitation is that, although the constraint derived from the closed-path theorem has been added to the numerical optimization framework so that the local self-intersections are prevented, there is no constraint for avoiding global self-intersections on the computed LPFB. Therefore, in the measurements for classification, the term $S(\Omega)$ is needed. The method for preventing global self-intersections is still under investigation. The angle expansion method for self-intersections in [41] seems to be a good candidate.

The third limitation comes from finding the thresholds for the classifier in the segmentation algorithm, which is by no means an easy job. Wang [52] determines them by supervised learning: samples of nearly developable surfaces and non-developable surfaces are first selected to train the classifiers. The values of $L^2(\Omega)$ and $A(\Omega)$ are computed on LPFB + IP of the sample patches. Then, the thresholds are drawn so that most samples can be classified into a correct category. Therefore, the thresholds depend on the samples employed to train the classifiers that are not robust enough.

The last limitation of the method its inability to give smooth boundaries that may be desired by toy fabrication and texture applications. Therefore, the second step of this segmentation algorithm needs to be replaced by a new method that can generate more smooth boundaries.

5.5 WireWarping: Surface Flattening with Length-Preserved Feature Curves

Study from industrial cases (for example [56, 63]) shows that the property of length-preservation on feature curves is very important to industrial applications for controlling the shape and dimension of products fabricated from planar pieces. This section presents an approach (named as *WireWarping*) for computing a flattened planar piece with length-preserved feature curves from a 3D piecewise linear surface patch. WireWarping simulates warping a given 3D surface patch onto a plane with the feature curves as tendon wires to preserve the length of their edges. The boundaries and feature curves are referred to as *wires* for the rest of this section. During warping, the surface-angle variations between edges on wires are minimized so that the shape of a planar piece is similar to its corresponding 3D patch.

The problem of surface flattening (or parameterization) is usually formulated under a constrained optimization framework. Its resultant 3D patch generally is not a developable surface, and length variations are always found in the flattening results. For an engineering application like the 3D garment design and manufacturing shown in Fig. 5.24, the length variations will lead to many problems. If length variations occur on the boundaries of two pieces that are going to be sewn together, unexpected wrinkles will form on the fabricated product. If length variations happen in the interior region of a patch, it will destroy the designed fit (for example, if the length of chest girth varies on the flattened patterns of the shirt shown in Fig. 5.24, the products made by these patterns may be too tight or too loose). A good garment shape and fit (i.e., without unexpected wrinkles) are two necessary criteria to evaluate whether a suit is a high-end garment product. This is also true for other industrial applications (like the shoe industry and the furniture industry). Therefore, the designers in these industries desire a surface flattening tool, which can preserve the length of boundaries and feature curves (such as the black and green curves in Fig. 5.24) on a 2D piece according to its 3D surface patch.

Problem Definition. Given a piecewise linear surface patch in $\Re^3$, its counterpart pattern in $\Re^2$ is computed to make the length of edges on the boundaries and feature curves (named as *wires*) optimally invariant; meanwhile, the angles between neighboring edges on the wires in $\Re^2$ are optimized to preserve their values on the given 3D surface patch.

5.5.1 Preliminary Definitions

Necessary definitions and preliminaries are given below.

Definition 5.4 *Feature curves* are the piecewise linear curves formed by polygonal edges on the given piecewise linear surface P to be flattened, where every segment on a feature curve is required to have the same length on P and on the flattened piece D.

Definition 5.5 For a feature curve, if its planar shape on the flattened piece D has been predefined, it is named as *a key feature curve*; other feature curves are called *accessory feature curves*, whose planar shapes are determined by minimizing the variation between the surface angle and the planar angle at each endpoint of segments.

Definition 5.6 *Darts* are the curves defined on the piecewise linear surface P, which specify places to be cut out.

For different products, different sets of feature curves are defined by industrial designers. Feature curves are adopted in the 3D design to control the shape of final products, and the role of these feature curves is like control curves for surface lofting in geometric modeling systems. However, for products that are fabricated from 2D sheet materials, an important requirement of the feature curves is that they should be length-invariant during flattening (in other words, work like tendon wires). The red and green curves shown in Fig. 5.24 are feature curves. In all examples of this section, the key feature curves are colored in red, the accessory feature curves are displayed in green, and the boundary curves are shown in black. Boundary curves are classified into accessory feature curves if no special explanation is given. Darts are illustrated in blue in all examples, which are converted into boundary curves at the beginning of flattening by iteratively introducing duplicate edges on those edges belonging to the dart curves. All feature curves are in general called *wires*.

Definition 5.7 Each region circled by feature curves on the given surface P is defined as a *wire-patch*.

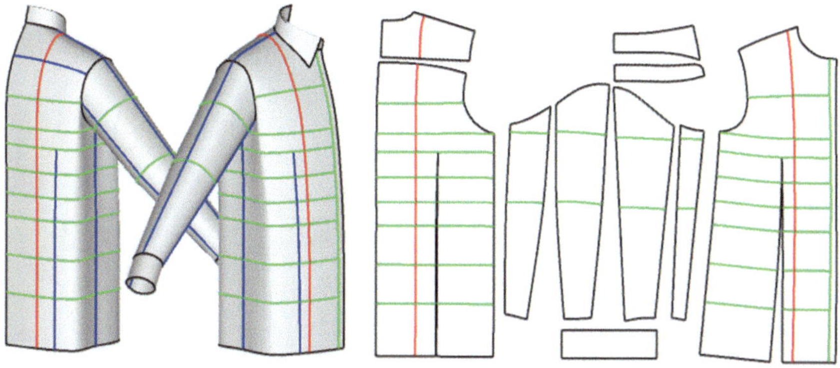

Fig. 5.24 An example of surface flattening on a 3D shirt (*front* and *back views*), where the *black curves* are the boundaries of 3D surface patches, the *blue curves* are the darts that will be cut out (so that become *black curves*), the *red* ones are the key feature curves, and the *green* ones are the accessory feature curves—the definitions about key feature curves and accessory feature curves are given in Sect. 5.5.1 (Image taken from [56] with permission)

For a given surface patch P, it can be segmented into several wire-patches. Figure 5.25a illustrates the wire-patches on a shirt with different colors. The boundary of a wire-patch is recorded by a list of wire-nodes, where each wire-node coincides with a vertex on the wires of the surface patch P. A vertex v on a feature curve may have more than one wire-nodes attached (for example, the vertex circled by blue curves in Fig. 5.25b). The number of wire-nodes associated with a surface vertex v is determined by the number of wire-patches adjacent to v.

Definition 5.8 A *wire-node* is denoted by q^i_j where the superscript refers to the index of the wire-patch P_i holding it and the subscript represents its index in P_i (ordered anticlockwise); $v(q^i_j)$ represents the vertex holding q^i_j and is named as the *host vertex* of q^i_j.

For three neighboring wire-nodes q^i_{j-1}, q^i_j and q^i_{j+1} on the same wire-patch, α^i_j is employed to represent the surface angle on P_i formed by them. The value of α^i_j can be evaluated by summing up the angles of polygons f_k with $f_k \in P_i$ at the host vertex $v(q^i_j)$. The 2D angle formed by these three wire-nodes after flattening is denoted by θ^i_j—details about how to compute θ^i_j are addressed below. A data-structure entity is developed for wire-nodes so that the host vertex $v(q^i_j)$ of a wire-node q^i_j can be easily found in a constant time complexity. The wire-nodes in other adjacent wire-patches at $v(q^i_j)$ are also stored in this entity. In addition, in order to travel among neighboring wire-patches, the following wire-curve is defined.

Definition 5.9 A *wire-curve* is defined by an ordered list of directional edges on P which separates two neighboring wire-patches.

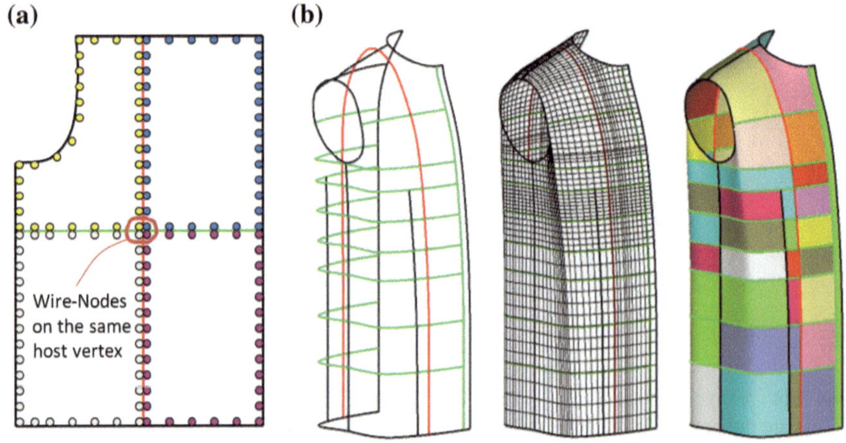

Fig. 5.25 An illustration of wire-patches. **a** The wire-nodes belonging to different wire-patches are shown in different *colors*. **b** *Left* the wires, *middle* the given piecewise linear surface, and *right* the wire-patches that are visualized in different *colors* (Image taken from [56] with permission)

It is obvious that wire-curves coincide with feature curves, and the boundary of a wire-patch consists of several wire-curves in general. A wire-patch stores a collection of its wire-curves and a wire-curve entity records its left or right wire-patches in the data structure.

5.5.2 Formulation

Length-preserved warping of a single wire-patch can be formulated in a constrained optimization framework as LPFB in Sect. 5.5 (specifically, (Eq. 5.66)). To compute a flattening of wire-patches by means of global warping, the subsystems of the constrained optimization are integrated into a consistent system to warp all wire-patches together. In addition to the closed-path constraint and the position coincident constraints, a compatibility constraint is introduced so that the sum of 2D angles of the wire-nodes associated with an interior host vertex v is 2π.

Without loss of generality, if there are totally m wire-patches constructed on a given surface P, there are $\sum_{p=1}^{m} n_p$ wire-nodes where n_p represents the number of wire-nodes for the wire-patch P_p whose index is p. As all wire-patches are warped together, every wire-node has one local index in the wire-patch and another global index. To simplify the expression, a permutation function $\Gamma_p(b)$ is defined for returning the global index of a wire-node on the wire patch P_p with the local index b, and its inverse function $\Gamma_p^{-1}(j)$ gives the local index of a wire-node q_j on the wire-patch P_p. The goal of global warping is to find constrained optimal 2D angles for wire-nodes so that the global distortion of flattening is minimized, which can be formulated as follows.

$$\begin{aligned}
&\min_{\theta_i} \sum_i \frac{1}{2}(\theta_i - \alpha_i)^2 \\
&\text{s.t. } n_p \pi - \sum_{b=1}^{n_p} \theta_{\Gamma_p(b)} \equiv 2\pi \qquad (\forall p = 1, ..., m) \\
&\sum_{b=1}^{n_p} l_b \cos \phi_b \equiv 0, \ \sum_{b=1}^{n_p} l_b \sin \phi_b \equiv 0 \ (\forall p = 1, ..., m) \\
&\sum_{q_k \in v} \theta_k \equiv 2\pi \qquad (\forall v \in \Phi)
\end{aligned} \tag{5.88}$$

where Φ represents the collection of interior vertices on accessory feature curves, θ_i is the 2D angle associated with the wire-node q_i, α_i represents its 3D surface angle, and l_b denotes the length of an edge on wires. In summary, if there are l wire-nodes with their 2D angles locked by the key feature curves, the number of variables for the above problem is

$$n_{\text{var}} = (\textstyle\sum_{p=1}^{m} n_p) - l,$$

and the set of free wire-nodes is defined as Q_{act}. If there are r interior vertices on the accessory feature curves, the total number of constraints is

$$n_{\text{con}} = 3m + r.$$

Using the Lagrange multipliers [34], the constrained optimization in Eq. (5.88) can be converted into an augmented objective function that

$$\begin{aligned}
J = &\sum_{i \in Q_{act}} \frac{1}{2}(\theta_i - \alpha_i)^2 \\
&+ \sum_p \left[\lambda_{\theta_p}((n_p - 2)\pi - \sum_{b=1}^{n_p} \theta_{\Gamma_p(b)}) + \lambda_{x_p}(\sum_{b=1}^{n_p} l_b \cos \phi_b) + \lambda_{y_p}(\sum_{b=1}^{n_p} l_b \sin \phi_b) \right]. \\
&+ \sum_{v \in \Phi} \lambda_v (2\pi - \sum_{q_k \in v} \theta_k)
\end{aligned} \tag{5.89}$$

The objective function can be minimized by Newton's method with the sequential linearly constrained programming [56].

5.5.3 Laying Out Feature Curves and Interior Mesh Vertices

After computing the optimal 2D angles on all wire-nodes, the wire-patches and their interior mesh vertices in $\Re^2$ need to be laid out. The feature curves are placed first, and the positions of interior mesh vertices are computed with the fixed boundary of wire-patches.

For placing feature curves, wire-patches are reordered by propagation and stored in a list Ψ. Starting from a seed wire-patch P_s, after inserting P_s into Ψ all neighboring wire-patches of P_s are checked. If any neighbor P_r not in Ψ is found, P_r is inserted into Ψ and the neighbors of P_r are checked recursively. With the wire-patches ordered in Ψ, the host vertices of wire-nodes can be placed patch by patch. The wire-nodes in a wire-patch are classified into two types: *fixed nodes*—those with their host vertices' 2D positions known, and *free nodes*—those whose host vertices have not been placed yet. The wire-nodes in a wire-patch P_i are searched in an anticlockwise direction and the first free node q_e, which is next to a fixed node q_f, is found. The position of q_e's host vertex $\mathbf{v}(q_e)$ can be determined by the positions of $\mathbf{v}(q_f)$ and its previous fixed node $\mathbf{v}(q_{f-})$, the edge length of $\mathbf{v}(q_f)\mathbf{v}(q_e)$ in $\Re^3$, and the optimal 2D angle $\theta(q_f)$ at q_f. Similarly, the latter free nodes on this wire-patch P_i can be placed consecutively. No fixed wire-node can be found on the first wire-patch in Ψ. Wang [56] randomly chooses two neighboring wire-nodes and fixes them in $\Re^2$ by reserving the distance between them. In this way, all feature curves (i.e., the boundaries of all wire-patches) can be placed appropriately in $\Re^2$.

The mesh vertices not associated with any wire-node (in other words, the interior mesh vertices of wire-patches) finally need to be positioned in $\Re^2$ to generate a correct mesh surface representation. Every vertex $\mathbf{v}_i$ is first placed at the average position of the boundary vertices of the wire-patch holding it. Next, the positions of $\mathbf{v}_i$ are moved iteratively by the operator

$$\mathbf{v}_i^p \leftarrow \frac{1}{w(\mathbf{v}_i)} \sum_{j \in N(\mathbf{v}_i)} \|\mathbf{v}_i \mathbf{v}_j\|^{-1} \mathbf{v}_j^p, \tag{5.90}$$

where $\|...\|$ denotes the distance between two vertices on the given surface, the superscript p means the planar coordinates of a vertex, $N(\mathbf{v}_i)$ represents the one-ring neighbors of the vertex $\mathbf{v}_i$, and $w(\mathbf{v}_i)$ is the summed weights as

$$w(\mathbf{v}_i) = \sum_{j \in N(\mathbf{v}_i)} \|\mathbf{v}_i \mathbf{v}_j\|^{-1}.$$

This is, in fact, an iterative version to solve a Laplacian-like system [31], which has been proven to be stable. The iteration stops when the movement of all vertices are smaller than 10^{-5}, and the number of iterations is in the range of 10–100. To further speed up the computation, we introduce a relaxing factor $\tau = 1.5$ (like the improvement of convergency using relaxation for the Gauss–Seidel solver in [6]) to let

$$\mathbf{v}_i^p \leftarrow \mathbf{v}_i^p + \tau \left(\left(\frac{1}{w(\mathbf{v}_i)} \sum_{j \in N(\mathbf{v}_i)} \|\mathbf{v}_i \mathbf{v}_j\|^{-1} \mathbf{v}_j \right) - \mathbf{v}_i \right), \tag{5.91}$$

so that the number of iteration steps can be reduced by about two third in most examples. An alternative to determine the planar coordinates of interior vertices is to apply the intrinsic parameterization [10] by considering each wire-patch as a surface path with a fixed boundary.

5.5.4 Numerical Analysis and Least-Norm Solution

The Newton's method solves a constrained optimization problem by converting the objective function and the constraints into an augmented objective function $J(X)$ with X as the variable vector. Then, the update vector δ in each iteration step is computed by the linear system, $\nabla^2 J(X)\delta = -\nabla J(X)$, which is formed by the Hessian matrix $\nabla^2 J(X)$ and the gradient $\nabla J(X)$. However, the Newton's algorithm has no control over the magnitude of δ. Therefore, vibration is easily generated when the status variable X is near optimum. In some extreme cases, such vibration may move the system to a status that can hardly converge. Figure 5.26a shows an example of such vibration when using Newton's method to compute the global *WireWarping*.

To make the Newton's method more robust, the soft-line-search strategy [29] is always employed to determine the actual update step size $\alpha\delta$ $(0 < \alpha \leq 1)$ (for example, [27]). However, such a line-search introduces additional sub-routine of iterations, so it actually slows down the computation. A least-norm solution is introduced below to increase the robustness of optimization while retaining the same efficiency as Newton's method in each iteration step.

As stated by Zayer et al. [62], carefully selecting alternative variables could make the linearization more accurate so that the computation converges faster than the Newton steps. Equation (5.88) can be reformulated by changing the variables from θ_i to the angle estimation error $e_i = \theta_i - \beta_i$, where β_i is the current angle at wire-node i and θ_i is the optimal angle to be computed.

$$\begin{aligned}
&\min_{e_i} \sum_i e_i^2 \\
&\text{s.t. } n_p\pi - \sum_{b=1}^{n_p} (\beta_{\Gamma_p(b)} + e_{\Gamma_p(b)}) \equiv 2\pi \quad (\forall p = 1, ..., m) \\
&\sum_{b=1}^{n_p} l_b \cos\phi_b \equiv 0, \ \sum_{b=1}^{n_p} l_b \sin\phi_b \equiv 0 \ (\forall p = 1, ..., m) \\
&\sum_{q_k \in v} (\beta_k + e_k) \equiv 2\pi \qquad\qquad (\forall v \in \Phi)
\end{aligned} \tag{5.92}$$

For the constraints with ϕ_b, as $\phi_i = \pi - (e_i + \beta_i) + \phi_{i-1}$ and $\phi_1 = \pi - (e_i + \beta_1)$ according to Eq. (5.64), ϕ_i can be expressed as

$$\phi_i = i\pi - \sum_{h=1}^{i} (e_h + \beta_h) = \Delta_i + \xi_i$$

with $\Delta_i = i\pi - \sum_{h=1}^{i} \beta_h$ and $\xi_i = -\sum_{h=1}^{i} e_h$. Using Taylor expansion

$$\begin{aligned}
\cos(\Delta_i + \xi_i) &= \cos\Delta_i + (-\sin\Delta_i)\xi_i + (-\tfrac{1}{2}\cos\Delta_i)\xi_i^2 + \cdots \\
\sin(\Delta_i + \xi_i) &= \sin\Delta_i + (\cos\Delta_i)\xi_i + (-\tfrac{1}{2}\sin\Delta_i)\xi_i^2 + \cdots
\end{aligned}$$

the series can then be truncated by retaining the linear terms only (i.e., with the approximation error $O(\xi_i^2)$). Therefore, the non-linear constraints $\sum_{b=1}^{n_p} l_b \cos\phi_b \equiv 0$ and $\sum_{b=1}^{n_p} l_b \sin\phi_b \equiv 0$ are linearized into

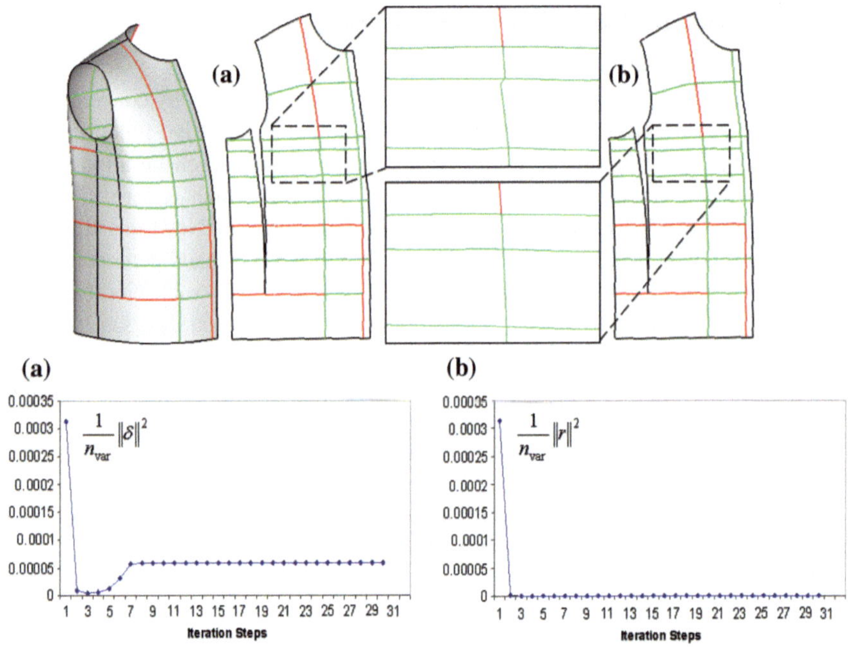

Fig. 5.26 Numerical vibration occurs when using Newton's method to solve the global *WireWarping* for flattening the front piece of a shirt—as shown in the zoom-window of (**a**), unwanted curve distortion is generated. The reason is that the numerical computation vibrates—see the chart of $\|\delta\|^2$. Following [56], the *green* and *red lines* here represent the key feature curves and the accessory feature curves respectively. Applying the least-norm scheme [57] can relieve the numerical oscillation—see the result shown in (**b**) (Image taken from [57] with permission)

$$\sum_{b=1}^{n_p} (e_{\Gamma_p(b)} \sum_{i=b}^{n_p} l_i \sin \Delta_i) = -\sum_{i=1}^{n_p} l_i \cos \Delta_i \tag{5.93}$$

and

$$\sum_{b=1}^{n_p} (e_{\Gamma_p(b)} \sum_{i=b}^{n_p} l_i \cos \Delta_i) = \sum_{i=1}^{n_p} l_i \sin \Delta_i \tag{5.94}$$

respectively. In summary, Eq. (5.92) is converted into

$$\min \|\mathbf{r}\|^2 \text{ s.t. } \mathbf{Cr} = \mathbf{b} \tag{5.95}$$

where $\mathbf{C}$ is a $n_{\text{con}} \times n_{\text{var}}$ matrix. As discussed in [56], if there are l wire-nodes with their 2D angles locked by the key feature curves, the number of variables for the above problem is $n_{\text{var}} = (\sum_{p=1}^{m} n_p) - l$. If there are r interior vertices on the accessory feature curves, the total number of constraints is $n_{\text{con}} = 3m + r$.

In general, $n_{\text{con}} < n_{\text{var}}$, there are multiple solutions for $\mathbf{Cr} = \mathbf{b}$. Among them, the one that leads to a minimal norm $\|\mathbf{r}\|^2$ on the variable vector $\mathbf{r}$ is usually demanded. This is a least-norm problem. For a full rank coefficient matrix $\mathbf{C}$, the least-norm problem has a unique solution (c.f. [49])

$$\mathbf{r} = \mathbf{C}^T(\mathbf{C}\mathbf{C}^T)^{-1}\mathbf{b}. \tag{5.96}$$

The value of $\mathbf{r}$ can be solved by finding a solution to the normal equation $(\mathbf{C}\mathbf{C}^T)\mathbf{x} = \mathbf{b}$ following by a substitution that $\mathbf{r} = \mathbf{C}^T\mathbf{x}$. The matrix $\mathbf{C}$ has full rank as the constraints in Eq. (5.88) are independent.

Starting from letting $\beta_i = \beta_i^0$, we iteratively update the value of β_i by solving e_i in Eq. (5.95) and updating its value with $\beta_i = \beta_i + e_i$ in each step. The iteration is stopped when $\frac{1}{n_{\text{var}}}\|\mathbf{r}\|^2 < 10^{-8}$ is satisfied. The resultant optimal angle for each wire-node is then determined. Why is such a least-norm solution in each iteration step more robust? The major reason is that among all possible solutions, the one with the minimal estimation error is adopted. While the Newton's update just moves the system variables along the optimal direction but not determine an optimal step size when not conducting the soft-line-search strategy, the least-norm solution actually mimics the soft-line-search. Another minor benefit of the least-norm solution is that the second derivative of $J(X)$ does not need to be computed.

Although the case of $n_{\text{con}} \geq n_{\text{var}}$ is never found in Wang's tests [56, 57], the linearization of global warping problem as above equations gives a possible solution to compute the update vector $\mathbf{r}$ by

$$\mathbf{C}^T\mathbf{C}\mathbf{r} = \mathbf{C}^T\mathbf{b} \tag{5.97}$$

which is in fact a least-square solution.

Same as other optimization techniques, the least-norm solution still relies on good initial angle values on wire-nodes.

- For those surfaces without key feature curves determined, the warping of feature curves are flexible. Thus, the Least-Square Conformal Map presented in [24] is used to pre-flatten the surface into plane, and the 2D angle at each wire-node is then adopted as the initial value of iteration.
- For those surfaces with the shape of key feature curves specified (for example, the perpendicular key feature curves are specified in Fig. 5.26), the initial value β_i^0 at a wire-node q_i is determined as

$$\beta_i^0 = \begin{cases} \alpha_i(2\pi - \sum_{q_j \in v} \alpha_j^L) / \sum_{q_k \in v} \alpha_k & q_i \in v,\ v \in \Phi \\ \alpha_i & \text{otherwise} \end{cases} \tag{5.98}$$

where q_k and q_j are the wire-nodes associated with the same vertex as q_i. q_j is the node on key feature curves with its 2D angle specified as α_j^L, q_k is on an accessory feature curve, and Φ is the set of interior vertices on accessory feature curves.

The initial values of angles determined by the above methods ensure that the angle compatible constraint—$\sum_{q_k \in v} \theta_k \equiv 2\pi \ (\forall v \in \Phi)$ has been satisfied at the beginning, which makes the computation easier to converge.

The least-norm solution of the global *WireWarping* introduced in this section has been tested on several examples. The first test is given to the shirt model shown in Fig. 5.26a, and the result by using the least-norm solution is given in Fig. 5.26b. It is easy to find that the computation converges quickly (that is, $\frac{1}{n_{\text{var}}} \|\mathbf{r}\|^2 \rightarrow 0$ after two steps of iteration).

The least-norm solution solves a linear equation system with dimension $n_{\text{con}} \times n_{\text{con}}$ (in other words, $\mathbf{CC}^T$) in each iteration plus one step of substitution. For the numerical computation of Newton's approach proposed in [56], the update vector in each step is also determined by computing a $(3m + r) \times (3m + r)$ linear system followed by substitution. As $n_{\text{con}} = 3m + r$, the computations in each step by Wang's method [56] and the least-norm solution are the same. Therefore, the computation speed of least-norm solution is faster as it usually needs fewer steps than Newton's method does to converge. More comparisons can be found in [57].

5.6 WireWarping++: Surface Flattening with Length Control

The WireWarping method introduced above gives a good solution to flatten a 3D mesh surface into planar pieces by simulating the warping of a given 3D surface into a 2D one with the boundaries and feature curves serving as tendon. The lengths of boundaries and feature curves can be preserved strictly. To make the shape of a 2D piece similar to its corresponding 3D surface, WireWarping minimizes the variations in surface angles on feature curves during warping. However, when feature curves pass through highly non-developable 3D regions, strictly preventing their lengths induces unstable distortion in 2D (see Fig. 5.27 for the quarter-sphere example with two feature curves from [56]). Moreover, among all feature curves in practical applications, only a few curves, whose lengths are quite important for the shape control of final fabricated products, need to be strictly preserved. Such feature curves are denoted as *rigid feature curves* in the rest of this section. Except these rigid feature curves, the lengths of all other feature curves can be varied within a controlled range, and these curves are named as *elastic feature curves*. As shown in Fig. 5.27, an improved 2D shape can be obtained by allowing length variations on elastic feature curves using an extension of WireWarping—*WireWarping++* [63].

Furthermore, during the industrial tests on WireWarping, it was found that two special types of topology on the connected feature curves may fail the tests. The first one occurs in a network of feature curves with hinged feature curves. As the angle of a wire node on the end of such curves is not defined, the numerical system in WireWarping becomes crispy. The second one is that, when the 3D surface patch to be flattened has separate boundary loops, poor flattening results are generated. These problems are solved by the proposed approach through a topology processing procedure.

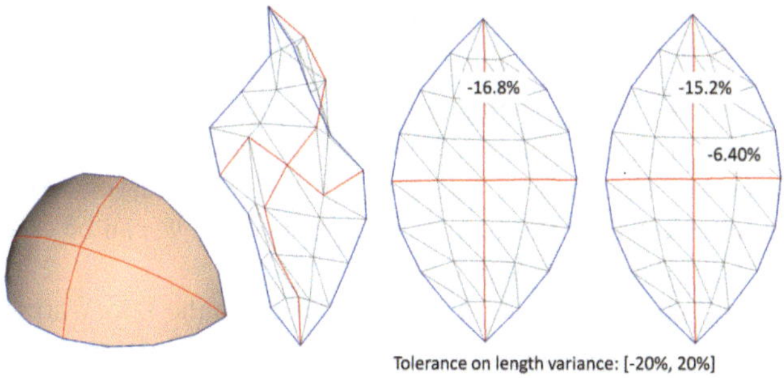

Fig. 5.27 Surface flattening on a highly non-developable surface—quarter-sphere. (*Left*) The given mesh model with two orthogonal feature curves defined in *red*. (*Middle-left*) The flattening result by WireWarping [56] (i.e., keeping the lengths of two feature curves and the boundaries invariant). (*Middle-right*) The result obtained using the WireWarping++ approach with the vertical feature curve being *elastic*. (*Right*) Another result obtained using WireWarping++ with two elastic feature curves. The sign of the length variance represents the status of shrinkage (−) or elongation (+). In all these flattening examples, the lengths of boundaries are kept unchanged (Image taken from [63] with permission ©2011 IEEE)

5.6.1 Multi-Loop Optimization Framework

In WireWarping++, the surface flattening results are computed under a multi-loop optimization framework. A shape error function is defined in the outer-loop to find an optimal length variation on the elastic feature curves. The selection of shape error function is quite important to numerical stability and convergence in the outer-loop optimization and is discussed in the next section.

From the knowledge of differential geometry, only a developable surface can retain the isometric property between any two points on it. For a given patch P which is far from developable, a length-preserved WireWarping flattening may cause large distortion on the flattened 2D pattern if the feature curves lie on those highly non-developable regions. In this case, the stretch energy can hardly be released as the feature curves are strictly constrained. Other flattening algorithms focusing on minimizing angle error (or surface stretch) can improve the shape in these cases but has no length control on feature curves and boundaries. The method introduced here optimizes the shape with controlled length variation in a defined range on the elastic feature curves while still preserving the invariant lengths on rigid feature curves.

A new variable vector $\mathbf{h}$ is introduced, where the dimension $n_{\mathbf{h}}$ is the number of elastic feature curves. $h_i \in \mathbf{h}$ represents the ratio of length variation on the ith elastic feature curve F_i^e. The length variation range for F_i^e is specified by a positive coefficient ϵ_i as $h_i \in [-\epsilon_i, \epsilon_i]$. When using l_i^0 to denote the original length of feature curves and l_i to denote the varied length, h_i can be represented as

$$h_i = \frac{l_i - l_i^0}{l_i^0} \quad (\forall i = 1, ..., n),$$

with $-\epsilon_i \leq h_i \leq \epsilon_i$. In other words,

$$l_i = (1 + h_i) l_i^0. \tag{5.99}$$

By a fixed vector $\mathbf{h}$, the new lengths, l_is, can be substituted into the least-norm solution of WireWarping to obtain a 2D pattern according to the varied lengths of elastic feature curves. Notice that, the lengths of rigid feature curves are kept invariant. As the flattening result Ψ of this approach depends on $\mathbf{h}$, Ψ can be considered as a function of $\mathbf{h}$ and is denoted as $\Psi(\mathbf{h})$.

By defining a shape error function $E(\Psi(\mathbf{h}))$ based on the WireWarping flattening $\Psi(\mathbf{h})$, how to find a good flattening now becomes a constrained optimization problem as

$$\begin{aligned} &\min_{\mathbf{h}}\{E(\Psi(\mathbf{h}))\} \\ &\text{s.t.} \; -\epsilon_i \le h_i \le \epsilon_i. \end{aligned} \tag{5.100}$$

For each inequality, $-\epsilon_i \le h_i \le \epsilon_i$, it can be rewritten into two inequalities as $h_i + \epsilon_i \ge 0$ and $-h_i + \epsilon_i \ge 0$. Therefore, there are totally $2n_{\mathbf{h}}$ inequality constraints. The active set method (see [29] for reference) is adopted to introduce them into the optimization. Briefly speaking, the inequality constraints are partitioned into an active set and an inactive set—only the constraints in the active set are added into the numerical system as equality constraints, which can be solved by the quasi-Newton method [34].

When using the quasi-Newton method to solve a constrained optimization problem, the objective function $E(\Psi(\mathbf{h}))$ should be second-order continuous with respect to the variable $\mathbf{h}$. It should also be noticed that, the selection of a shape error metric $E(\cdots)$ will affect the optimized 2D shape and the convergence of numerical computation. This problem is discussed in detail in the next section.

In summary, the surface flattening algorithm, which allows controlled length variation on elastic feature curves, is, in fact, under a multi-loop optimization framework. In the inner-loop, a WireWarping flattening in terms of length variation $\mathbf{h}$ is computed by the least-norm approach [57]. In the outer-loop, a quasi-Newton method iteratively minimizes $E(\mathbf{h})$ under the active constraints, which ensures that the optimized $\mathbf{h}$ is not beyond the tolerance of length variation. The optimization starts from an initial guess $\mathbf{h}^0$, and ends when a minimized $E(\mathbf{h})$ is achieved. In the quasi-Newton approach, an approximate Hessian matrix is updated in each iteration by using the *Broyden fletcher goldfarb shanmo* (BFGS) method. In addition, a line-search algorithm with constraints as penalty terms [29] is employed to make the outer-loop optimization more numerically stable. The pseudo-code of the optimization algorithm is shown in Algorithm 6.

Notice that the active set method cannot completely ensure that the resultant value of h_i falls in the range of $[-\epsilon_i, \epsilon_i]$. Therefore, a post-verification step is employed to see if the resultant h_i is in the range. If not, h_i must be projected back into the interval to guarantee length control.

Algorithm 6 MultiloopOptmFlattening

Initialize $\mathbf{h}^0$ of a given mesh patch;
Compute $E(\Psi(\mathbf{h}))$ and $\nabla_{\mathbf{h}} E$, and let the Hessian matrix be I;
Let $\delta_{\mathbf{h}}^0 = -\nabla_{\mathbf{h}} E(\mathbf{h}^0)$, and empty the set of active constraints Ω_{act};
for $i = 1$ to m **do**
  Find the update vector $\delta_{\mathbf{h}}^i$ of $\mathbf{h}$ and the Lagrange multiplier λ_c^i by the quasi-Newton method;
  Using the *penalty line-search* algorithm [29] to find a proper scalar $\alpha \in [0, 1]$ to let $E(\Psi(\mathbf{h}^{i-1} + \alpha\delta_{\mathbf{h}}^i)) < E(\Psi(\mathbf{h}^{i-1}))$;
  $\mathbf{h}^i = \mathbf{h}^{i-1} + \alpha\delta_{\mathbf{h}}^i$;
  if $\|\delta_{\mathbf{h}}^i\| < \tau_1$ or $\|E(\Psi(\mathbf{h}^i)) - E(\Psi(\mathbf{h}^{i-1}))\| < \tau_2$ **then**
    return;
  end if
  Update the set of active constraints by checking whether every inequality constraint is satisfied at $\mathbf{h}^i$;
  Update the approximate Hessian matrix by the *BFGS* method;
end for
Get the final 2D pattern by using the length-preserved WireWarping with the optimal $\mathbf{h}$;
return;

5.6.2 Shape Error Function

The shape error function $E(\Psi(\mathbf{h}))$ measures the distortion on flattened planar patterns generated by WireWarping with a given length variation $\mathbf{h}$ on the elastic feature curves. Several functions, such as angle variation, edge length variation, area difference, and signed area difference, are studied. However, none of them is sensitive enough to the global shape distortion on the planar pattern. In other words, the converging speed of optimization is slow.

As reported by Zhang and Wang [63], the best shape error function found in their tests is the as-rigid-as-possible (ARAP) metric defined in [26]. The function measures the distortion on all triangles as

$$E = \frac{1}{2}\sum_{t=1}^{T}\sum_{i=0}^{2}\cot(\theta_t^i)\|(\mathbf{u}_t^i - \mathbf{u}_t^{i+1}) - L_t(\mathbf{v}_t^i - \mathbf{v}_t^{i+1})\|^2, \tag{5.101}$$

where $\mathbf{v}_t^i$ and $\mathbf{u}_t^i$ are the 3D and the 2D coordinates of the ith vertex in a triangle t, and θ_t^i is the angle opposite to the edge $\mathbf{v}_t^i\mathbf{v}_t^{i+1}$ in 3D. $L_t(\cdots)$ is a rigid transformation matrix to map the triangle t onto a plane. Notice that, not only the vertices on feature curves but also other vertices are evaluated here.

Another matter of the optimization approach is how to determine the initial guess $\mathbf{h}^0$. Zhang and Wang [63] apply the ARAP mesh parameterization [26] to obtain a flattening D for a given patch P. The length variations between D and P on the elastic feature curves are then employed as the initial value $\mathbf{h}^0$.

We also need to verify the continuity of the shape error function employed in this approach to see if its second order differentiation (in terms of length variation vector **h**) around the initial guess is continuous. Although it is hard to prove that analytically, the trial tests conducted on all kinds of surface patches verify that the shape error function presented in Eq. (5.101) satisfies this requirement. The continuity analysis on several examples has been given in [63], and all examples give a smooth function on the first-order and second-order continuities. To conclude, the function in Eq. (5.101) reflects the shape error quite well and is suitable for the optimization.

5.6.3 Topology Processing

Using the WireWarping++ method to flatten mesh surfaces has certain requirements on the topology of the network of wires (i.e., the feature curves and boundaries). There are two types of topology that make the numerical system of WireWarping++ unstable: (1) the network with hinged feature curves and (2) the patch with separate boundary loops. To enable the flattening of surface patches with such a topology, the following two algorithms are developed to process the topology of feature curves network Υ.

5.6.3.1 Processing on Hinged Feature Curves

The definition of hinged wires is first given below, and then the processing method used to eliminate them is presented.

Definition 5.10 For a feature curve c_f defined in the network Υ, if the left and right regions of any portion of the curve belong to the same wire-patch, such a portion of c_f is defined as a *hinged wire*.

Vertices located on the hinged wires are called *hinged wire-nodes*. When using WireWarping++ to compute flattened surfaces, the surface angles on wire-nodes formed by adjacent edges on the wires are computed in the inner loop of optimization. However, such surface angles are not defined on the hinged wire-nodes, thus crashing the numerical solver. Notice that hinged wires are different from darts, which are parts of boundaries of the given patch P and have well-defined surface angle. Hinged wires must be processed to vanish the hinged wire-nodes.

Definition 5.11 For the network of feature curves Υ defined on a given surface patch P, if there is a vertex $\mathbf{v} \in P$ that has only one adjacent edge on Υ, the vertex $\mathbf{v} \in P$ is defined as a *tail-node*.

The elimination of hinged wires can be achieved by extending the hinged wires starting from the tail-nodes. The steps of topology processing on hinged features are detailed as follows.

1. First, all the tail nodes are detected on the network of feature curves, Υ. To eliminate the hinged wires, a surface curve path, which starts from the tail nodes

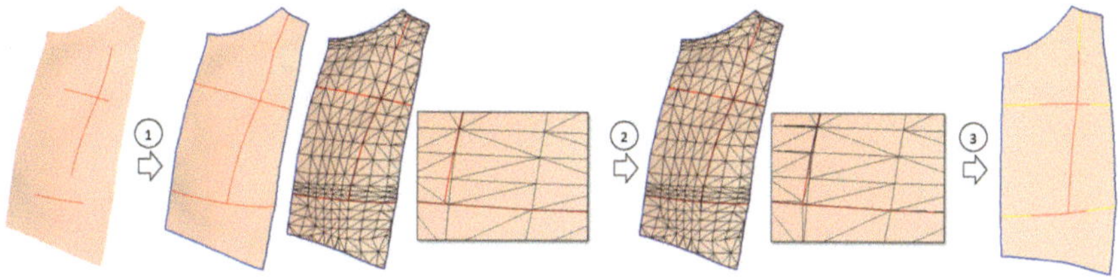

Fig. 5.28 Topology processing on hinged feature curves. Step (1) Extending the hinged curves along their endpoints' tangent directions by discrete surface geodesic curves. Step (2) Applying the CDT to convert surface curves into the edges of *triangles*. Step (3) Flattening the surface by WireWarping++ with the newly added feature curves (in *yellow*) which are 'super' elastic

and finally intersects another feature curve or boundary, needs to be found. The regions on the left and right sides of a hinged wire are then separated into two wire-patches. The surface path that extends the hinged feature curve along its tangent vector at the tail node and goes along the geodesic direction on P is a good choice. For a piecewise linear surface P, the piecewise linear geodesic curve on it along a given direction can be incrementally computed by ensuring that the curve always has the equal left and right surface angles on P. Details can be found in [51] and [19]. Figure 5.28 shows an example of such an extension.

2. Second, the topology of the original mesh surface needs to be modified according to the extended geodesic curve paths found in the previous step. A Constrained delaunay triangulation (CDT) is employed to carry out the triangulation to make the newly added surface curve into triangle edges (see Step 2 in Fig. 5.28).
3. Last, every newly added feature curve c_f is set as a super elastic feature curve with uncontrolled variant length. The range of length variation h_f is not controlled by the range $[-\epsilon_f, \epsilon_f]$ any more. Instead, $1 + h_f > 0$ is set as the corresponding constraint to ensure that the length of an elastic feature curve is positive.

After fixing all hinged wires, WireWarping++ can be applied to flatten the surface patch. The flattening result of a patch after fixing all hinged wires is shown in Fig. 5.28.

5.6.3.2 Connecting Separate Boundary Loops

WireWarping++ works well on surfaces with a disk-like topology. However, it has problems when flattening a surface with separate boundary loops.

Definition 5.12 A *boundary loop* of a surface patch P is a set of connected boundary vertices where each pair of neighboring vertices is connected by a boundary edge.

Definition 5.13 Two boundary loops are defined as *separate* if there is no feature curve linking them.

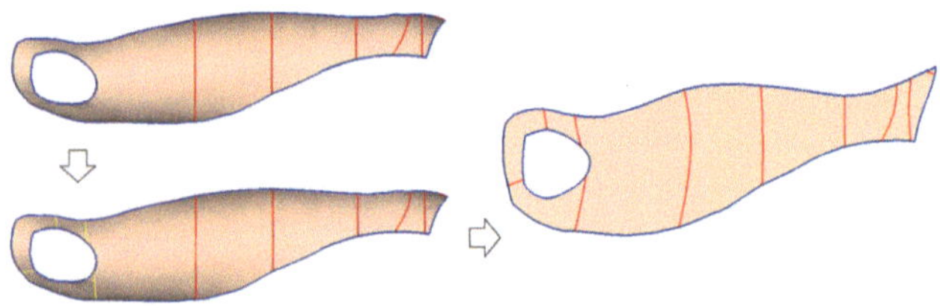

Fig. 5.29 Topology processing on separate boundary loops, where the newly added feature curves are set as 'super' elastic

As there is no feature curve (i.e., wire) linking them, the two separate boundary loops (wires) are decoupled in the numerical system of WireWarping, which leads to a degenerate result. An automatic connecting method, which is akin to the virtual cutting scheme in Sect. 5.4 and [52], is introduced below. The algorithm consists of three steps:

1. After detecting all separate boundary loops, the one with no feature curve connected, L_j, is selected. The shortest path from the vertices on L_j to the vertices on other boundary loops passing through the edges of triangles is determined by the multi-source Dijkstra's algorithm [8]. The shortest zigzag path is further smoothed by an iterative refinement procedure to approximate a geodesic curve [51].
2. If the above curve starts from $\mathbf{v}_s$ on L_j, the farthest vertex from $\mathbf{v}_s$ on L_j, $\mathbf{v}_f$, is found. After that, the shortest path from $\mathbf{v}_f$ to the boundary vertices on other boundary loops is generated by the single source Dijkstra's algorithm [8] and refined by Wang's method [51]. To get better results, two more such curves can be added starting from the vertices between $\mathbf{v}_s$ and $\mathbf{v}_f$ or those between $\mathbf{v}_f$ and $\mathbf{v}_s$. Therefore, a total of four curves are added.
3. Repeat the above two steps until all boundary loops join more than one feature curves. After that, the newly added feature curves are triangulated into edges using CDT, and are assigned as super elastic feature curves.

When applying the WireWarping++ algorithm to such processed surface patches with separate boundary loops, satisfactory flattening results can be obtained. Figure 5.29 shows an example of flattening a surface patch with separate boundary loops.

5.6.4 Case Study: Design and Manufacturing of Jeans

The proposed WireWarping++ method has been studied in a real application of user-customized design of jeans. After designing a 3D model for a user as the shape shown in the top-left corner of Fig. 5.30, the corresponding 2D patterns for fabrication need to be computed. The results of WireWarping++ were compared with those generated by the state of the art in literature (i.e., the results generated by the angle based

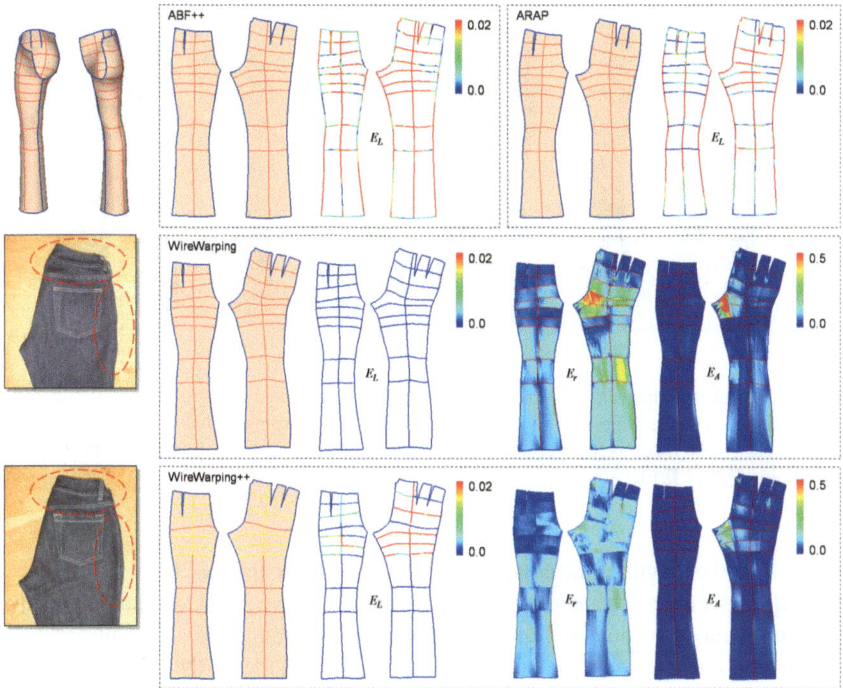

Fig. 5.30 Surface flattening of a pair of jeans. (*Top row*) The *leftmost* one is the 3D given surface with *feature lines* and *cutting lines* (boundaries) defined. The flattening results obtained from the angle-based flattening approach (ABF++) [40] is shown in the *middle*, and the result generated by the ARAP parameterization [26] is shown in the *right*. The *colors* on the curves indicate the length variation on the *feature curves* and boundaries. (*Middle row*) The surface flattening results generated by WireWarping [56, 57], where the lengths of all *feature curves* are invariant. The *color* maps show the distribution of the aspect ratio E_r (in the *left*) and the (ARAP) energy E_A on all *triangles*. (*Bottom row*) The resultant patterns generated by WireWarping++ are shown, where length variations occur only on the specified elastic *feature curves* (in *yellow*), and both the aspect ratio E_r and the (ARAP) energy E_A on the 2D patterns are reduced compared with the results obtained from WireWarping. The maximum length variation range is set to 5.0 %. The jeans fabricated from the 2D patterns generated by WireWarping and WireWarping++ are also shown in the *middle* and *bottom* rows respectively (Image taken from [63] with permission © 2011 IEEE)

flattening (ABF++) [40], the ARAP [26], and WireWarping [56, 57]). Notice that ABF++ does not preserve the scale of a flattening, so the flattening result is scaled by the longest edge length in 3D.

Several metrics are adopted to verify the distortion of flattening results.

- *Edge-length variation.* The length variation of each edge e on a feature curve is measured by

$$E_L = \frac{\|l_e^0 - l_e\|}{l_e^0}, \tag{5.102}$$

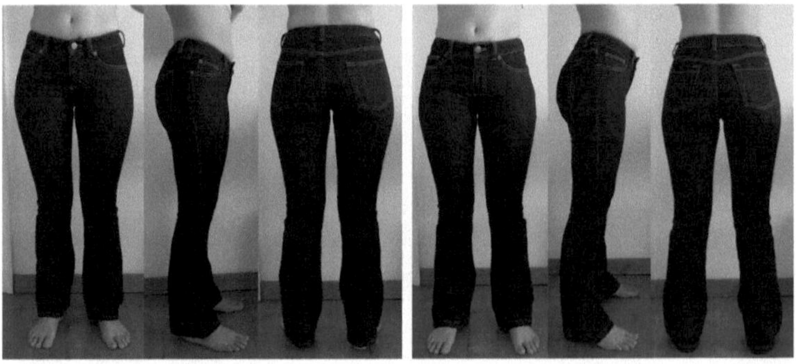

Fig. 5.31 Photos of the jeans fabricated from the patterns generated by WireWarping versus those generated by WireWarping++. The one generated by WireWarping++ fits the back waist band and the back yoke much better—commented by a fashion specialist (Image taken from [63] with permission ©2011 IEEE)

where l_e^0 is the length of the edge e in 3D, and l_e is its length in 2D.
Global aspect ratio. The aspect ratio E_r was proposed by Azariadis and Sapidis [3] to measure the distortion on the results of surface flattening. The ideal value of E_r is one, which is shown only on isometric mappings. In the tests, the color map of the aspect ratio is displayed on every face.

- *ARAP energy*. The ARAP energy defined by Liu et al. [26] below is also computed on every triangular face.

$$E_A = \frac{1}{2}\sum_{i=0}^{2} \cot(\theta_t^i)\|(\mathbf{u}_t^i - \mathbf{u}_t^{i+1}) - L_t(\mathbf{v}_t^i - \mathbf{v}_t^{i+1})\|^2 \tag{5.103}$$

The value of this function measures the stretch from 3D surface to 2D pattern. A color map in terms of E_A is also used to illustrate the distortion per face in the experimental results.

From Fig. 5.30, it is obvious that ABF++ and ARAP cannot preserve the lengths of feature curves and boundaries. This is a significant disadvantage when applying them to the sheet manufacturing industries. Although the WireWarping method can preserve the lengths of all feature curves and boundaries, it gives large distortion on some triangles (see the color maps of E_r and E_A in the second row of Fig. 5.30). The WireWarping++ approach proposed in this section gives superior results than WireWarping in the aspect ratio E_r and the ARAP stretch metric E_A. The shape of the jeans pants fabricated from the patterns generated by WireWarping++ is also better (see Figs. 5.30 and 5.31).

References

1. Aono, M., Breen, D.E., Wozny, M.J.: Modeling methods for the design of 3d broadcloth composite parts. Comput. Aided Des. **33**, 989–1007 (2001)
2. Azariadis, P., Aspragathos, N.: Design of plane developments of doubly curved surfaces. Comput. Aided Des. **29**, 675–685 (1997)
3. Azariadis, P., Sapidis, N.: Planar development of free-form surfaces: quality evaluation and visual inspection. Computing **72**(1–2), 13–27 (2004)
4. Botsch, M., Kobbelt, L.: A remeshing approach to multiresolution modeling. In: SGP '04: Proceedings of the 2004 Eurographics/ACM SIGGRAPH Symposium on Geometry Processing, pp. 185–192 (2004)
5. do Carmo, M.P.: Differential Geometry of Curves and Surfaces. Prentice-Hall, Englewood Cliffs (1976)
6. Chapra, S.C., Canale, R.P.: Numerical Methods for Engineers: With Software and Programming Applications. McGraw-Hill Higher Education, New York (2003)
7. Cohen-Steiner, D., Alliez, P., Desbrun, M.: Variational shape approximation. ACM Trans. Graph. **23**(3), 905–914 (2004)
8. Cormen, T.H., Leiserson, C.E., Rivest, R.L., Stein, C.: Introduction to Algorithms, 2nd edn. MIT Press, Cambridge (2001)
9. Decaudin, P., Julius, D., Wither, J., Boissieux, L., Sheffer, A., Cani, M.P.: Forum virtual garments: a fully geometric approach for clothing design. Comput. Graph **25**, 625–634 (2006)
10. Desbrun, M., Meyer, M., Alliez, P.: Intrinsic parameterizations of surface meshes. Comput. Graph. Forum **21**, 209–218 (2002)
11. Desbrun, M., Meyer, M., Schröder, P., Barr, A.H.: Implicit fairing of irregular meshes using diffusion and curvature flow. In: Proceedings of ACM SIGGRAPH 99, pp. 317–324 (1999)
12. Duda, R., Hart, P., Stork, D.: Pattern Classification, 2nd edn. Wiley, New York (2001)
13. Floater, M.S., Hormann, K.: Surface parameterization: a tutorial and survey. In: Dodgson, N.A., Floater, M.S., Sabin, M.A. (eds.) Advances in Multiresolution for Geometric Modelling, p. 186. Springer, Heidelberg (2005)
14. Julius, D., Kraevoy, V., Sheffer, A.: D-charts: quasi-developable mesh segmentation. Comput. Graph. Forum **24**, 581–590 (2005)
15. Karni, Z., Gotsman, C.: Spectral compression of mesh geometry. In: Proceedings of ACM SIGGRAPH 2000, pp. 279–286 (2000)
16. Karni, Z., Gotsman, C., Gortler, S.J.: Free-boundary linear parameterization of 3D meshes in the presence of constraints. In: Proceedings of Shape Modeling and Applications 2005, pp. 268–277 (2005)
17. Katz, S., Tal, A.: Hierarchical mesh decomposition using fuzzy clustering and cuts. ACM Trans. Graph. **22**(3), 954–961 (2003)
18. Kilian, M., Flöry, S., Chen, Z., Mitra, N., Sheffer, A., Pottmann, H.: Curved folding. ACM Trans. Graph. **27**, 75:1–75:9 (2008)
19. Kimmel, R., Sethian, J.: Computing geodesic paths on manifolds. Proc. Natl. Acad. Sci. USA **95**, 8431–8435 (1998)
20. Kobbelt, L.: $\sqrt{3}$-subdivision. In: Proceedings of SIGGRAPH 2000, pp. 103–112 (2000)
21. Kobbelt, L., Schröder, P.: A multiresolution framework for variational subdivision. ACM Trans. Graph. **17**, 209–237 (1998)
22. Larsen, E., Gottschalk, S., Lin, M.C., Manocha, D.: Fast proximity queries with swept sphere volumes. In: Proceedings of International Conference on Robotics and Automation, pp. 3719–3726 (2000)
23. Lévy, B.: OpenNL. http://www.loria.fr/levy/software/ (2005)
24. Levy, B., Petitjean, S., Ray, N., Maillot, J.: Least squares conformal maps for automatic texture atlas generation. In: Proceedings of SIGGRAPH '02, pp. 362–371 (2002)
25. Li, S., Demmel, J., Gilbert, J.: SuperLU. http://crd.lbl.gov/xiaoye/SuperLU/ (2006)
26. Liu, L., Zhang, L., Xu, Y., Gotsman, C., Gortler, S.J.: A local/global approach to mesh parameterization. Comput. Graph. Forum **27**, 1495–1504 (2008)

27. Liu, Y., Pottmann, H., Wallner, J., Yang, Y.L., Wang, W.: Geometric modeling with conical meshes and developable surfaces. ACM Trans. Graph. **25**, 681–689 (2006)
28. McCartney, J., Hinds, B.K., Chong, K.W.: Pattern flattening for orthotropic materials. Comput. Aided Des. **37**, 631–644 (2005)
29. Madsen, K., Nielsen, H.B., Tingleff, O.: Optimization with Constraints. Course Notes, Technical University of Denmark (2004)
30. Metaxas, D.N.: Physics-based deformable models. Kluwer Academic, Dordrecht (1997)
31. Meyer, M., Desbrun, M., Schröder, P., Barr, A.: Discrete differential-geometry operators for triangulated 2-manifolds. In: Hege, H.-C., Polthier, K. (eds.) Visualization and Mathematics, vol. III, pp. 35–58. Springer, Heidelberg (2003)
32. Mortenson, M.E.: Geometric Modeling. Wiley, New York (1997)
33. Nealen, A., Igarashi, T., Sorkine, O., Alexa, M.: Fibermesh: designing freeform surfaces with 3D curves. ACM Trans. Graph. **26**(3), 41 (2007)
34. Nocedal, J., Wright, S.J.: Numerical Optimization. Springer, New York (1999)
35. Press, W.H., Flannery, B.P., Teukolsky, S.A., Vetterling, W.T.: Numerical Recipes in C: The Art of Scientific Computing. University Press, Cambridge (1995)
36. Provot, X.: Deformation constraints in a mass-spring model to describe rigid cloth behavior. In: Proceedings of Graphics Interface 95, pp. 147–154 (1995)
37. Sander, P.V., Snyder, J., Gortler, S.J., Hoppe, H.: Texture mapping progressive meshes. In: Proceedings of the 28th Annual Conference on Computer Graphics and Interactive Techniques, SIGGRAPH '01, pp. 409–416 (2001)
38. Shatz, I., Tal, A., Leifman, G.: Paper craft models from meshes. Vis. Comput., Special Issue of Pacific Graphics 2006 **22**(9–11), 825–834 (2006)
39. Sheffer, A.: Spanning tree seams for reducing parameterization distortion of triangulated surfaces. In: Proceedings of the Shape Modeling International, 2002 (SMI'02), SMI'02, pp. 61–66 (2002)
40. Sheffer, A., Levy, B., Mogilnitsky, M., Bogomyakov, A.: ABF++: fast and robust angle based flattening. ACM Trans. Graph. **24**, 311–330 (2005)
41. Sheffer, A., de Sturler, E.: Parameterization of faceted surfaces for meshing using angle based flattening. Eng. Comput. **17**(3), 326–337 (2001)
42. Sorkine, O.: Differential representations for mesh processing. Comput. Graph. Forum **25**, 789–807 (2006)
43. Sorkine, O., Cohen-Or, D.: Least-squares meshes. In: Proceedings of Shape Modeling International 2004, pp. 191–199 (2004)
44. Sorkine, O., Cohen-Or, D., Irony, D., Toledo, S.: Geometry-aware bases for shape approximation. IEEE Trans. Vis. Comput. Graph. **11**, 171–180 (2005)
45. Sorkine, O., Cohen-Or, D., Lipman, Y., Alexa, M., Rössl, C., Seidel, H.P.: Laplacian surface editing. In: Eurographics/ACM SIGGRAPH Symposium on Geometry Processing 2004, pp. 179–188 (2004)
46. Tang, K., Wang, C.C.L.: Modeling developable folds on a strip. ASME J. Comput. Inf. Sci. Eng. **5**, 35–47 (2005)
47. Taubin, G.: A signal processing approach to fair surface design. In: Proceedings of ACM SIGGRAPH 95, pp. 351–358 (1995)
48. Taubin, G.: Linear anisotropic mesh filtering. In: Technical Report of IBM Research, TR-RC2213 (2001)
49. Vandenberghe, L.: Applied Numerical Computing (course-notes). http://www.ee.ucla.edu/vandenbe/ (2007)
50. Volino, P., Magnenat-Thalmann, N.: Accurate collision response on polygon mehses. In: Proceedings of Computer Animation 2000, pp. 137–144 (2000)
51. Wang, C.C.L.: Cyber tape: an interactive measurement tool on polyhedral surface. Comput. Graph. **28**(5), 731–745 (2004)
52. Wang, C.C.L.: Computing length-preserved free boundary for quasi-developable mesh segmentation. IEEE Trans. Vis. Comp. Graph. **14**(1), 25–36 (2008)

53. Wang, C.C.L.: Flattenable mesh surface fitting on boundary curves. ASME Trans. J. Comput. Inf. Sci. Eng. **8**, 21006 (2008)
54. Wang, C.C.L.: A least-norm approach to flattenable mesh surface processing. In: Proceedings of IEEE International Conference on Shape Modeling and Applications 2008, pp. 131–138 (2008)
55. Wang, C.C.L.: Towards flattenable mesh surfaces. Comput. Aided Des. **40**, 109–122 (2008)
56. Wang, C.C.L.: WireWarping: a fast surface flattening approach with length-preserved feature curves. Comput. Aided Des. **40**(3), 381–395 (2008)
57. Wang, C.C.L.: A note on least-norm solution of global WireWarping. Comput. Aided Des. **41**(9), 695–698 (2009)
58. Wang, C.C.L., Smith, S.S.F., Yuen, M.M.F.: Surface flattening based on energy model. Comput. Aided Des. **34**, 823–833 (2002)
59. Wang, C.C.L., Tang, K.: Achieving developability of a polygonal surface by minimum deformation: a study of global and local optimization approaches. Vis. Comput. **20**, 521–539 (2004)
60. Wang, C.C.L., Tang, K., Yeung, B.M.L.: Freeform surface flattening based on fitting a woven mesh model. Comput. Aided Des. **37**, 799–814 (2005)
61. Yamauchi, H., Gumhold, S., Zayer, R., Seidel, H.P.: Mesh segmentation driven by gaussian curvature. Vis. Comput., Special Issue of Pacific Graphics 2005 **21**(8–10), 659–668 (2005).
62. Zayer, R., Lévy, B., Seidel, H.P.: Linear angle based parameterization. In: Proceedings of Eurographics Symposium on Geometry Processing, pp. 135–141 (2007)
63. Zhang, Y., Wang, C.C.L.: WireWarping++: robust and flexible surface flattening with length control. IEEE Trans. Autom. Sci. Eng. **8**(1), 205–215 (2011)
64. Zorin, D., Schröder, P., Sweldens, W.: Interpolating subdivision for meshes with arbitrary topology. In: Proceedings of SIGGRAPH 96, pp. 189–192 (1996)

Chapter 6
Compression Products Fabricated by Extensible Materials

6.1 Elastic Medical Braces

Reports have shown that more than seven million physician office visits per year in the United States are related to problems with wrist joints suffering from repetitive strain injury [1]. If other types of joint injury (say, sports-related ones) were included, the number would become much bigger. The dramatic increase in the use of computers and various kinds of automatic equipment is a major contributor to this end. Elastic braces, whose purpose is to restrict the motion of an injured joint so that it will eventually heal by itself, are the most commonly used assistive medical devices for joint injuries. A brace will exert different biomechanical effects on different individuals since they have different joint shapes. Therefore, for better and faster treatment of patients, more and more physicians now request custom-made braces specifically designed for individual patients rather than choosing from off-the-shelf or mass-produced ones.

A brace is made of a piece of elastic fabric with certain material characteristics. To restrict the motion of a joint, the brace should be in a *positive tensile* state—it must be stretched so that normal pressure can be generated upon the joint. The design task of a brace is to find a planar pattern/geometry of the brace that is at the *rest or relaxed state* meaning that neither stretch nor compression occurs at this time, so that when worn, the brace generates a desired normal pressure distribution. In most cases (though not every), the planar pattern is sewn at two matching seams so as to form a relaxed closed cylindrical shape.

Currently, the flattening procedure is determined manually through a tedious and inefficient trial-and-error process: markers are placed on the sample points in an initial guess pattern; the corresponding brace is then worn by the patient or put on a cast model of the joint; normal pressure is measured at some key sample points, and the boundaries of the pattern (the matching seams) are adjusted in an ad hoc manner by looking at the movements of the markers and the errors of the normal pressure. The research presented by Wang and Tang [16] was motivated by this and a computer system was developed to automate the design process of customized elastic

C. C. L. Wang, *Geometric Modeling and Reasoning of Human-Centered Freeform Products*, DOI: 10.1007/978-1-4471-4360-4_6, © Springer-Verlag London 2013

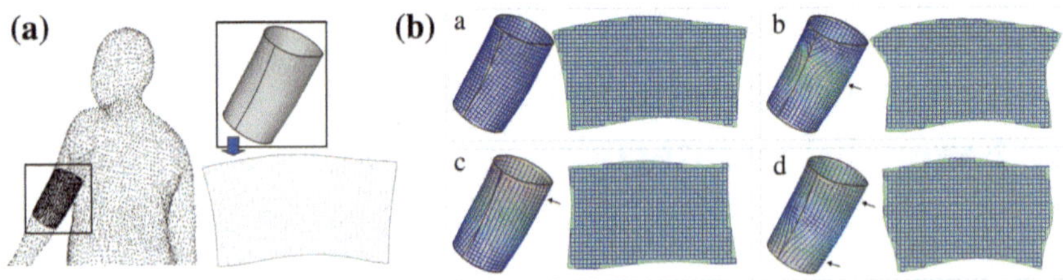

Fig. 6.1 Geometric design of a customized assistive medical brace: **a** the 3D mesh surface of the brace is acquired from a scanned human model, and the initial planar pattern is obtained by the parameterization algorithm of Desbrun et al. [2], and **b** different planar patterns are computed by fitting a woven model on the 3D brace surface while satisfying different user-specified normal pressure requirements—different colors on the 3D brace represent different strain levels, where bluc dcnotcs zero red denotes the highest value (0.25). The places pointed by arrows are with large normal pressures specified (Image taken from [16] with permission)

brace. Specifically, given the material characteristics of the brace, the geometry of the local body shape near the joint where the brace is to be worn, and the desired normal pressure at some designated points on the joint, the computer program outputs the correct corresponding flattened planar pattern of the brace. The approach is introduced in this section below.

A woven-model-based fitting and flattening algorithm for 3D freeform surfaces is used in this approach because, in most cases, braces are made of woven-like materials. Figure 6.1 shows an example that illustrates the flattening algorithm. After obtaining the scanned 3D triangular mesh model of the surface on which the brace is to be put, an initial 2D pattern of the brace can be determined by the intrinsic mesh parameterization algorithm [2] (see Fig. 6.1a). Based on this initial pattern, a woven model is then fitted onto the 3D mesh surface of the brace (shown in the top left of Fig. 6.1b). The next step, which is the main part of the algorithm, is to redistribute the woven nodes onto the 3D mesh surface (including possible insertion and removal of nodes), so that required normal pressures are given at specified points. To accomplish this, the correlation between the normal pressure and the strains of the woven model on the surface is first established, and then the node-distribution problem is converted into a strain-energy minimization problem and solved through a diffusion process. Figure 6.1b shows several resultant patterns of a given 3D brace shape based on different user-assigned normal pressures.

6.1.1 Woven Model for Elastic Brace

The geometric design of a user-customized elastic medical brace is, in fact, based on the simulation of stretching a woven model onto a freeform polygonal mesh surface M which represents the 3D shape of a joint. Therefore, the woven model is first defined below.

A brace can be regarded as a ply of woven fabric composed of horizontal and vertical threads interwoven in a specific fashion (for example, the one shown in

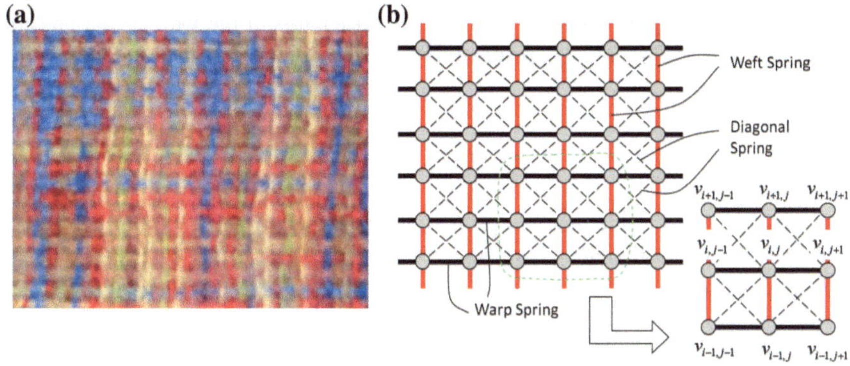

Fig. 6.2 Woven model: **a** piece of real woven fabric; **b** the spring mesh representation for a piece of woven fabric (Image taken from [16] with permission)

Fig. 6.2a). From the study of materials [4], the following definitions and assumptions of the ply are adopted for generating an elastic brace.

Assumption 6.1 All the weft threads are fabricated with the same type of materials, and hence have a constant tensile stiffness coefficient. The same is true for all the warp threads.

Assumption 6.2 A ply generally has a strong tensile-strain resistance in the thread direction and a much weaker shear-strain resistance.

Assumption 6.3 No slippage occurs at the crossings of weft (vertical) and warp (horizontal) threads.

Assumption 6.4 To simplify the physical model, frictions—both the friction between threads and the friction between the woven fabric and the surface wearing the brace—are ignored.

Based on the above assumptions, the woven fabric is modeled by a spring mesh Γ. An example spring mesh model is shown in Fig. 6.2b. There are three components in this model, that is, weft (vertical) springs, warp (horizontal) springs, and diagonal springs. For real woven fabrics, there is no diagonal thread in general. The reason for adding diagonal springs is to simulate the shear deformation resistance. Each of these three types of springs has its own initial length at which the spring attains zero energy. A *woven node* is an intersection between springs whose position determines the deformation of the springs connected to it. Each node is indexed by $v_{i,j}$, where i, j are integers representing the indices of row and column respectively. For a mesh node $v_{i,j}$, its valence is the number of springs connecting to it. If the valence is eight, $v_{i,j}$ is an *internal node*; otherwise, $v_{i,j}$ is called a *boundary node*. For a mesh node $v_{i,j}$, if there is another mesh node $v_{i+a,j+b}$ satisfying: (1) $a, b \in \{-1, 0, 1\}$ and (2) $|a|+|b| > 0$, $v_{i+a,j+b}$ is called a neighboring node of $v_{i,j}$. For a woven mesh in 2D, all the weft springs are aligned in one direction and all the warp springs are aligned in another direction. In the formulation below, their directions are orthogonal to each other although they do not have to. If the initial lengths of the weft spring and the warp spring are l_{weft} and l_{warp} respectively, the initial length of the diagonal spring is given by

$$l_{\text{diag}} = \sqrt{l_{\text{weft}}^2 + l_{\text{warp}}^2}.$$

Remark 6.1 Anisotropic material properties can be achieved by assigning different spring stiffness coefficients, k_{weft}, k_{warp} and k_{diag}, to the weft, the warp, and the diagonal springs.

Notice that by Assumption 6.1, all weft springs should have the same stiffness coefficient, and so do the warp and the diagonal springs. In addition, by Assumption 6.2, the remark below can be obtained.

Remark 6.2 In general, $k_{\text{diag}} \ll k_{\text{weft}}$ and $k_{\text{diag}} \ll k_{\text{warp}}$.

Since no slippage occurs at a woven node (by Assumption 6.3) and frictions are neglected (by Assumption 6.4), the forces generated on the two warp springs linked to a particular woven node should be equal to each other—this is also true for the weft springs. Based on Remark 6.2, the forces contributed by diagonal springs are extremely small, so they can be ignored in the simulation model when formulating the correlation between normal pressures and tensile strains. However, the function of diagonal springs cannot be ignored when releasing elastic energy on the woven model, where they prevent the woven mesh from being overlapped. Thus, the following remark can be highlighted.

Remark 6.3 When in equilibrium, the strain on a single weft or warp thread is a constant.

A planar woven mesh Γ can be easily fitted onto a 3D freeform surface M if a planar parameterization Ω of M has been determined. For a woven node $v_{i,j} \in \Gamma$, suppose it falls in a triangle $T_k \in M$ on Ω, the barycentric coordinates of $v_{i,j}$ in T_k on Ω can be computed. By applying the same barycentric coordinates to the same triangle T_k in 3D, $v_{i,j}$ is mapped onto the 3D freeform surface M. With this mapping method, every woven node can be located on the freeform surface precisely. However, the directions and lengths of different kinds of springs may not be preserved as compared with their 2D counterparts. This leads to strains—so normal pressure is generated on M. The objective of the approach introduced below is to find a correct Ω that satisfies the requirements of normal pressures at specified points.

6.1.2 Methodology

The main methodology for computing the planar pattern of a brace that exerts prescribed normal pressures at certain specified points is presented below.

6.1.2.1 Normal Pressure and Tensile Strain

The purpose of an assistive medical brace is to exert normal pressure on a joint. This is jointly accomplished by the tensile strain (stretch) of the brace and the curvature of the joint surface.[1] To a physician, only the normal pressure is meaningful, which though must be converted to tensile strain for the computation. Motivated by the theories of solid mechanics [5], Wang and Tang [16] stipulated that it can be reasonably modeled as

$$P_n = f\left(\int_0^{2\pi} g(\kappa(\theta))\sigma(\theta)\mathrm{d}\theta\right)$$

where both $f(\cdots)$ and $g(\cdots)$ are some positive and monotonic scalar functions, $\kappa(\theta)$ represents the normal curvature in the direction of θ on the tangent plane at the surface point, and $\sigma(\theta)$ denotes the normal stress in θ. Without loss of generality, the value of the normal pressure at a woven node v on M can be computed by the formula

$$P_v = s\int_0^{2\pi} \kappa_n(\theta)\sigma(\theta)\mathrm{d}\theta$$

where s is a coefficient related to the thickness and the material of fabrics of the brace that can be determined through some material tests. Without material tests, s is set to one in the numerical tests conducted by Wang and Tang [16]. The equation is then further simplified to

$$P_v \approx s\int_0^{2\pi} \kappa_H\sigma(\theta)\mathrm{d}\theta \tag{6.1}$$

by replacing $\kappa_n(\theta)$ with the mean curvature κ_H at the surface point on M. The mean curvature at a point on the polygonal mesh surface can be computed by the method of Meyer et al. [9]. Wang and Tang [16] adopt the following formula to approximate the function of $\sigma(\theta)$:

$$\sigma(\theta) = a\cos\theta + b\sin\theta + c\cos^2 + \mathrm{d}\sin^2 .$$

For a woven node $v_{i,j}$, the stress function is assumed to interpolate the four stresses shown on its adjacent warp and weft springs (see Fig. 6.3). That is,

$$\sigma(0) = \sigma_1, \sigma\left(\frac{\pi}{2}\right) = \sigma_2, \sigma(\pi) = \sigma_3, \sigma\left(\frac{3\pi}{2}\right) = \sigma_4.$$

---

[1] Notice that no normal pressure can be generated upon a flat surface no matter how hard the brace is stretched.

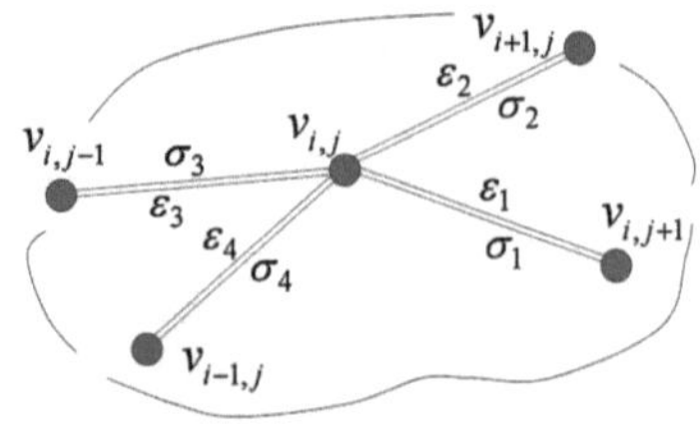

Fig. 6.3 Stresses and strains on woven springs around a woven node $v_{i,j}$ (Image taken from [16] with permission)

This interpolation simplification is valid if the density of the woven grid is high, which is assumed to be true in our case. By Remark 6.2, here simply neglect the diagonal springs, since their stiffness is generally much smaller than that of the warp and weft springs. Therefore, one can have

$$a = \frac{1}{2}(\sigma_1 - \sigma_3), b = \frac{1}{2}(\sigma_2 - \sigma_4), c = \frac{1}{2}(\sigma_1 + \sigma_3), d = \frac{1}{2}(\sigma_2 + \sigma_4)$$

which yields

$$P_v \approx s \int_0^{2\pi} \kappa_H \sigma(\theta) \mathrm{d}\theta = \frac{s\pi\kappa_H}{2}(\sigma_1 + \sigma_2 + \sigma_3 + \sigma_4).$$

Since

$$\sigma_1 = k_{\text{warp}}\varepsilon_1, \sigma_2 = k_{\text{weft}}\varepsilon_2, \sigma_3 = k_{\text{warp}}\varepsilon_3, \sigma_4 = k_{\text{weft}}\varepsilon_4,$$

the normal surface pressure at v can be expressed as

$$P_v \approx \frac{s\pi\kappa_H}{2}(k_{\text{warp}}(\varepsilon_1 + \varepsilon_3) + k_{\text{weft}}(\varepsilon_2 + \varepsilon_4)) \tag{6.2}$$

with

$$\varepsilon_1 = \frac{\|\mathbf{p}_{i,j}\mathbf{p}_{i,j+1}\| - L_{\text{warp}}}{L_{\text{warp}}}, \varepsilon_2 = \frac{\|\mathbf{p}_{i,j}\mathbf{p}_{i+1,j}\| - L_{\text{weft}}}{L_{\text{weft}}},$$
$$\varepsilon_3 = \frac{\|\mathbf{p}_{i,j}\mathbf{p}_{i,j-1}\| - L_{\text{warp}}}{L_{\text{warp}}}, \varepsilon_4 = \frac{\|\mathbf{p}_{i,j}\mathbf{p}_{i-1,j}\| - L_{\text{weft}}}{L_{\text{weft}}},$$

where $\mathbf{p}_{i,j}$ denotes the 3D position of a woven node $v_{i,j}$ on the mesh surface M. As mentioned above, L_{warp} and L_{weft} are the initial (relaxed) lengths of the warp and weft springs, respectively.

6.1.2.2 Strains on Threads

The next step is to further relate the normal pressures to the tensile strains on woven springs since they directly relate to the positions of nodes on the surface. As the woven fitting result must be in an equilibrium state, Remark 6.3 indicates that $\varepsilon_1 = \varepsilon_3 = \varepsilon_{\text{warp}}$ and $\varepsilon_2 = \varepsilon_4 = \varepsilon_{\text{weft}}$. Thus, Eq. (6.2) becomes

$$P_v \approx s\pi\kappa_H(k_{\text{warp}}\varepsilon_{\text{warp}} + k_{\text{weft}}\varepsilon_{\text{weft}}). \tag{6.3}$$

For a given P_v, if $\varepsilon_{\text{weft}}$ is known, the above equation gives the required tensile strain on the warp springs at v. From preliminary physical experiments, Wang and Tang [16] found that the weft strains on medical assistive braces are usually very small. Limited by the geometry of various medical assistive braces, the strains on weft threads are mainly generated by the friction between a brace and the human skin. Since the friction is neglected (due to Assumption 6.4), the following remark can be obtained.

Remark 6.4 When in equilibrium, the strains on all weft springs are assumed to be a near-zero constant, $\varepsilon_{\text{weft}} \approx 0$.

For any row of woven warp springs on Γ, if there are totally m constraints (normal pressures) assigned to this thread, the desired tensile strain $\varepsilon_{\text{warp}}$ can be determined by the least-square fitting with the objective function

$$J_{\varepsilon_{\text{warp}}} = \sum_{v=1}^{m}(s\pi\kappa_{H_v}(k_{\text{warp}}\varepsilon_{\text{warp}} + k_{\text{weft}}\varepsilon_{\text{weft}}) - P_v^0)^2.$$

Letting $\partial J_{\varepsilon_{\text{warp}}}/\partial\varepsilon_{\text{warp}} \equiv 0$ gives

$$\varepsilon_{\text{warp}} = \sum_{v=1}^{m}(P_v^0\kappa_{H_v} - s\pi\kappa_{H_v}^2 k_{\text{weft}}\varepsilon_{\text{weft}}) / \sum_{v=1}^{m}(s\pi\kappa_{H_v}^2 k_{\text{warp}}) \tag{6.4}$$

where κ_{H_v} is the mean curvature of the surface M at the position of the woven node v, and P_v^0 is the user-specified normal pressure.

There are two critical problems here. First, Eq. (6.4) works only when the positions of all the nodes of the warp thread on M are known. Second, if a warp thread does not pass through any point on M with a prescribed normal pressure, its tensile strain $\varepsilon_{\text{warp}}$ still needs to be calculated. For the first problem, one can perform a strain-energy minimization to place all the warp threads on M. The second problem can be solved by introducing a smooth curve $\varepsilon(t)$, called the warp strain distribution function, to interpolate those tensile warp strains determined by Eq. (6.4), where t is the parameter in the range [0, 1] corresponding to the row indices of the threads. The warp strains on those threads without normal pressure constraints are given by $\varepsilon(t)$. Besides meeting the prescribed normal pressure requirement, the function

$\varepsilon(t)$ should also satisfy certain endpoint constraints (for example, $\varepsilon'(0) = 0$ and $\varepsilon'(1) = 0$).

6.1.2.3 Elastic Energy due to Strains

Based on Eqs. (6.3) and (6.4), warp threads must be placed properly on M so that the prescribed normal pressure can be achieved. This placement procedure is formulated under the framework of elastic energy minimization. For a given woven model Γ, the embedded elastic energy is formulated as

$$J_E = \sum_{e \in \Gamma} \frac{1}{2} k_e (\|\mathbf{p}_\alpha \mathbf{p}_\beta\| - L_e)^2 \tag{6.5}$$

where $\mathbf{p}_\alpha$ and $\mathbf{p}_\beta$ are the 3D positions of the woven nodes v_α and v_β, respectively, for a spring $e \in \Gamma$. The constant k_e is one of the spring constants, k_{warp}, k_{weft}, or k_{diag}, depending on the type of spring e, and L_e is the rest length of e. The value of J_E depends on the positions of woven nodes on M. When randomly moving a woven node on the given surface M, J_E will be changed. When $J_E = 0$, it means that every spring $e \in \Gamma$ now satisfies its rest length L_e. However, due to some constraints (for example, the given geometry shape), zero J_E usually cannot be achieved. Therefore, a J_E with a minimal value is desired. As aforementioned, the initial 3D fitting of the woven model Γ on M can be obtained by any mesh parameterization (such as [2]) of the given 3D mesh surface M. In general, the lengths of springs on M are not the same as their rest lengths in 2D (that is, L_{warp}, L_{weft}, or L_{diag}). As a result, J_E can be minimized by moving the woven nodes on M through a diffusion process (details are presented in Sect. 6.1.3). The upper-left figure in Fig. 6.1b shows an example of stretch-free energy minimization. However, it is imperative to notice that, when $L_e = L_{\text{warp}}$, L_{weft} or L_{diag}, minimizing the elastic energy J_E will not yield a woven fitting preserving the desired strains as given in Eq. (6.4). Therefore, the energy function needs to be modified. For a spring e with a prescribed tensile strain ε_e, the desired length of e on M is changed to

$$L_e = (1 + \varepsilon_e) L_{\text{warp}} \tag{6.6}$$

for a warp spring, and

$$L_e = (1 + \varepsilon_e) L_{\text{weft}} \tag{6.7}$$

for a weft spring. For a diagonal spring linking two warp springs and two weft springs, the expected 3D length of the diagonal spring is computed by

$$L_e = ((1 + \bar{\varepsilon}_{\text{warp}})^2 L_{\text{warp}}^2 + (1 + \bar{\varepsilon}_{\text{weft}})^2 L_{\text{weft}}^2)^{1/2} \tag{6.8}$$

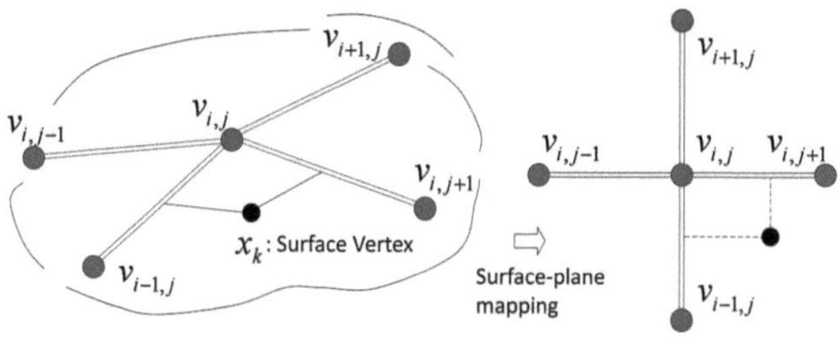

Fig. 6.4 Interpolation scheme for the surface-plane mapping (Image taken from [16] with permission)

where $\bar{\varepsilon}_{\text{warp}}$ is the average of the strains on the two warp springs, and $\bar{\varepsilon}_{\text{weft}}$ is the average of the strains on the two weft springs. Substituting Eqs. (6.6)–(6.8) into (6.5), the minimization of J_E will give rise to the distribution of the woven nodes on M satisfying the spring strains that will attain the prescribed normal pressure.

6.1.2.4 Surface-Plane Mapping

The result of minimizing J_E in Eq. (6.5) is a mapping between the nodes on surface M and their counterparts in the plane. This mapping is discrete since only nodes are mapped. Therefore, it needs to be interpolated so as to become continuous—every point on M has its counterpart defined in the plane.

An interpolation scheme akin to the vertex mapping method proposed by Wang et al. [18] is employed here. Briefly, for any vertex $\mathbf{x}_k \in M$, its closest woven node $v_{i,j}$ (with position $\mathbf{p}_{i,j}$) on Γ can be determined by a local search if every polygonal face of M contains pointers to the woven nodes lying on the face. After that, a unit weft vector $\mathbf{t}_{\text{weft}}$ is formed by either $\mathbf{p}_{i,j}\mathbf{p}_{i+1,j}$ or $\mathbf{p}_{i,j}\mathbf{p}_{i-1,j}$—in other words, $\mathbf{p}_{i,j}\mathbf{p}_{i+1,j}$ is chosen if the projection of $\mathbf{p}_{i,j}\mathbf{x}_k$ on $\mathbf{p}_{i,j}\mathbf{p}_{i+1,j}$ is positive; otherwise, $\mathbf{p}_{i,j}\mathbf{p}_{i-1,j}$ is used. In a similar manner, a unit warp vector is formed by either $\mathbf{p}_{i,j}\mathbf{p}_{i,j+1}$ or $\mathbf{p}_{i,j}\mathbf{p}_{i,j-1}$. The projections of $\mathbf{p}_{i,j}\mathbf{x}_k$ on $\mathbf{t}_{\text{warp}}$ and $\mathbf{t}_{\text{weft}}$ are $< \mathbf{p}_{i,j}\mathbf{x}_k, \mathbf{t}_{\text{warp}} >$ and $< \mathbf{p}_{i,j}\mathbf{x}_k, \mathbf{t}_{\text{weft}} >$, respectively. Therefore, by keeping the same ratio of their lengths in 3D and 2D, the planar coordinates of $\mathbf{x}_k$ can be determined by

$$x_k = L_{\text{warp}} \left(j \pm < \mathbf{p}_{i,j}\mathbf{x}_k, \mathbf{t}_{\text{warp}} > / \|\mathbf{p}_{i,j}\mathbf{p}_{i,j\pm 1}\| \right) \tag{6.9}$$

$$y_k = L_{\text{weft}} \left(i \pm < \mathbf{p}_{i,j}\mathbf{x}_k, \mathbf{t}_{\text{weft}} > / \|\mathbf{p}_{i,j}\mathbf{p}_{i\pm 1,j}\| \right) \tag{6.10}$$

where the sign $\pm$ is determined by the direction of $\mathbf{p}_{i,j}\mathbf{x}_k$. An illustration of this interpolation is shown in Fig. 6.4.

6.1.3 Algorithmic Details

The exact placement of the warp threads on M is obtained through an energy diffusion process that minimizes the energy J_E of Eq. (6.5). The overall algorithm consists of three steps—as explained in **Algorithm** *WovenPlacement*.

Algorithm 1 WovenPlacement

```
Compute the initial planar shape Ω of the given mesh M by the intrinsic parameterization [2],
which gives the initial fitting of the woven model Γ;
repeat
  Determine the expected strain of every weft and warp spring on Γ;
  Move the woven nodes on the surface of M through energy diffusion;
until terminal condition is satisfied
```

The normal pressure assigned by a physician is specified only at some discrete points on the mesh surface M, which may not coincide exactly with the woven nodes during the diffusion. Therefore, at the beginning of the second step, if a point has a prescribed normal pressure, its closest woven node on M is located and assigned with the corresponding normal pressure. Then, Eq. (6.4) is invoked to calculate the strains on all warp threads which have at least one such located node. The remaining problem in strain calculation is to obtain the warp strain distribution function $\varepsilon(t)$ that interpolates for those warp threads not containing any such located nodes. The function $\varepsilon(t)$ should be smooth and flat at the two ends (that is, $\varepsilon'(0) = 0$ and $\varepsilon'(1) = 0$). In addition, only tensile strains should be appeared in the brace (that is, $\varepsilon(t) \geq 0$ ($\forall t \in [0, 1]$). This is a typical curve interpolation problem. A B-spline curve is used to fit the function, $\varepsilon(t)$. To ensure $\varepsilon(t) \geq 0$, the following approximation (instead of interpolation) is used.

$$\varepsilon(t) = \sum_{i=0}^{n-1} q_i^2 N_{i,3}(t) \tag{6.11}$$

where q_i^2 are control points with non-negative values, $N_{i,3}(t) \geq 0$ are B-spline basis functions of degree three, and the function $\varepsilon(t)$ can be obtained by determining the values of q_is. Details about B-spline curve fitting can be found in [10] or any texts on B-spline curves.

The movement of woven nodes in the third step of **Algorithm** *WovenPlacement* is governed by the elastic energy function defined in Eq. (6.5) together with Eqs. (6.6)–(6.8). To minimize J_E, one should let the position $\mathbf{p}$ of every node v in the woven model Γ satisfy

$$\frac{\partial J_E}{\partial \mathbf{p}} = \sum_{j \in N(v)} k_j \frac{\mathbf{p}\mathbf{p}_j}{\|\mathbf{p}\mathbf{p}_j\|} (\|\mathbf{p}\mathbf{p}_j\| - L_{e_j}) = 0 \tag{6.12}$$

where $N(v)$ represents the one-ring neighbors of v—in other words, those with a spring linked to v. A diffusion process similar to that of Desbrun et al. [3] is used to solve Eq. (6.12). Taking the current $\frac{\partial J_E}{\partial \mathbf{p}}$ as a force in a spring-mass system, the new position of v is computed by

$$\mathbf{p}_{\text{new}} = \mathbf{p} - \lambda \frac{\partial J_E}{\partial \mathbf{p}} \tag{6.13}$$

where λ is a damping factor to control the movement of v in every iteration. This diffusion process is similar to the quasi-Newton type numerical optimization scheme.

Here, the movement of a woven node v is along the geodesic path determined by $\frac{\partial J_E}{\partial \mathbf{p}}$ with a distance of $\lambda \| \frac{\partial J_E}{\partial \mathbf{p}} \|$. Choosing a small λ will lead to more accurate results but at the cost of a low diffusion speed, whereas using a large λ may make the system become unstable. Wang and Tang [16] use $\lambda = 0.125/\max\{k_{\text{warp}}, k_{\text{weft}}\}$ in all their tests and find it to work well. The method introduced in [6] is adopted here to compute the discrete geodesic path following a given direction vector $\mathbf{t}_0$ on a given mesh surface at a point $\mathbf{p}_0$. The algorithm is, in fact, a local incremental approach. At every point $\mathbf{p}_i$ on the mesh, the geodesic path locally coincides with the intersection curve of the plane formed by $\mathbf{p}_i$, $\mathbf{t}_i$ and $\mathbf{n}_i$, where $\mathbf{n}_i$ is the mesh surface normal at $\mathbf{p}_i$ and $\mathbf{t}_i$ is the tangent of the geodesic path at this point.

In the diffusion process, the overall orientation of the woven threads on M may become twisted in some sense. This is undesirable from a physician's point of view. As a solution to this problem, at the beginning of the diffusion, users are asked to specify two points on M as two orientation markers. After every iteration step, the corresponding points of the two markers mapped onto the plane are checked. If the two points do not lie on a vertical line, all the planar vertices are rotated to align the two corresponding points to a vertical line.

Moreover, during the diffusion, some nodes may move beyond the boundary of M; if so, these nodes together with the springs linked to them are removed from Γ. If some gap emerges near the boundary of M, new nodes will often need to be inserted into M.

Another issue related to the diffusion of elastic energy is the selection of initial values. Similar to most other numerical optimization techniques, the result of energy diffusion relies heavily on the given initial values. From investigations, Wang and Tang [16] find that, similar to other spring systems, it is easier for the system to converge when the springs are deformed from the tensile state back to the rest state compared with the opposite—in other words, from the compressed state to the rest state. Therefore, in the first step of the overall algorithm, after obtaining the intrinsic parameterization (which is close to a stretch-free state), the planar coordinates of all the vertices are scaled down with a ratio $r < 1$ in both vertical and horizontal directions. As a result, most springs in the woven model after the first step are in the tensile state. In [16], Wang and Tang set r to 0.75 in the warp direction and to 1.0 in the weft direction so that the woven threads in the weft direction are not affected.

The iterative diffusion algorithm needs certain error measurement to evaluate the convergence of the iteration. Wang and Tang [16] measure the *squared strain difference* (SSD) on the fitted woven model. For all warp and weft springs (diagonal springs are excluded as their contribution to the normal pressure is ignored due to Lemma 5.2), the SSD is defined as

$$E_{\mathrm{SSD}} = \sum_{e} (\varepsilon_e - \varepsilon_e^0)^2 \tag{6.14}$$

where ε_e is either $\varepsilon_{\mathrm{warp}}$ or $\varepsilon_{\mathrm{weft}}$ depending on the type of the spring and ε_e^0 is the desired strain on the spring. An empirical threshold E_{SSD}^0 is used to determine the termination of the iteration: the diffusion will continue if $E_{\mathrm{SSD}} > E_{\mathrm{SSD}}^0$, unless a prescribed maximum number of iterations is reached.

As shown in Fig. 6.1, this approach can be used to generate planar patterns for different normal pressure configurations. Four configurations are tested and shown in Fig. 6.1b:

- *Configuration A*: No normal pressure is assigned and the fitting result minimizes the stretch (see top left of Fig. 6.1b); this is often called stretch-free although it is not really 'stretch-free' in general.
- *Configuration B*: A large normal pressure is specified in the middle of the brace while a small normal pressure is assigned to the two ends of the brace; the resultant pattern leads to a tighter stretch in the middle of the brace (see top right of Fig. 6.1b).
- *Configuration C*: A large normal pressure is assigned to one end of the brace while a small one is specified at the other end; the final brace has a large stretch on one side while it is almost stretch-free on the other (see bottom left of Fig. 6.1b).
- *Configuration D*: A large normal pressure is assigned to the two ends of the brace; the result is depicted at the bottom right of Fig. 6.1b, which shows almost no stretch in the middle of the brace.

The spring stiffness coefficients adopted in this test are $k_{\mathrm{warp}} = 0.5$, $k_{\mathrm{weft}} = 0.5$, and $k_{\mathrm{diag}} = 0.05$, which simulate a woven fabric with similar strengths in both directions—*isotropic* case ($k_{\mathrm{warp}} \approx k_{\mathrm{weft}}$). Moreover, *anisotropic* case ($k_{\mathrm{warp}} \ll k_{\mathrm{weft}}$) can also be simulated by using different coefficients for k_{warp} and k_{weft} (for example, $k_{\mathrm{warp}} = 0.5$, $k_{\mathrm{weft}} = 5.0$, and $k_{\mathrm{diag}} = 0.05$).

Since this is an initial study, many issues and questions do still remain (see the discussions in [16]) and can be stated specifically as follows:

- Only the elasticity in two directions—warp and weft—is considered; therefore, when in equilibrium, the strain (and stress) on a single weft or warp thread is a constant. The result is that only very simple patterns of strain distribution can be simulated by the woven model.
- The computation of redistributing the woven nodes on the surface is based on the knowledge of a strain distribution. However, computing such a strain distribution in general is not straightforward. Thus, again, the woven-model-based approach in [16] can only mimic very simple and limited patterns of strain distribution.

- The strain on the springs around a point with a user-specified normal pressure are calculated by fitting a quadratic polynomial. Such an approximation is not accurate enough.
- The boundary of the woven model is zigzag, which brings great difficulty to model the physical interactions between the 2D pieces that will be sewn together.
- There is no guarantee that all the threads will be in the tensile state, which is though strongly required by a compression garment.

These deficiencies are overcome by a more complicated physical/geometric approach presented in the following section.

6.2 A Physical/Geometric Approach for Patterns of Compression Garment

A new physical/geometric approach, which is able to model more complicated elastic behaviors of fabrics and determine the 2D patterns of a given 3D mesh surface satisfying the given strain and/or normal pressure distribution, is presented in this section. In this approach, every triangle edge is simulated by a tensile truss bar, and the equilibrium equation is established at every vertex on the given mesh surface. The correlation between the normal pressure and the stress on the truss bars is carefully analyzed. Moreover, critical geometric constraints are introduced to ensure that the network of truss bars in the relaxed state can be flattened without significant stretches. The sewing behavior between assembled patches is also modeled. After assembling all these modules, the strain distribution on the truss bars is finally determined with a least-square based iterative minimization.

6.2.1 Physical Model

Without loss of generality, the input 3D surface M is assumed to be represented by a triangular mesh. To simulate the physical behavior of elastic fabrics on M, every triangle edge is considered to be a linear truss bar. The human body that supports the elastic fabrics is defined by a given 3D shape H. The following definitions and assumptions are assigned to the physical model of compression garments.

Definition 6.1 The given triangular mesh surface M is a piece of elastic fabric in its final shape when it is worn on H.

Definition 6.2 The elastic fabric is represented as a network of linear truss bars in connectivity on M.

Assumption 6.5 All the linear truss bars are of the same material, and have the same cross-section.

From the study of elastic materials [4], the strain of a linear truss bar e is

$$\varepsilon_e = (l_e - l_e^0)/l_e^0 \tag{6.15}$$

where l_e is the current length of e and l_e^0 is the length in the relaxed state. The correlation between the strain ε_e and the stress σ_e of e is

$$\sigma_e = k\varepsilon_e. \tag{6.16}$$

The corresponding force of this stress is its integral over the entire cross-section of the truss bar. By Assumption 6.5, the stiffness coefficient k in Eq. (6.16) is constant over the entire fabric, and the correlation between the strain and the force is linear. This simplifies the subsequent formulation in this section; otherwise, the equilibrium equation introduced below would become nonlinear and make the problem much more difficult to be solved.

Definition 6.3 For a compression garment, the strain ε_e on every truss bar e should satisfy $\varepsilon_e \geq 0$.

Definition 6.4 For a linear truss bar e, its generated force f_e is proportional to its strain ε_e with a constant ratio.

Assumption 6.6 The nodes linking the truss bars coincide with the vertices of M when the whole truss structure is in equilibrium.

Assumption 6.7 The friction between the elastic fabric and the surface of human body is negligible.

Assumption 6.8 The human body H under compression agrees with the shape of M, and the deformation of H is neglected.

The forces at every joint node of the truss structure should be in equilibrium so that the whole system is in a stable state. Based on this assumption, the following lemma can be derived.

Lemma 6.1 *The projection of forces in the tangent plane, Ω_v, to a node $v \in M$ as generated by the truss bars around v, is in equilibrium.*

Proof Let, $\mathbf{f}_e$ be the force on a truss bar $e \in E(v)$ with $E(v)$ denoting the collection of edges linking to v. $\mathbf{f}_e$ can be decomposed into the component perpendicular to Ω_v and that in Ω_v. By Assumptions 6.6 and 6.7, the perpendicular component is balanced by the support from the human body and the friction is ignored. Therefore, one can have the following equilibrium equation in the plane Ω_v:

$$\sum_{e \in E(v)} \Omega_v(\mathbf{f}_e) \equiv 0 \tag{6.17}$$

where $\Omega_v(\cdots)$ stands for the projection of a vector onto the plane Ω_v. Figure 6.5 gives an illustration of this.

Q.E.D.

Motivated by the strain–stress theories in solid mechanics [5], the correlation between the tensile strain and the normal pressure can be modeled as

$$P_v = s \int_0^{2\pi} \kappa_n(\theta)\sigma(\theta)d\theta \tag{6.18}$$

where $\kappa_n(\theta)$ is the normal curvature in direction θ on the tangent plane at the surface point, $\sigma(\theta)$ denotes the normal stress in θ, and s is a parameter to reflect the effect on different materials. Wang and Tang [17] further simplify the formula into a finite sum on the truss structure model as

$$P_v = \frac{s}{2\pi\kappa_H} \sum_{e \in E(v)} (-\mathbf{t}_e \cdot \mathbf{n}_v)\theta_e \varepsilon_e \tag{6.19}$$

where $\mathbf{t}_e$ is the unit vector of truss bar e pointing outwards from v, $\mathbf{n}_v$ is the unit normal vector to surface M at v, κ_H is the mean curvature at v, and θ_e reflects the weight of the truss bar e contributing to the normal pressure. Here, the value of θ_e is the average angle of e's left and right triangles at the vertex v on Ω_v. As illustrated in Fig. 6.5, $\theta_e = 0.5(\alpha_1 + \alpha_2)$.

In practice, the 2D patterns of a compression garment are sewn together by suture (see Fig. 6.6a). The physical effect of this sewing operation needs to be modeled as well. Generally speaking, during sewing, the boundaries of fabrics are in the relaxed state. If some distortion is introduced after the pieces are sewn together, unwanted wrinkles might occur and thus violate the original design intent—this is one critical criterion to evaluate whether a garment is well designed and fabricated. Since the suture in general is not elastic, the boundaries of fabrics locked by suture have almost no freedom of deformation. Therefore, the following definition and assumption are introduced to the physical model.

Definition 6.5 The boundary of a given mesh surface M is defined as a *sewn boundary* if it is linked to other pieces by suture; otherwise, it is defined as a *free boundary*.

Assumption 6.9 The strains of the truss bars overlapping the boundary of a fabric piece should be zero when the system of truss structure is in equilibrium.

Notice that the above assumption is only for sewn boundaries; the strain may not be zero for those free boundaries (for example, the free boundary of elastic braces shown in Fig. 6.2). Moreover, the input mesh surfaces to be assembled are assumed to be stitched together by making the boundary vertices coincident (for example, as shown in Fig. 6.6b, three pieces—in different colors—are stitched together by linking the three orange vertices). The reaction forces among them would resist this stitching. Therefore, the equilibrium at the sewing nodes relies on the forces from the

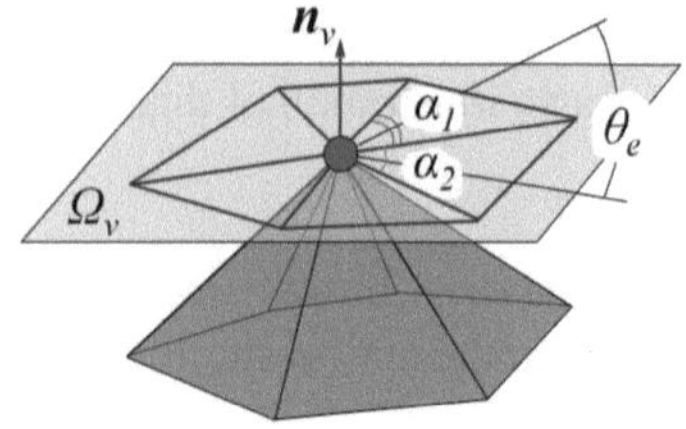

Fig. 6.5 An illustration of the physical model at an interior vertex (Image taken from [17] with permission)

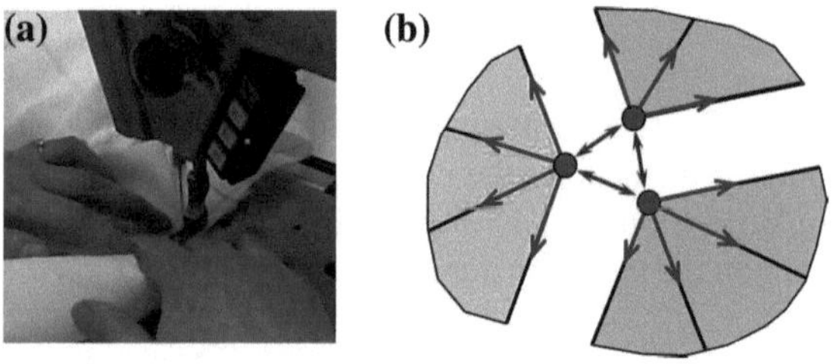

Fig. 6.6 An illustration of the sewing model—the *blue* arrows are forces generated by the tensile truss bars to the nodes, and the *red* ones are the reaction forces between sewing nodes (Image taken from [17] with permission)

truss bars linked to all these nodes. More specifically, Eq. (6.17) given above needs to be modified as

$$\sum_{e \in E(v)} \Omega_v(\mathbf{f}_e) + \sum_{v_s \in S(v)} \sum_{e \in E(v_s)} \Omega_v(\mathbf{f}_e) \equiv 0 \tag{6.20}$$

where $S(v)$ denotes the collection of all nodes that should be sewn together with node v during the fabrication.

The physical model developed in this section works together with the geometric constraints derived next to satisfy the physical properties of compression garment on the 3D shape warped from the computed 2D patterns.

6.2.2 Geometric Constraints

The behavior of elastic fabric pieces on a compression garment should not only be governed by the equilibrium equations (that is, Eq. (6.17) and (6.20)) described in the previous section, but should also have correct geometric constraints to ensure that the 3D shape can really be fabricated from 2D patterns. This leads to the following developability constraints imposed on every interior mesh vertex.

From differential geometry, it is well known that a surface can be developed into a plane without local stretch only if the Gaussian curvature is zero everywhere on it. When dealing with a piecewise linear mesh surface, the Gaussian curvature is converted to the form in terms of the angles of triangles at the involved vertex, which has been successfully used in several existing systems (for example, [11, 13, 15]).

Therefore, a similar formula is used below to constrain the local geometry of a fabric piece for compression garment.

For a vertex v on M, using $\alpha_f(v)$ to symbolize the vertex angle of a triangular face f at v, the condition

$$\sum_{f \in F(v)} \alpha_f(v) \equiv 2\pi$$

should be satisfied if all the faces around v are to be flattened without any cracks and/or overlaps, where $F(v)$ is the set of triangles coincided at v. Notice that, in this problem setting, the triangles in 3D have already been distorted in order to generate compression. Therefore, the angle of face f in its original relaxed state must be used here to replace $\alpha_f(v)$.

Constraint 6.1 Let $\alpha_f^0(v)$ represent the angle of the relaxed triangle f at v, the geometric constraint at an interior vertex v is defined as

$$\sum_{f \in F(v)} \alpha_f^0(v) \equiv 2\pi. \tag{6.21}$$

Notice that this constraint is only set for interior vertices. For boundary vertices, the constraint in terms of angles should be

$$\sum_{f \in F(v)} \alpha_f^0(v) < 2\pi. \tag{6.22}$$

The computation of planar patterns is converted to a problem of determining the optimal original length of each truss bar so that the above geometric constraint and the equilibrium equations can be satisfied (in the minimum energy sense). The *triangle constraint* defined below is also important and necessary as it ensures the validity of the length of a truss bar.

Constraint 6.2 Let, l_1^0, l_2^0 and l_3^0 represent the relaxed lengths of three bars e_1, e_2 and e_3, respectively, in a triangle in M; they must satisfy

$$l_i^0 + l_j^0 - l_k^0 > 0 \tag{6.23}$$

if $l_k^0 \geq l_i^0$ and $l_k^0 \geq l_j^0$ with $i, j, k \in \{1, 2, 3\}$.

The inequality constraint in Eq. (6.23) prevents the case that the summed length of two edges in a triangle is even shorter than the third triangle edge.

6.2.3 Numerical Solution

This section presents a numerical scheme that computes the optimal tensile strain on every linear truss bar so that the desired normal pressure distribution on M can be realized (in the minimum energy sense). Nodes are not moved during the computation. Users are allowed to specify the desired normal pressure at some designated vertices on M (referred to as *anchor points*). As a direct result of the tensile strains on M, the corresponding 2D patterns of M are also obtained. The optimal strains should ensure the force equilibrium at all the vertices (that, joint nodes of the truss structure), realize the desired normal pressures, and satisfy the geometric constraints.

The intuitive variables in the numerical system are the strains ε_e on the truss bars. To safeguard the $\varepsilon_e \geq 0$ requirement (as in Definition 6.3), a new variable x_e is introduced as

$$\varepsilon_e = x_e^2. \tag{6.24}$$

According to Lemma 6.1, the projection of forces of the truss bars around a vertex v onto its tangent plane should be in equilibrium. According to Assumption 6.5, the forces are proportional to the strains on the truss bars. Therefore, the equilibrium equation (Eq. (6.17)) can be derived to become

$$\sum e \in E(v)(\mathbf{t}_{xv} \cdot \mathbf{t}_e)\mathbf{f}_e = \sum e \in E(v)(\mathbf{t}_{xv} \cdot \mathbf{t}_e)\sigma_e = \sum e \in E(v)(\mathbf{t}_{xv} \cdot \mathbf{t}_e)k\varepsilon_e = 0$$
$$\sum e \in E(v)(\mathbf{t}_{yv} \cdot \mathbf{t}_e)\mathbf{f}_e = \sum e \in E(v)(\mathbf{t}_{yv} \cdot \mathbf{t}_e)\sigma_e = \sum e \in E(v)(\mathbf{t}_{yv} \cdot \mathbf{t}_e)k\varepsilon_e = 0$$

with $\mathbf{f}_e$ and σ_e being the tensile force and stress on the truss bar e, and

$$\sum_{e \in E(v)} (\mathbf{t}_{xv} \cdot \mathbf{t}_e)x_e^2 = 0, \quad \sum_{e \in E(v)} (\mathbf{t}_{yv} \cdot \mathbf{t}_e)x_e^2 = 0, \tag{6.25}$$

where $\mathbf{t}_{xv}$ and $\mathbf{t}_{yv}$ are the unit vectors of the local frame established at v such that $n_v = \mathbf{t}_{xv} \times \mathbf{t}_{yv}$ (that is, $\mathbf{n}_v$, $\mathbf{t}_{xv}$ and $\mathbf{t}_{yv}$ are orthogonal to each other). Similar formulae can be derived from Eq. (6.20) for nodes that are sewn together on the boundaries. These equations should be satisfied at all vertices of M. In addition, by Assumption 6.9, $x_e = 0$ should be enforced for those truss bars on the sewn boundary.

The strain–pressure conversion equation derived from Eq. (6.19) is defined on every anchor point by replacing ε_e in Eq. (6.19) with x_e^2. The equations for geometric constraints are also reformulated in terms of x_e, and the relaxed length l_e^0 is converted into $l_e^0 = l_e/(1 + x_e^2)$ with the length of truss bar e, l_e, on the given mesh surface M. The inequality constraints are introduced to the numerical system by the active set method. In short, the inequality constraints are partitioned into an active set and an inactive one—only the constraints in the active set are added to the system during computation (see [8] for reference).

6.2.3.1 Linearization and Least-Square Solution

Almost all the equations listed above are nonlinear in nature. To derive a numerical solution for them, one can first linearize the equations and then utilize the least-square method to iteratively update the value of x_e until a satisfactory solution is obtained.

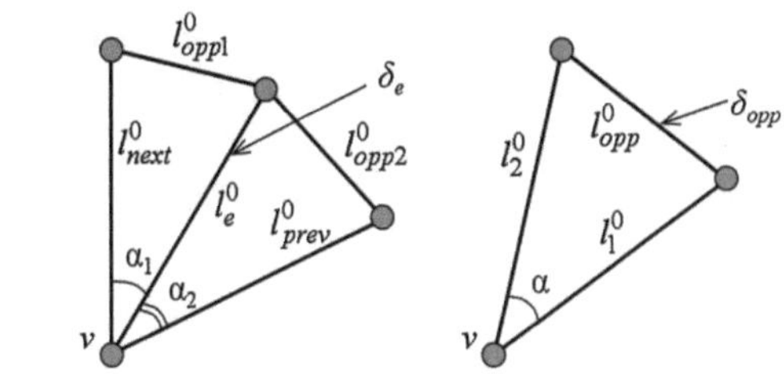

Fig. 6.7 The derivative of vertex angle in terms of strain. *Left* The variable δ_e is for the bar adjacent to the vertex v. *Right* The variable δ_{opp} is for the bar opposite to the vertex (Image taken from [17] with permission)

The linearization is obtained by the first-order Taylor expansion. Explicitly, Eq. (6.25) is linearized to

$$\sum_{e \in E(v)} (\mathbf{t}_{xv} \cdot \mathbf{t}_e) x_e^2 + \sum_{e \in E(v)} 2(\mathbf{t}_{xv} \cdot \mathbf{t}_e) x_e \delta_e \approx 0 \tag{6.26}$$

$$\sum_{e \in E(v)} (\mathbf{t}_{yv} \cdot \mathbf{t}_e) x_e^2 + \sum_{e \in E(v)} 2(\mathbf{t}_{yv} \cdot \mathbf{t}_e) x_e \delta_e \approx 0. \tag{6.27}$$

where $\delta_e = x_e^{\text{new}} - x_e$ and x_e is the current value of the truss bar e. The formulae have an approximation error $O(\delta_e^2)$. For the normal pressures assigned to the anchor points, Eq. (6.19) can be converted into a linear form in terms of δ_e as

$$p_v \approx s \sum_{e \in E(v)} a_e x_e^2 + 2s \sum_{e \in E(v)} a_e x_e \delta_e \tag{6.28}$$

where $a_e = (-\mathbf{t}_e \cdot \mathbf{n}_v)\theta_e / 2\pi\kappa_H$ and the approximation error is $O(\delta_e^2)$.

As for the geometric constraint (Eq. (6.21)), the value of $\alpha_f^0(v)$ depends not only on the strains of bars coincident at v but also on the stains of the bars opposite to v (see Fig. 6.7). Therefore, applying the Taylor expansion to Eq. (6.21) leads to

$$\alpha_v + \sum \frac{\partial \alpha_v}{\partial l_e^0} \frac{\partial l_e^0}{\partial x_e} \delta_e + \sum \frac{\partial \alpha_v}{\partial l_{\text{opp}}^0} \frac{\partial l_{\text{opp}}^0}{\partial x_{\text{opp}}} \delta_{\text{opp}} \approx 0 \tag{6.29}$$

where $\alpha_v = \sum_{f \in F(v)} \alpha_f^0(v)$ is the current vertex angle at v, l_e^0 is the relaxed length of the truss bars adjacent to v, and l_{opp}^0 is the relaxed length of bars opposite to v.

Regarding the derivatives (see Fig. 6.7), one can have $\partial \alpha_v / \partial l_e^0 = \partial(\alpha_1 + \alpha_2 + const) / \partial l_e^0$, where α_1 and α_2 are functions of $(l_e^0, l_{\text{next}}^0, l_{\text{opp1}}^0)$ and $(l_e^0, l_{\text{last}}^0, l_{\text{opp2}}^0)$, respectively and $\partial \alpha_v / \partial l_{\text{opp}}^0 = \partial \alpha / \partial l_{\text{opp}}^0$, with α depending on $(l_1^0, l_2^0, l_{\text{opp}}^0)$ by the $\arccos(\cdots)$ function. Notice that all the angles here are evaluated by the relaxed lengths of truss bars (that is, l_e^0) instead of their current lengths. A similar linearization can be applied to Eq. (6.22) for boundary vertices, which leads to the same formula as Eq. (6.29)—notice that, this linear equation for boundary vertices is added to the numerical system only when Eq. (6.22) is not satisfied.

For the triangle constraint, when Eq. (6.23) is disobeyed, the following linearization of it is inserted into the numerical system.

$$l_i^0 + l_j^0 - l_k^0 + \frac{\partial l_i^0}{\partial x_i}\delta_i + \frac{\partial l_j^0}{\partial x_j}\delta_j - \frac{\partial l_k^0}{\partial x_k}\delta_k \approx 0 \tag{6.30}$$

To satisfy the sewing constraint (Eq. (6.20)), it can be linearized in a way similar to that of Eqs. (6.26) and (6.27). Then, owing to Assumption 6.9, one can let

$$\delta_e + x_e = 0 \tag{6.31}$$

for any truss bar e on a sewn boundary. Furthermore, in order to add damping factors to make the numerical system more stable, the update of x_e for all non-sewn boundary bars should be small in each iteration step. Therefore, the following damping equation is introduced to the numerical system for all truss bars except the ones on the sewn boundary.

$$\delta_e = 0 \tag{6.32}$$

Notice that as the least-square solution is used to determine the value of δ_e, the value of δ_e on interior bars is a value balanced from all constraints in Eqs. (6.26)–(6.31) but *not* zero.

Integrating all the above linear equations (Eqs. (6.26)–(6.31)) can have

$$\begin{bmatrix} \mathbf{A} \\ \mathbf{I} \end{bmatrix} [\delta_e] = \begin{bmatrix} \mathbf{b}_a \\ \mathbf{b}_i \end{bmatrix} \tag{6.33}$$

where $\mathbf{A}$ and $\mathbf{b}_a$ are derived from Eqs. (6.26)–(6.29), and $\mathbf{b}_i$ is from Eqs. (6.30) and (6.31). This is an over-determined linear equation system, which can be solved by the least-square solution as

$$\begin{bmatrix} \mathbf{A}^T & \mathbf{I} \end{bmatrix} \begin{bmatrix} \mathbf{A} \\ \mathbf{I} \end{bmatrix} [\delta_e] = \begin{bmatrix} \mathbf{A}^T & \mathbf{I} \end{bmatrix} \begin{bmatrix} \mathbf{b}_a \\ \mathbf{b}_i \end{bmatrix}. \tag{6.34}$$

In the implementation of Wang and Tang [17], after setting initial values to $\mathbf{x} = [x_e]$, they solve Eq.(6.34) and then update the values of $\mathbf{x}$ by $\mathbf{x} \leftarrow \mathbf{x} + \delta$ iteratively. The iteration continues until either the zero-norm condition $\|\delta\|^2 < 10^{-5}$ is met or the maximum number of iterations (for example, 500) is reached. To different rows in the linear system of Eq.(6.33), different weights may be adopted. In all their tests, both sides of Eq.(6.29) are multiplied with a weight of 5.0 (as compared with a weight of 1.0 for all the other rows) so as to highlight the developability constraint, and the weights on anchor points are set as a ratio of the number of vertices on M to the number of anchor points.

6.2.3.2 Computing 2D Shape

After the optimal values of x_e are determined, the relaxed lengths of all the truss bars can be calculated. The only task left is to determine the 2D shape of each pattern. To do so, Wang and Tang [17] first compute the angles in all triangles by the relaxed lengths of triangle edges. Then, the angle-based least-square formulation [11] is used to determine the 2D coordinates of each vertex. Using the method proposed by Sheffer et al. [11], the positions of two vertices on the longest edge are fixed and a system of linear equations relating angles to the planar coordinates is solved. The reason for not adopting the intuitive greedy reconstruction is that, while the flattening of each triangle alone generates a very small numerical error, these errors accumulate at the advancing front. Thus, when the involved mesh surface has many triangles, the accumulated error in the end could be very large. The least-square reconstruction somewhat balances out the cumulative error. The 2D mesh obtained can be further optimized by the method of Wang et al. [14].

6.2.4 Experimental Results and Verification

Several error measurement terms can be defined for the evaluation of experimental results.

The first error term is the deviation between the prespecified desired normal pressure and the computed normal pressure M, as

$$E_p = \frac{1}{m_a} \sum_v |P_v - P_v^0| \tag{6.35}$$

where m_a is the number of anchor points, P_v^0 is the desired normal pressure, and P_v is as defined in Eq.(6.19). In the tests conducted by Wang and Tang [17], the value of s in Eq.(6.19) is simply set as 1.0. More realistic values depending on the material properties can be determined by calibration procedures, which beyond the scope of this book.

The second error term measures the level of equilibrium at the vertices in the final M, as

$$E_F = \sum_e \|(\mathbf{t}_e \cdot \mathbf{t}_x)\varepsilon_e\|^2 + \sum_e \|(\mathbf{t}_e \cdot \mathbf{t}_y)\varepsilon_e\|^2. \tag{6.36}$$

While these two error items relate to the physical property of the system, the following two equations

$$E_{\theta_{\text{mean}}} = \frac{1}{m_{\text{int}}} \sum_v |2\pi - \sum_{f \in F(v)} \alpha_f^0(v)| \tag{6.37}$$

Table 6.1 Statistics of errors

Example	E_P	E_F	$E_{\theta_{mean}}$	$E_{\theta_{max}}$	E_L
Brace (free)	N/A	2.7×10^{-4}	3.4×10^{-4}	0.036	5.7×10^{-3}
Brace A	6.0×10^{-6}	1.6×10^{-3}	7.4×10^{-4}	0.038	1.7×10^{-2}
Brace B	1.6×10^{-4}	3.2×10^{-3}	2.8×10^{-3}	0.037	1.5×10^{-2}
Brace C	7.0×10^{-5}	4.2×10^{-4}	7.8×10^{-4}	0.038	2.4×10^{-2}

$$E_{\theta_{\max}} = \max\{|2\pi - \sum_{f \in F(v)} \alpha_f^0(v)|\} \tag{6.38}$$

quantify the developability of the final M, that is, how well it can be flattened. m_{int} is the number of non-boundary vertices on M. Notice that $\alpha_f^0(v)$ is computed with the relaxed lengths of the truss bars in the triangle f. The final error i.e., E_L, is used to measure the length change on the flattened patterns from the original relaxed value l_e^0, as

$$E_L = \frac{1}{m_e} \sum_e |l_e^0 - l_e^*| / l_e^0. \tag{6.39}$$

where l_e^* is the length of edge e after flattening and m_e is the number of truss bars (i.e., edges) on M.

In Fig. 6.8, this method is tested on a medical elastic brace example with four different configurations of normal pressure assignment. Figure 6.8a–d display the results of strain distribution and planar patterns under different normal pressure assignments. It is worth mentioning that the configuration in Fig. 6.8d cannot be simulated by the method in Sect. 6.1. For comparison, the 2D patterns generated by the least squares conformal map (LSCM) [7] and the length-preserved free boundary (LPFB) [12] are given in Figs. 6.8e, f. Table 6.1 lists the results of error measurement of the four configurations, all of which are small, indicating that the numerical iterations converged for all the four configurations in the tests.

In order to verify this physical/geometric approach, Wang and Tang [17] compared the results computed by this scheme with the simulation results from finite element analysis (FEA). Two models under different compressions are tested by a commercial FEA software. Their conclusion is that the results computed by this approach are close to the results from FEA.

6.3 Material-Related Coefficients

In the two approached introduced above, the computation usually involves material-related coefficients. To evaluate the material-related coefficients and verify the simulation results, a test bed needs to be built. This section introduces such a test bed.

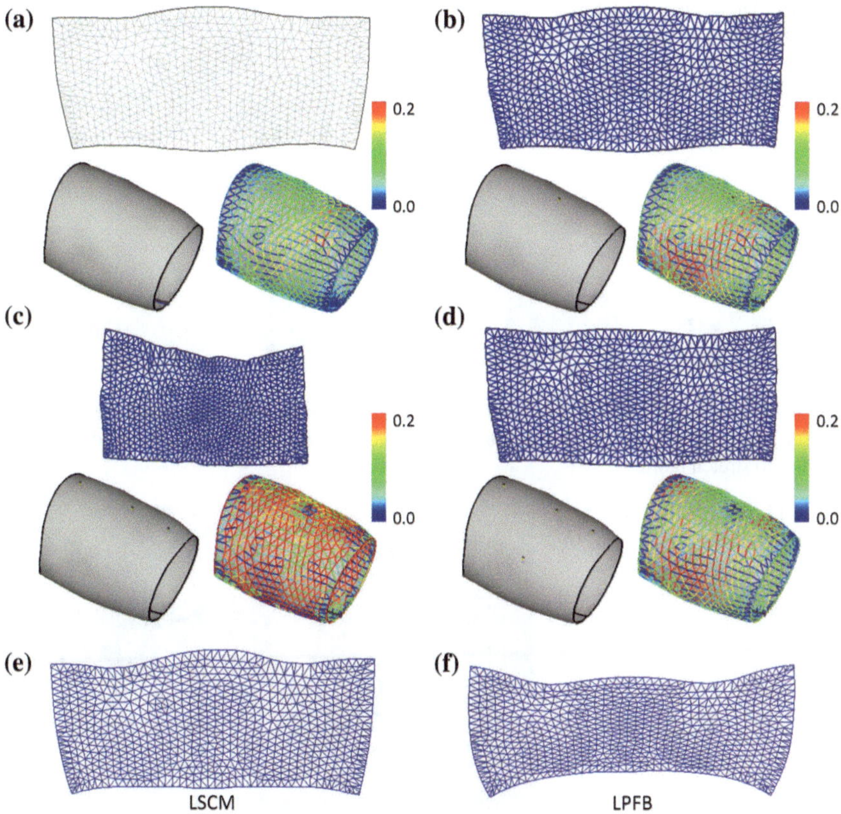

Fig. 6.8 The medical brace example with different normal pressures specified at the anchor points (in *black*): **a** the brace is free—no specific normal pressure is assigned, **b** brace A—with one anchor point, **c** brace B—with three anchor points, and **d** brace C—with four anchor point. The color maps show different strain distributions corresponding to the four different normal pressure configurations. The final 2D patterns (in *red*) that are superimposed with the one from the free brace in (a), are also given for comparison. **e** and **f** give the planar meshes generated by the least squares conformal map (LSCM) and the length-preserved free boundary (LPFB) respectively, which are different from the patterns generated by this approach (Image taken from [17] with permission)

The test bed consists of several cylindrical, conical, and spherical models in various dimensions. Compression pressure sensors are integrated onto the surface of the models. When warping different fabrics onto these models, the value of compression pressure can be evaluated. Meanwhile, several markers are printed on the fabric pieces which can be identified by an imaging system to report the corresponding strains. The pressure and strain values can be used to calibrate the material-related coefficients of the above physical/geometric models.

As shown in Fig. 6.9, three sets of models with cylindrical, conical (which, in fact, is a cone without head), and ellipsoidal shapes are made using rapid prototyping (RP) to imitate different parts of human body, such as upper arm, forearm, lower leg, and

Fig. 6.9 Models that are fabricated by rapid prototyping are used in the test bed to imitate different parts of human body

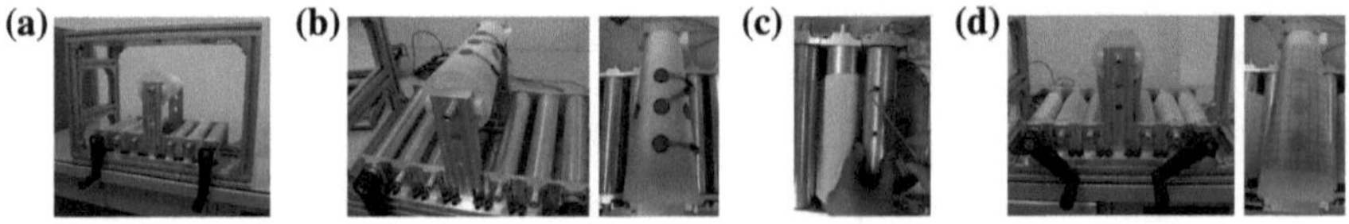

Fig. 6.10 Test bed system: **a** the machine with a model inserted, **b** five sensors are adhered to the surface of RP models, **c** a sample fabric piece placed zigzag onto the shafts, and **d** rolling the handles to stretch fabrics

Fig. 6.11 The test bed with a camera mounted

thigh. Force sensors are used to measure the force applied to the geometric RP models at various locations during stretching. Since the force sensors are placed onto the curved surface of the models, they are preferably thin and flexible. In view of these, the force sensing resistors (FSR - Part No. 402) produced by Interlink Electronics are adopted as they match the requirements well. The testing machine mainly comprises five shafts, two locked shafts, 12 bearings with holders, two handles, one fixed holder and one aluminum frame as shown in Fig. 6.10.

During the test, a sample fabric piece is inserted in zigzag manner over each bearing with holder and each end of the fabric piece is locked on the locked shaft. In the middle of the machine is a shaft on a fixed holder. There is no bearing on this holder as it is used to hold the RP models for observation of fabric extension while the remaining shafts and locked shafts are on holders with bearings so that they can be easily rotated. The aluminum frame is used to hold all these components tightly. After the fabric piece is locked, one can rotate the handles. The fabric piece is then stretched harder and harder. Five sensors are stuck onto the surface of the RP geometric models using adhesive tape as shown in Fig. 6.10b. The data of the force applied to these five sensors are then transferred to a computer and saved for subsequent data analysis. About 300 grids with dimensions of 1cm by 1cm are drawn onto the surface of the sample fabric piece to measure the extension under stretching.

A program for data analysis can be developed, which allows to use a camera mounted on top of the aluminum frame as displayed in Fig. 6.11 to observe fabric extensions. By looking down at the surface of the shaft on the fixed holder, a 2D line can be observed which can then be calculated into a 3D one using our software with the 3D geometry of RP models input. The strain of the fabric piece is then resulted. By this setup, the stress–strain relationship of different testing fabrics on different geometric shapes can be measured.

References

1. Atroshi, I., Gummesson, C.: Prevalence of carpal tunnel syndrome in a general population. J. Am. Med. Assoc. **282**, 153–158 (1999)
2. Desbrun, M., Meyer, M., Alliez, P.: Intrinsic parameterizations of surface meshes. Comput. Graphics Forum **21**, 209–218 (2002)
3. Desbrun, M., Meyer, M., Schröder, P., Barr, A.H.: Implicit fairing of irregular meshes using diffusion and curvature flow. In: Proceedings of ACM SIGGRAPH 99, pp. 317–324 (1999)
4. Dowling, N.E.: Mechanical behavior of materials. Prentice-Hall, New York (1993)
5. Fung, Y.C.: Foundations of solid mechanics. Prentice-Hall, Englewood Cliffs (1965)
6. Kumar, G.R., Srinivasan, P., Holla, V.D., Shastry, K., Prakash, B.: Geodesic curve computations on surfaces. Comput. Aided Geom. Des. **20**, 119–133 (2003)
7. Levy, B., Petitjean, S., Ray, N., Maillot, J.: Least squares conformal maps for automatic texture atlas generation. In: Proceedings of SIGGRAPH '02, pp. 362–371 (2002)
8. Madsen, K., Nielsen, H.B., Tingleff, O.: Optimization with Constraints. Technical University of Denmark, Course Notes (2004)
9. Meyer, M., Desbrun, M., Schröder, P., Barr, A.: Discrete differential-geometry operators for triangulated 2-manifolds. In: Visualization and Mathematics III, pp. 35–58 (2003)
10. Piegl, L., Tiller, W.: The NURBS Book 2nd edn. Springer, Berlin (1997)
11. Sheffer, A., Levy, B., Mogilnitsky, M., Bogomyakov, A.: ABF++: Fast and robust angle based flattening. ACM Trans. Graph. **24**, 311–330 (2005)
12. Wang, C.C.L.: Computing length-preserved free boundary for quasi-developable mesh segmentation. IEEE Trans. Vis. Comp. Graph. **14**(1), 25–36 (2008)
13. Wang, C.C.L.: Towards flattenable mesh surfaces. Comput. Aided Des. **40**, 109–122 (2008)
14. Wang, C.C.L., Smith, S.S.F., Yuen, M.M.F.: Surface flattening based on energy model. Comput-Aided Des **34**, 823–833 (2002)
15. Wang, C.C.L., Tang, K.: Achieving developability of a polygonal surface by minimum deformation: a study of global and local optimization approaches. Vis. Comput. **20**, 521–539 (2004)
16. Wang, C.C.L., Tang, K.: Woven model based geometric design of elastic medical braces. Comput. Aided Des. **39**, 69–79 (2007)
17. Wang, C.C.L., Tang, K.: Pattern computation for compression garment by a physical/geometric approach. Comput. Aided Des. **42**(2), 78–86 (2010)
18. Wang, C.C.L., Tang, K., Yeung, B.M.L.: Freeform surface flattening based on fitting a woven mesh model. Comput. Aided Des. **37**, 799–814 (2005)

Index

MIX
Papier aus verantwortungsvollen Quellen
Paper from responsible sources
FSC® C105338

If you have any concerns about our products,
you can contact us on
ProductSafety@springernature.com

In case Publisher is established outside the EU,
the EU authorized representative is:
Springer Nature Customer Service Center GmbH
Europaplatz 3, 69115 Heidelberg, Germany

Printed by Libri Plureos GmbH
in Hamburg, Germany